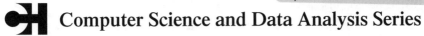
Computer Science and Data Analysis Series

Exploratory Data Analysis
with MATLAB®

Chapman & Hall/CRC
Series in Computer Science and Data Analysis

The interface between the computer and statistical sciences is increasing, as each discipline seeks to harness the power and resources of the other. This series aims to foster the integration between the computer sciences and statistical, numerical and probabilistic methods by publishing a broad range of reference works, textbooks and handbooks.

SERIES EDITORS

John Lafferty, Carnegie Mellon University
David Madigan, Rutgers University
Fionn Murtagh, Queen's University Belfast
Padhraic Smyth, University of California Irvine

Proposals for the series should be sent directly to one of the series editors above, or submitted to:

Chapman & Hall/CRC Press UK

23-25 Blades Court
London SW15 2NU
UK

Published Titles

Bayesian Artificial Intelligence
Kevin B. Korb and Ann E. Nicholson

Exploratory Data Analysis with MATLAB®
Wendy L. Martinez and Angel R. Martinez

Forthcoming Titles

Correspondence Analysis and Data Coding with JAVA and R
Fionn Murtagh

R Graphics
Paul Murrell

Nonlinear Dimensionality Reduction
Vin de Silva and Carrie Grimes

Computer Science and Data Analysis Series

Exploratory Data Analysis
with MATLAB®

Wendy L. Martinez
Angel R. Martinez

CHAPMAN & HALL/CRC

A CRC Press Company

Boca Raton   London   New York   Washington, D.C.

Library of Congress Cataloging-in-Publication Data

Martinez, Wendy L.
   Exploratory data analysis with MATLAB / Wendy L. Martinez, Angel R. Martinez.
      p. cm.
   Includes bibliographical references and index.
   ISBN 1-58488-366-9 (alk. paper)
      1. Multivariate analysis. 2. MATLAB. 3. Mathematical statistics. I. Martinez, Angel R.
   II. Title.

   QA278.M3735 2004
   519.5'35--dc22   2004058245

Visit the CRC Press Web site at www.crcpress.com

© 2005 by Chapman & Hall/CRC Press

No claim to original U.S. Government works
International Standard Book Number 1-58488-366-9
Library of Congress Card Number 2004058245
Printed in the United States of America   2 3 4 5 6 7 8 9 0
Printed on acid-free paper

This book is dedicated to our children:

Angel and Ochida

Deborah and Nataniel

Jeff and Lynn

and

Lisa (Principessa)

Table of Contents

Part I
Introduction to Exploratory Data Analysis

Chapter 1
Introduction to Exploratory Data Analysis

Part II
EDA as Pattern Discovery

Chapter 2
Dimensionality Reduction - Linear Methods

Chapter 9
Distribution Shapes

Chapter 10
Multivariate Visualization

Appendix A
Proximity Measures

Appendix B
Software Resources for EDA

Appendix C

Appendix D
Introduction to MATLAB

Appendix E
MATLAB Functions

Preface

One of the goals of our first book, *Computational Statistics Handbook with MATLAB®* [2002], was to show some of the key concepts and methods of computational statistics and how they can be implemented in MATLAB.[1] A core component of computational statistics is the discipline known as exploratory data analysis or EDA. Thus, we see this book as a complement to the first one with similar goals: *to make exploratory data analysis techniques available to a wide range of users.*

Exploratory data analysis is an area of statistics and data analysis, where the idea is to first explore the data set, often using methods from descriptive statistics, scientific visualization, data tours, dimensionality reduction, and others. This exploration is done without any (hopefully!) pre-conceived notions or hypotheses. Indeed, the idea is to use the results of the exploration to guide and to develop the subsequent hypothesis tests, models, etc. It is closely related to the field of data mining, and many of the EDA tools discussed in this book are part of the toolkit for knowledge discovery and data mining.

This book is intended for a wide audience that includes scientists, statisticians, data miners, engineers, computer scientists, biostatisticians, social scientists, and any other discipline that must deal with the analysis of raw data. We also hope this book can be useful in a classroom setting at the senior undergraduate or graduate level. Exercises are included with each chapter, making it suitable as a textbook or supplemental text for a course in exploratory data analysis, data mining, computational statistics, machine learning, and others. Readers are encouraged to look over the exercises, because new concepts are sometimes introduced in them. Exercises are computational and exploratory in nature, so there is often no unique answer!

As for the background required for this book, we assume that the reader has an understanding of basic linear algebra. For example, one should have a familiarity with the notation of linear algebra, array multiplication, a matrix inverse, determinants, an array transpose, etc. We also assume that the reader has had introductory probability and statistics courses. Here one should know about random variables, probability distributions and density functions, basic descriptive measures, regression, etc.

In a spirit similar to the first book, this text is *not* focused on the theoretical aspects of the methods. Rather, the main focus of this book is on the *use* of the

---

[1] MATLAB® and Handle Graphics® are registered trademarks of The MathWorks, Inc.

EDA methods. Implementation of the methods is secondary, but where feasible, we show students and practitioners the implementation through algorithms, procedures, and MATLAB code. Many of the methods are complicated, and the details of the MATLAB implementation are not important. In these instances, we show how to use the functions and techniques. The interested reader (or programmer) can consult the M-files for more information. Thus, readers who prefer to use some other programming language should be able to implement the algorithms on their own.

While we do not delve into the theory, we would like to emphasize that the methods described in the book have a theoretical basis. Therefore, at the end of each chapter, we provide additional references and resources, so those readers who would like to know more about the underlying theory will know where to find the information.

MATLAB code in the form of an Exploratory Data Analysis Toolbox is provided with the text. This includes the functions, GUIs, and data sets that are described in the book. This is available for download at

http://lib.stat.cmu.edu

and

http://www.infinityassociates.com

Please review the **readme** file for installation instructions and information on any changes. M-files that contain the MATLAB commands for the exercises are also available for download.

We also make the disclaimer that our MATLAB code is not necessarily the most efficient way to accomplish the task. In many cases, we sacrificed efficiency for clarity. Please refer to the example M-files for alternative MATLAB code, courtesy of Tom Lane of The MathWorks, Inc.

We describe the EDA Toolbox in greater detail in Appendix B. We also provide website information for other tools that are available for download (at no cost). Some of these toolboxes and functions are used in the book and others are provided for informational purposes. Where possible and appropriate, we include some of this free MATLAB code with the EDA Toolbox to make it easier for the reader to follow along with the examples and exercises.

We assume that the reader has the Statistics Toolbox (Version 4 or higher) from The MathWorks, Inc. Where appropriate, we specify whether the function we are using is in the main MATLAB software package, Statistics Toolbox, or the EDA Toolbox. The development of the EDA Toolbox was mostly accomplished with MATLAB Version 6.5 (Statistics Toolbox, Version 4), so the code should work if this is what you have. However, a new release of MATLAB and the Statistics Toolbox was introduced in the middle of writing this book, so we also incorporate information about new functionality provided in these versions.

We would like to acknowledge the invaluable help of the reviewers: Chris Fraley, David Johannsen, Catherine Loader, Tom Lane, David Marchette, and Jeff Solka. Their many helpful comments and suggestions resulted in a better book. Any shortcomings are the sole responsibility of the authors. We owe a special thanks to Jeff Solka for programming assistance with finite mixtures and to Richard Johnson for allowing us to use his Data Visualization Toolbox and updating his functions. We would also like to acknowledge all of those researchers who wrote MATLAB code for methods described in this book and also made it available for free. We thank the editors of the book series in Computer Science and Data Analysis for including this text. We greatly appreciate the help and patience of those at CRC press: Bob Stern, Rob Calver, Jessica Vakili, and Andrea Demby. Finally, we are indebted to Naomi Fernandes and Tom Lane at The MathWorks, Inc. for their special assistance with MATLAB.

Disclaimers

1. Any MATLAB programs and data sets that are included with the book are provided in good faith. The authors, publishers, or distributors do not guarantee their accuracy and are not responsible for the consequences of their use.

2. Some of the MATLAB functions provided with the EDA Toolbox were written by other researchers, and they retain the copyright. References are given in Appendix B and in the **help** section of each function. Unless otherwise specified, the EDA Toolbox is provided under the GNU license specifications:

`http://www.gnu.org/copyleft/gpl.html`

3. The views expressed in this book are those of the authors and do not necessarily represent the views of the United States Department of Defense or its components.

Wendy L. and Angel R. Martinez
October 2004

Part I

Introduction to Exploratory Data Analysis

Chapter 1

Introduction to Exploratory Data Analysis

We shall not cease from exploration
And the end of all our exploring
Will be to arrive where we started
And know the place for the first time.

T. S. Eliot, "Little Gidding" (the last of his *Four Quartets*)

The purpose of this chapter is to provide some introductory and background information. First, we cover the philosophy of exploratory data analysis and discuss how this fits in with other data analysis techniques and objectives. This is followed by an overview of the text, which includes the software that will be used and the background necessary to understand the methods. We then present several data sets that will be employed throughout the book to illustrate the concepts and ideas. Finally, we conclude the chapter with some information on data transforms, which will be important in some of the methods presented in the text.

1.1 What is Exploratory Data Analysis

John W. Tukey [1977] was one of the first statisticians to provide a detailed description of *exploratory data analysis* (EDA). He defined it as "detective work - numerical detective work - or counting detective work - or graphical detective work." [Tukey, 1977, page 1] It is mostly a philosophy of data analysis where the researcher examines the data without any pre-conceived ideas in order to discover what the data can tell him about the phenomena being studied. Tukey contrasts this with **confirmatory data analysis** (CDA), an area of data analysis that is mostly concerned with statistical hypothesis testing, confidence intervals, estimation, etc. Tukey [1977] states that "Confirmatory data analysis is judicial or quasi-judicial in character." CDA methods typically involve the process of making inferences about or estimates of some population characteristic and then trying to evaluate the precision associated with the results. EDA and CDA should not be used separately from each other, but rather they should be used in a complementary way. The analyst explores the data looking for patterns and structure that leads to hypotheses and models.

Tukey's book on EDA was written at a time when computers were not widely available and the data sets tended to be somewhat small, especially by today's standards. So, Tukey developed methods that could be accomplished using pencil and paper, such as the familiar box-and-whisker plots (also known as boxplots) and the stem-and-leaf. He also included discussions of data transformation, smoothing, slicing, and others. Since this book is written at a time when computers are widely available, we go beyond what Tukey used in EDA and present computationally intensive methods for pattern discovery and statistical visualization. However, our philosophy of EDA is the same - that those engaged in it are *data detectives*.

Tukey [1980], expanding on his ideas of how exploratory and confirmatory data analysis fit together, presents a typical straight-line methodology for CDA; its steps follow:

1. State the question(s) to be investigated.
2. Design an experiment to address the questions.
3. Collect data according to the designed experiment.
4. Perform a statistical analysis of the data.
5. Produce an answer.

This procedure is the heart of the usual confirmatory process. To incorporate EDA, Tukey revises the first two steps as follows:

1. Start with some idea.
2. Iterate between asking a question and creating a design.

Forming the question involves issues such as: What can or should be asked? What designs are possible? How likely is it that a design will give a useful answer? The ideas and methods of EDA play a role in this process. In conclusion, Tukey states that EDA is an attitude, a flexibility, and some graph paper.

A small, easily read book on EDA written from a social science perspective is the one by Hartwig and Dearing [1979]. They describe the CDA mode as one that answers questions such as "Do the data confirm hypothesis XYZ?" Whereas, EDA tends to ask "What can the data tell me about relationship XYZ?" Hartwig and Dearing specify two principles for EDA: *skepticism* and *openness*. This might involve visualization of the data to look for anomalies or patterns, the use of resistant statistics to summarize the data, openness to the transformation of the data to gain better insights, and the generation of models.

Some of the ideas of EDA and their importance to teaching statistics were discussed by Chatfield [1985]. He called the topic *initial data analysis* or IDA. While Chatfield agrees with the EDA emphasis on starting with the noninferential approach in data analysis, he also stresses the need for looking at how the data were collected, what are the objectives of the analysis, and the use of EDA/IDA as part of an integrated approach to statistical inference.

Hoaglin [1982] provides a summary of EDA in the *Encyclopedia of Statistical Sciences*. He describes EDA as the "flexible searching for clues and evidence" and confirmatory data analysis as "evaluating the available evidence." In his summary, he states that EDA encompasses four themes: resistance, residuals, re-expression and display.

Resistant data analysis pertains to those methods where an arbitrary change in a data point or small subset of the data yields a small change in the result. A related idea is *robustness*, which has to do with how sensitive an analysis is to departures from the assumptions of an underlying probabilistic model.

Residuals are what we have left over after a summary or fitted model has been subtracted out. We can write this as

$$residual = data - fit.$$

The idea of examining residuals is common practice today. Residuals should be looked at carefully for lack of fit, heteroscedasticity (nonconstant variance), nonadditivity, and other interesting characteristics of the data.

Re-expression has to do with the transformation of the data to some other scale that might make the variance constant, might yield symmetric residuals, could linearize the data or add some other effect. The goal of re-expression for EDA is to facilitate the search for structure, patterns, or other information.

Finally, we have the importance of *displays* or *visualization* techniques for EDA. As we described previously, the displays used most often by early practitioners of EDA included the stem-and-leaf plots and boxplots. The use of scientific and statistical visualization is fundamental to EDA, because often the only way to discover patterns, structure or to generate hypotheses is by visual transformations of the data.

Given the increased capabilities of computing and data storage, where massive amounts of data are collected and stored simply because we can do so and not because of some designed experiment, questions are often generated *after* the data have been collected [Hand, Mannila and Smyth, 2001; Wegman, 1988]. Perhaps there is an evolution of the concept of EDA in the making and the need for a new philosophy of data analysis.

1.2 Overview of the Text

This book is divided into two main sections: pattern discovery and graphical EDA. We first cover linear and nonlinear dimensionality reduction because sometimes structure is discovered or can only be discovered with fewer dimensions or features. We include some classical techniques such as principal component analysis, factor analysis, and multidimensional scaling, as well as some of the more recent computationally intensive methods like self-organizing maps, locally linear embedding, isometric feature mapping, and generative topographic maps.

Searching the data for insights and information is fundamental to EDA. So, we describe several methods that 'tour' the data looking for interesting structure (holes, outliers, clusters, etc.). These are variants of the grand tour and projection pursuit that try to look at the data set in many 2-D or 3-D views in the hope of discovering something interesting and informative.

Clustering or unsupervised learning is a standard tool in EDA and data mining. These methods look for groups or clusters, and some of the issues that must be addressed involve determining the number of clusters and the validity or strength of the clusters. Here we cover some of the classical methods such as hierarchical clustering and k-means. We also devote an entire chapter to a newer technique called model-based clustering that includes a way to determine the number of clusters and to assess the resulting clusters.

Evaluating the relationship between variables is an important subject in data analysis. We do not cover the standard regression methodology; it is assumed that the reader already understands that subject. Instead, we include a chapter on scatterplot smoothing techniques such as loess.

The second section of the book discusses many of the standard techniques of visualization for EDA. The reader will note, however, that graphical techniques, by necessity, are used throughout the book to illustrate ideas and concepts.

In this section, we provide some classic, as well as some novel ways of visualizing the results of the cluster process, such as dendrograms, treemaps, rectangle plots, and ReClus. These visualization techniques can be used to assess the output from the various clustering algorithms that were covered in the first section of the book. Distribution shapes can tell us important things about the underlying phenomena that produced the data. We will look at ways to determine the shape of the distribution by using boxplots, bagplots, q-q plots, histograms, and others.

Finally, we present ways to visualize multivariate data. These include parallel coordinate plots, scatterplot matrices, glyph plots, coplots, dot charts, and Andrews' curves. The ability to interact with the plot to uncover structure or patterns is important, and we present some of the standard methods such as linking and brushing. We also connect both sections by revisiting the idea of the grand tour and show how that can be implemented with Andrews' curves and parallel coordinate plots.

We realize that other topics can be considered part of EDA, such as descriptive statistics, outlier detection, robust data analysis, probability density estimation, and residual analysis. However, these topics are beyond the scope of this book. Descriptive statistics are covered in introductory statistics texts, and since we assume that readers are familiar with this subject matter, there is no need to provide explanations here. Similarly, we do not emphasize residual analysis as a stand-alone subject, mostly because this is widely discussed in other books on regression and multivariate analysis.

We do cover some density estimation, such as model-based clustering (Chapter 6) and histograms (Chapter 9). The reader is referred to Scott [1992] for an excellent treatment of the theory and methods of multivariate density estimation in general or Silverman [1986] for kernel density estimation. For more information on MATLAB implementations of density estimation the reader can refer to Martinez and Martinez [2002]. Finally, we will likely encounter outlier detection as we go along in the text, but this topic, along with robust statistics, will not be covered as a stand-alone subject. There are several books on outlier detection and robust statistics. These include Hoaglin, Mosteller and Tukey [1983], Huber [1981], and Rousseeuw and Leroy [1987]. A rather dated paper on the topic is Hogg [1974].

We use MATLAB® throughout the book to illustrate the ideas and to show how they can be implemented in software. Much of the code used in the examples and to create the figures is freely available, either as part of the downloadable toolbox included with the book or on other internet sites. This information will be discussed in more detail in Appendix B. For MATLAB product information, please contact:

> The MathWorks, Inc.
> 3 Apple Hill Drive
> Natick, MA, 01760-2098 USA
> Tel: 508-647-7000
> Fax: 508-647-7101
> E-mail: info@mathworks.com
> Web: www.mathworks.com

It is important for the reader to understand what versions of the software or what toolboxes are used with this text. The book was written using MATLAB Versions 6.5 and 7. We made some use of the MATLAB Statistics Toolbox, Versions 4 and 5. We will refer to the Curve Fitting Toolbox in Chapter 7, where we discuss smoothing. However, this particular toolbox is not needed to use the examples in the book.

To get the most out of this book, readers should have a basic understanding of matrix algebra. For example, one should be familiar with determinants, a matrix transpose, the trace of a matrix, etc. We recommend Strang [1988,

1993] for those who need to refresh their memories on the topic. We do not use any calculus in this book, but a solid understanding of algebra is always useful in any situation. We expect readers to have knowledge of the basic concepts in probability and statistics, such as random samples, probability distributions, hypothesis testing, and regression.

1.3 A Few Words About Notation

In this section, we explain our notation and font conventions. MATLAB code will be in Courier New bold font such as this: **function**. To make the book more readable, we will indent MATLAB code when we have several lines of code, and this can always be typed in as you see it in the book.

For the most part, we follow the convention that a vector is arranged as a column, so it has dimensions $p \times 1$.[1] Our data sets will always be arranged in a matrix of dimension $n \times p$, which is denoted as $\mathbf{X}$. Here n represents the number of observations we have in our sample, and p is the number of variables or dimensions. Thus, each row corresponds to a p-dimensional observation or data point. The ij-th element of $\mathbf{X}$ will be represented by x_{ij}. For the most part, the subscript i refers to a row in a matrix or an observation, and a subscript j references a column in a matrix or a variable. What is meant by this will be clear from the text.

In many cases, we might need to center our observations before we analyze them. To make the notation somewhat simpler later on, we will use the matrix $\mathbf{X}_c$ to represent our centered data matrix, where each row is now centered at the origin. We calculate this matrix by first finding the mean of each column of $\mathbf{X}$ and then subtracting it from each row. The following code will calculate this in MATLAB:

```
% Find the mean of each column.
[n,p] = size(X);
xbar = mean(X);
% Create a matrix where each row is the mean
% and subtract from X to center at origin.
Xc = X - repmat(xbar,n,1);
```

---

[1] The notation $m \times n$ is read "m by n," and it means that we have m rows and n columns in an array. It will be clear from the context whether this indicates matrix dimensions or multiplication.

1.4 Data Sets Used in the Book

In this section, we describe the main data sets that will be used throughout the text. Other data sets will be used in the exercises and in some of the examples. This section can be set aside and read as needed without any loss of continuity. Please see Appendix C for detailed information on all data sets included with the text.

1.4.1 Unstructured Text Documents

The ability to analyze free-form text documents (e.g., Internet documents, intelligence reports, news stories, etc.) is an important application in computational statistics. We must first encode the documents in some numeric form in order to apply computational methods. The usual way this is accomplished is via a term-document matrix, where each row of the matrix corresponds to a word in the lexicon, and each column represents a document. The elements of the term-document matrix contain the number of times the i-th word appears in j-th document [Manning and Schütze, 2000; Charniak, 1996]. One of the drawbacks to this type of encoding is that the order of the words is lost, resulting in a loss of information [Hand, Mannila and Smyth, 2001].

We now present a new method for encoding unstructured text documents where the order of the words is accounted for. The resulting structure is called the bigram proximity matrix (BPM).

Bigram Proximity Matrices

The **bigram proximity matrix** (BPM) is a nonsymmetric matrix that captures the number of times word pairs occur in a section of text [Martinez and Wegman, 2002a; 2002b]. The BPM is a square matrix whose column and row headings are the alphabetically ordered entries of the lexicon. Each element of the BPM is the number of times word i appears immediately before word j in the unit of text. The size of the BPM is determined by the size of the lexicon created by alphabetically listing the unique occurrences of the words in the corpus. In order to assess the usefulness of the BPM encoding we had to determine whether or not the representation preserves enough of the semantic content to make them separable from BPMs of other thematically unrelated collections of documents.

We must make some comments about the lexicon and the pre-processing of the documents before proceeding with more information on the BPM and the data provided with this book. All punctuation *within* a sentence, such as commas, semi-colons, colons, etc., were removed. All end-of-sentence punctuation, other than a period, such as question marks and exclamation points were converted to a period. The period is used in the lexicon as a word, and it is placed at the beginning of the alphabetized lexicon.

Other pre-processing issues involve the removal of noise words and stemming. Many natural language processing applications use a shorter version of the lexicon by excluding words often used in the language [Kimbrell, 1988; Salton, Buckley and Smith, 1990; Frakes and Baeza-Yates, 1992; Berry and Browne, 1999]. These words, usually called *stop words*, are said to have low informational content and thus, in the name of computational efficiency, are deleted. Not all agree with this approach [Witten, Moffat and Bell, 1994].

Taking the denoising idea one step further, one could also stem the words in the denoised text. The idea is to reduce words to their stem or root to increase the frequency of key words and thus enhance the discriminatory capability of the features. Stemming is routinely applied in the area of information retrieval (IR). In this application of text processing, stemming is used to enhance the performance of the IR system, as well as to reduce the total number of unique words and save on computational resources. The stemmer we used to pre-process the text documents is the Porter stemmer [Baeza-Yates and Ribero-Neto, 1999; Porter, 1980]. The Porter stemmer is simple; however, its performance is comparable with older established stemmers.

We are now ready to give an example of the BPM. The BPM for the sentence or text stream,

> *"The wise young man sought his father in the crowd."*

is shown in Table 1.1. We see that the matrix element located in the third row (*his*) and the fifth column (*father*) has a value of one. This means that the pair of words *his father* occurs once in this unit of text. It should be noted that in most cases, depending on the size of the lexicon and the size of the text stream, the BPM will be very sparse.

TABLE 1.1

Example of a BPM

	.	crowd	his	in	father	man	sought	the	wise	young
.										
crowd	1									
his					1					
in								1		
father				1						
man							1			
sought			1							
the		1							1	
wise										1
young						1				

Note that the zeros are left out for ease of reading.

By preserving the word ordering of the discourse stream, the BPM captures a substantial amount of information about meaning. Also, by obtaining the individual counts of word co-occurrences, the BPM captures the 'intensity' of the discourse's theme. Both features make the BPM a suitable tool for capturing meaning and performing computations to identify semantic similarities among units of discourse (e.g., paragraphs, documents). Note that a BPM is created for each text unit.

One of the data sets included in this book, which was obtained from text documents, came from the Topic Detection and Tracking (TDT) Pilot Corpus (Linguistic Data Consortium, Philadelphia, PA):

http://www.ldc.upenn.edu/Projects/TDT-Pilot/.

The TDT corpus is comprised of close to 16,000 stories collected from July 1, 1994 to June 30, 1995 from the Reuters newswire service and CNN broadcast news transcripts. A set of 25 events are discussed in the complete TDT Pilot Corpus. These 25 topics were determined first, and then the stories were classified as either belonging to the topic, not belonging, or somewhat belonging (*Yes*, *No*, or *Brief*, respectively).

In order to meet the computational requirements of available computing resources, a subset of the TDT corpus was used. A total of 503 stories were chosen that includes 16 of the 25 events. See Table 1.2 for a list of topics. The 503 stories chosen contain only the *Yes* or *No* classifications. This choice stems from the need to demonstrate that the BPM captures enough meaning to make a correct or incorrect topic classification choice.

TABLE 1.2

List of 16 Topics

Topic Number	Topic Description	Number of Documents Used
4	Cessna on the White House	14
5	Clinic Murders (Salvi)	41
6	Comet into Jupiter	44
8	Death of N. Korean Leader	35
9	DNA in OJ Trial	29
11	Hall's Copter in N. Korea	74
12	Humble, TX Flooding	16
13	Justice-to-be Breyer	8
15	Kobe, Japan Quake	49
16	Lost in Iraq	30
17	NYC Subway Bombing	24
18	Oklahoma City Bombing	76
21	Serbians Down F-16	16
22	Serbs Violate Bihac	19
24	US Air 427 Crash	16
25	WTC Bombing Trial	12

There were 7,146 words in the lexicon after denoising and stemming, so each BPM has $7,146^2$ elements. This is very high dimensional data ($7,146^2$ dimensions). We can apply several EDA methods that require the interpoint distance matrix only and not the original data (i.e., BPMs). Thus, we only include the interpoint distance matrices for different measures of semantic distance: IRad, Ochiai, simple matching, and L_1. It should be noted that the match and Ochiai measures started out as similarities (large values mean the observations are similar), and were converted to distances for use in the text. See Appendix A for more information on these distances and Martinez [2002] for other choices, not included here. Table 1.3 gives a summary of the BPM data we will be using in subsequent chapters.

TABLE 1.3

Summary of the BPM Data

Distance	Name of File
IRad	`iradbpm`
Ochiai	`ochiaibpm`
Match	`matchbpm`
L_1 Norm	`L1bpm`

One of the issues we might want to explore with these data is dimensionality reduction so further processing can be accomplished, such as clustering or supervised learning. We would also be interested in visualizing the data in some manner to determine whether or not the observations exhibit some interesting structure. Finally, we might use these data with a clustering algorithm to see how many groups are found in the data, to find latent topics or sub-groups or to see if documents are clustered such that those in one group have the same meaning.

1.4.2 Gene Expression Data

The Human Genome Project completed a map (in draft form) of the human genetic blueprint in 2001 (**http://www.nature.com/genomics/human**), but much work remains to be done in understanding the functions of the genes and the role of proteins in a living system. The area of study called *functional genomics* addresses this problem, and one of its main tools is DNA *microarray* technology [Sebastiani, et al., 2003]. This technology allows data to be collected on multiple experiments and provides a view of the genetic activity (for thousands of genes) for an organism.

We now provide a brief introduction to the terminology used in this area. The reader is referred to Sebastiani, et al. [2003] or Griffiths, et al. [2000] for more detail on the unique statistical challenges and the underlying biological and technical foundation of genetic analysis. As most of us are aware from introductory biology, organisms are made up of cells, and the nucleus of each cell contains DNA (deoxyribonucleic acid). DNA instructs the cells to produce proteins and how much protein to produce. Proteins participate in most of the functions living things perform. Segments of DNA are called *genes*. The *genome* is the complete DNA for an organism, and it contains the genetic code needed to create a unique life. The process of gene activation is called *gene expression*, and the expression level provides a value indicating the number of intermediary molecules (messenger ribonucleic acid and transfer ribonucleic acid) created in this process.

Microarray technology can simultaneously measure the relative gene expression level of thousands of genes in tissue or cell samples. There are two main types of microarray technology: cDNA microarrays and synthetic oligonucleotide microarrays. In both of these methods, a target (extracted from tissue or cell) is hybridized to a probe (genes of known identity or small sequences of DNA). The target is tagged with fluorescent dye before being hybridized to the probe, and a digital image is formed of the chemical reaction. The intensity of the signal then has to be converted to a quantitative value from the image. As one might expect, this involves various image processing techniques, and it could be a major source of error.

A data set containing gene expression levels has information on genes (rows of the matrix) from several experiments (columns of the matrix). Typically, the columns correspond to patients, tumors, time steps, etc. We note that with the analysis of gene expression data, either the rows (genes) or columns (experiments/samples) could correspond to the dimensionality (or sample size), depending on the goal of the analysis. Some of the questions that might be addressed through this technology include:

- What genes are expressed (or not expressed) in a tumor cell versus a normal cell?
- Can we predict the best treatment for a cancer?
- Are there genes that characterize a specific tumor?
- Are we able to cluster cells based on their gene expression level?
- Can we discover sub-classes of cancer or tumors?

For more background information on gene expression data, we refer the reader to Schena, et al. [1995], Chee, et al. [1996], and Lander [1999]. Many gene expression data sets are freely available on the internet, and there are also many articles on the statistical analysis of this type of data. We refer the interested reader to a recent issue of *Statistical Science* (Volume 18, Number 1, February 2003) for a special section on microarray analysis. One can also go to the *Proceedings of the National Academy of Science* website (**http://www.pnas.org**) for articles, many of which have the data available for download. We include three gene expression data sets with this book, and we describe them below.

Yeast Data Set

This data set was originally described in Cho, et al. [1998], and it showed the gene expression levels of around 6000 genes over two cell cycles and five phases. The two cell cycles provide 17 time points (columns of the matrix). The subset of the data we provide was obtained by Yeung and Ruzzo [2001] and is available at

http://www.cs.washington.edu/homes/kayee/model.

A full description of the process they used to get the subset can also be found there. First, they extracted all genes that were found to peak in only one of the five phases; those that peaked in multiple phases were not used. Then they removed any rows with negative entries, yielding a total of 384 genes.

The data set is called **yeast.mat**, and it contains two variables: **data** and **classlabs**. The **data** matrix has 384 rows and 17 columns. The variable **classlabs** is a vector containing 384 class labels for the genes indicating whether the gene peaks in phase 1 through phase 5.

Leukemia Data Set

The **leukemia** data set was first discussed in Golub, et al., [1999], where the authors measured the gene expressions of human acute leukemia. Their study included prediction of the type of leukemia using supervised learning and the discovery of new classes of leukemia via unsupervised learning. The motivation for this work was to improve cancer treatment by distinguishing between sub-classes of cancer or tumors. The author's website

http://www.genome.wi.mit.edu/MPR

has more information about their methods and procedure, and the full data set is available there.

They first classified the leukemias into two groups: (1) those that arise from lymphoid precursors or (2) from myeloid precursors. The first one is called acute lymphoblastic leukemia (ALL), and the second is called acute myeloid leukemia (AML). The distinction between these two classes of leukemia is well known, but a single test to sufficiently establish a diagnosis does not exist [Golub, et al., 1999]. As one might imagine, a proper diagnosis is critical to successful treatment and to avoid unnecessary toxicities. The authors turned to microarray technology and statistical pattern recognition to address this problem.

Their initial data set had 38 bone marrow samples taken at the time of diagnosis; 27 came from patients with ALL, and 11 patients had AML. They used oligonucleotide microarrays containing probes for 6,817 human genes to obtain the gene expression information. Their first goal was to construct a classifier using the gene expression values that would predict the type of leukemia. So, one could consider this as building a classifier where the observations have 6,817 dimensions, and the sample size is 38. They had to reduce the dimensionality, so they chose the 50 genes that have the highest correlation with the class of leukemia. They used an independent test set of leukemia samples to evaluate the classifier. This set of data consists of 34 samples, where 24 of them came from bone marrow and 10 came from peripheral blood samples. It also included samples from children and from different laboratories using different protocols.

They also looked at class discovery or unsupervised learning, where they wanted to see if the patients could be clustered into two groups corresponding to the types of leukemia. They used the method called self-organizing maps (Chapter 3), employing the full set of 6,817 genes. Another aspect of class discovery is to look for subgroups within known classes. For example, the patients with ALL can be further subdivided into patients with B-cell or T-cell lineage.

We decided to include only the 50 genes, rather than the full set. The **leukemia.mat** file has four variables. The variable **leukemia** has 50 genes (rows) and 72 patients (columns). The first 38 columns correspond to the initial training set of patients, and the rest of the columns contain data for the independent testing set. The variables **btcell** and **cancertype** are cell arrays of strings containing the label for B-cell, T-cell, or NA and ALL or AML, respectively. Finally, the variable **geneinfo** is a cell array where the first column provides the gene description, and the second column contains the gene number.

Example 1.1

We show a plot of the 50 genes in Figure 1.1, but only the first 38 samples (i.e., columns) are shown. This is similar to Figure 3B in Golub, et al., [1999]. We standardized each gene, so the mean across each row is 0 and the standard deviation is 1. The first 27 columns of the picture correspond to ALL leukemia, and the last 11 columns pertain to the AML leukemia. We can see by the color that the first 25 genes tend to be more highly expressed in ALL, while the last 25 genes are highly expressed in AML. The MATLAB code to construct this plot is given below.

```
% First standardize the data such that each row
% has mean 0 and standard deviation 1.
load leukemia
x = leukemia(:,1:38);
[n,p] = size(x);
y = zeros(n,p);
for i = 1:n
    sig = std(x(i,:));
    mu = mean(x(i,:));
    y(i,:)= (x(i,:)-mu)/sig;
end
% Now do the image of the data.
pcolor(y)
colormap(gray(256))
```

```
colorbar
title('Gene Expression for Leukemia')
xlabel('ALL (1-27) or AML (28-38)')
ylabel('Gene')
```

The results shown in Figure 1.1 indicate that we might be able to distinguish between AML and ALL leukemia using these data.
❑

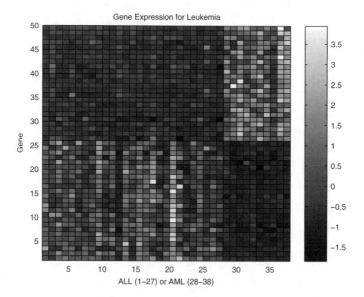

FIGURE 1.1

This shows the gene expression for the **leukemia** data set. Each row corresponds to a gene, and each column corresponds to a cancer sample. The rows have been standardized such that the mean is 0 and the standard deviation is 1. We can see that the ALL leukemia is highly expressed in the first set of 25 genes, and the AML leukemia is highly expressed in the second set of 25 genes.

Lung Data Set

Traditionally, the classification of lung cancer is based on clinicopathological features. An understanding of the molecular basis and a possible molecular classification of lung carcinomas could yield better therapies targeted to the type of cancer, superior prediction of patient treatment, and the identification of new targets for chemotherapy. We provide two data sets that were originally downloaded from **http://www.genome.mit.edu/MPR/lung** and described in Bhattacharjee, et al. [2001]. The authors applied hierarchical and probabilistic clustering to find subclasses of lung adenocarcinoma, and they showed the diagnostic potential of analyzing gene expression data by demonstrating the ability to separate primary lung adenocarcinomas from metastases of extra-pulmonary origin.

A preliminary classification of lung carcinomas comprises two groups: small-cell lung carcinomas (SCLC) or nonsmall-cell lung carcinomas (NSCLC). The NSCLC category can be further subdivided into 3 groups: adenocarcinomas (AD), squamous cell carcinomas (SQ), and large-cell carcinomas (COID). The most common type is adenocarcinomas. The data were obtained from 203 specimens, where 186 were cancerous and 17 were normal lung. The cancer samples contained 139 lung adenocarcinomas, 21 squamous cell lung carcinomas, 20 pulmonary carcinoids, and 6 small-cell lung carcinomas. This is called Dataset A in Bhattacharjee, et al. [2001]; the full data set included 12,600 genes. The authors reduced this to 3,312 by selecting the most variable genes, using a standard deviation threshold of 50 expression units. We provide these data in **lungA.mat**. This file includes two variables: **lungA** and **labA**. The variable **lungA** is a 3312 × 203 matrix, and **labA** is a vector containing the 203 class labels.

The authors also looked at adenocarcinomas separately trying to discover subclasses. To this end, they separated the 139 adenocarcinomas and the 17 normal samples and called it Dataset B. They also took fewer gene transcript sequences for this data set by selecting only 675 genes according to other statistical pre-processing steps. These data are provided in **lungB.mat**, which contains two variables: **lungB** (675 × 156) and **labB** (156 class labels). We summarize these data sets in Table 1.4.

TABLE 1.4

Description of Lung Cancer Data Set

Cancer Type	Label	Number of Data Points
*Dataset A (**lungA.mat**): 3,312 rows, 203 columns*		
Nonsmall cell lung carcinomas		
Adenocarcinomas	AD	139
Pulmonary carcinoids	COID	20
Squamous cell	SQ	21
Normal	NL	17
Small-cell lung carcinomas	SCLC	6
*Dataset B (**lungB.mat**): 675 rows, 156 columns*		
Adenocarcinomas	AD	139
Normal	NL	17

For those who need to analyze gene expression data, we recommend the Bioinformatics Toolbox from The MathWorks. The toolbox provides an integrated environment for solving problems in genomics and proteomics, genetic engineering, and biological research. Some capabilities include the ability to calculate the statistical characteristics of the data, to manipulate sequences, to construct models of biological sequences using Hidden Markov Models, and to visualize microarray data.

1.4.3 Oronsay Data Set

This data set consists of particle size measurements originally presented in Timmins [1981] and analyzed by Olbricht [1982], Fieller, Gilbertson & Olbricht [1984], and Fieller, Flenley and Olbricht [1992]. An extensive analysis from a graphical EDA point of view was conducted by Wilhelm, Wegman and Symanzik [1999]. The measurement and analysis of particle sizes is often used in archaeology, fuel technology (droplets of propellant), medicine (blood cells), and geology (grains of sand). The usual objective is to determine the distribution of particle sizes because this characterizes the environment where the measurements were taken or the process of interest.

The Oronsay particle size data were gathered for a geological application, where the goal was to discover different characteristics between dune sands and beach sands. This characterization would be used to determine whether or not midden sands were dune or beach. The middens were near places where prehistoric man lived, and geologists are interested in whether these middens were beach or dune because that would be an indication of how the coastline has shifted.

There are 226 samples of sand, with 77 belonging to an unknown type of sand (from the middens) and 149 samples of known type (beach or dune). The known samples were taken from *Cnoc Coig* (CC - 119 observations, 90 beach and 29 dune) and *Caisteal nan Gillean* (CG - 30 observations, 20 beach and 10 dune). See Wilhelm, Wegman and Symanzik [1999] for a map showing these sites on Oronsay island. This reference also shows a more detailed classification of the sands based on transects and levels of sand.

Each observation is obtained in the following manner. Approximately 60g or 70g of sand is put through a stack of 11 sieves of sizes 0.063mm, 0.09mm, 0.125mm, 0.18mm, 0.25mm, 0.355mm, 0.5mm, 0.71mm, 1.0mm, 1.4mm, and 2.0mm. The sand that remains on each of the sieves is weighed, along with the sand that went through completely. This yields 12 weight measurements, and each corresponds to a class of particle size. Note that there are two extreme classes: particle sizes less than 0.063mm (what went through the smallest sieve) and particle sizes larger than 2.0mm (what is in the largest sieve).

Flenley and Olbricht [1993] consider the classes as outlined above, and they apply various multivariate and exploratory data analysis techniques such as principal component analysis and projection pursuit. The **oronsay** data set was downloaded from:

http://www.galaxy.gmu.edu/papers/oronsay.html

More information on the original data set can be found at this website. We chose to label observations first with respect to midden, beach or dune (in variable **beachdune**):

- Class 0: midden (77 observations)
- Class 1: beach (110 observations)

- Class 2: dune (39 observations)

We then classify observations according to the sampling site (in variable **midden**), as follows

- Class 0: midden (77 observations)
- Class 1: *Cnoc Coig* - CC (119 observations)
- Class 2: *Caisteal nan Gillean* - CG (30 observations)

The data set is in the **oronsay.mat** file. The data are in a 226 × 12 matrix called **oronsay**, and the data are in raw format; i.e., untransformed and unstandardized. Also included is a cell array of strings called **labcol** that contains the names (i.e., sieve sizes) of the columns.

1.4.4 Software Inspection

The data described in this section were collected in response to efforts for process improvement in software testing. Many systems today rely on complex software that might consist of several modules programmed by different programmers, so ensuring that the software works correctly and as expected is important.

One way to test the software is by inspection, where software engineers inspect the code in a formal way. First they look for inconsistencies, logical errors, etc., and then they all meet with the programmer to discuss what they perceive as defects. The programmer is familiar with the code and can help determine whether or not it is a defect in the software.

The data are saved in a file called **software**. The variables are normalized by the size of the inspection (the number of pages or SLOC – single lines of code). The file **software.mat** contains the preparation time in minutes (**prepage, prepsloc**), the total work hours in minutes for the meeting (**mtgsloc**), and the number of defects found (**defpage, defsloc**). Software engineers and managers would be interested in understanding the relationship between the inspection time and the number of defects found. One of the goals might be to find an optimal time for inspection, where one gets the most payoff (number of defects found) for the amount of time spent reviewing the code. We show an example of these data in Figure 1.2. The defect types include compatibility, design, human-factors, standards, and others.

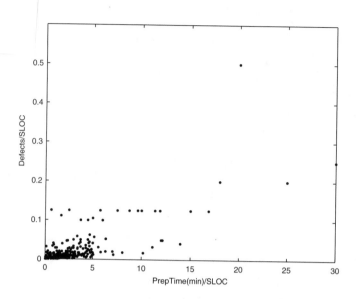

FIGURE 1.2

This is a scatterplot of the software inspection data. The relationship between the variables is difficult to see.

1.5 Transforming Data

In many real-world applications, the data analyst will have to deal with raw data that are not in the most convenient form. The data might need to be re-expressed to produce effective visualization or an easier, more informative analysis. Some of the types of problems that can arise include data that exhibit nonlinearity or asymmetry, contain outliers, change spread with different levels, etc. We can transform the data by applying a single mathematical function to all of the observations.

In the first sub-section below, we discuss the general power transformations that can be used to change the shape of the data distribution. This arises in situations when we are concerned with formal inference methods where the shape of the distribution is important (e.g., statistical hypothesis testing or confidence intervals). In EDA, we might want to change the shape to facilitate visualization, smoothing, and other analyses. Next we cover linear transformations of the data that leave the shape alone. These are typically changes in scale and origin and can be important in dimensionality reduction, clustering, and visualization.

1.5.1 Power Transformations

A *transformation* of a set of data points $x_1, x_2, ..., x_n$ is a function T that substitutes each observation x_i with a new value $T(x_i)$ [Emerson and Stoto, 1983]. Transformations should have the following desirable properties:

1. The order of the data is preserved by the transformation. Because of this, statistics based on order, such as medians are preserved; i.e., medians are transformed to medians.

2. They are continuous functions guaranteeing that points that are close together in raw form are also close together using their transformed values, relative to the scale used.

3. They are smooth functions that have derivatives of all orders, and they are specified by elementary functions.

Some common transformations include taking roots (square root, cube root, etc.), finding reciprocals, calculating logarithms, and raising variables to positive integral powers. These transformations provide adequate flexibility for most situations in data analysis.

Example 1.2

This example uses the software inspection data shown in Figure 1.2. We see that the data are skewed, and the relationship between the variables is difficult to understand. We apply a log transform to both variables using the following MATLAB code, and show the results in Figure 1.3.

```
load software
% First transform the data.
X = log(prepsloc);
Y = log(defsloc);
% Plot the transformed data.
plot(X,Y,'.')
xlabel('Log PrepTime/SLOC')
ylabel('Log Defects/SLOC')
```

We now have a better idea of the relationship between these two variables, which will be examined further in Chapter 7.
❑

Some transformations of the data may lead to insights or discovery of structures that we might not otherwise see. However, as with any analysis, we should be careful about creating something that is not really there, but is just an artifact of the processing. Thus, in any application of EDA, the analyst should go back to the subject area and consult domain experts to verify and help interpret the results.

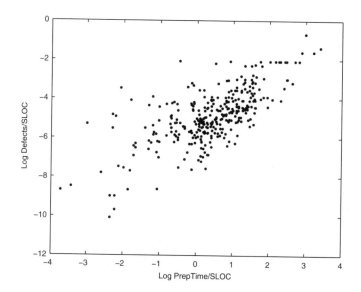

FIGURE 1.3

Each variate was transformed using the logarithm. The relationship between preparation time per SLOC and number of defects found per SLOC is now easier to see.

1.5.2 Standardization

If the variables are measurements along a different scale or if the standard deviations for the variables are different from one another, then one variable might dominate the distance (or some other similar calculation) used in the analysis. We will make extensive use of interpoint distances throughout the text in applications such as clustering, multidimensional scaling, and nonlinear dimensionality reduction. We discuss several 1-D standardization methods below. However, we note that in some multivariate contexts, the 1-D transformations may be applied to each variable (i.e., on the column of **X**) separately.

Transformation Using the Standard Deviation

The first standardization we discuss is called the sample *z-score*, and it should be familiar to most readers who have taken an introductory statistics class. The transformed variates are found using

$$z = \frac{(x - \bar{x})}{s}, \tag{1.1}$$

where x is the original observed data value, $\bar{x}$ is the sample mean, and s is the sample standard deviation. In this standardization, the new variate z will have a mean of zero and a variance of one.

When the z-score transformation is used in a clustering context, it is important that it be applied in a global manner across all observations. If standardization is done within clusters, then false and misleading clustering solutions can result [Milligan and Cooper, 1988].

If we do not center the data at zero by removing the sample mean, then we have the following

$$z = \frac{x}{s}. \qquad (1.2)$$

This transformed variable will have a variance of one and a transformed mean equal to $\bar{x}/s$. The standardizations in Equations 1.1 and 1.2 are linear functions of each other, so Euclidean distances (see Appendix A) calculated on data that have been transformed using the two formulas result in identical dissimilarity values.

For robust versions of Equations 1.1 and 1.2, we can substitute the median and the interquartile range for the sample mean and sample standard deviation respectively. This will be explored in the exercises.

Transformation Using the Range

Instead of dividing by the standard deviation, as above, we can use the range of the variable as the divisor. This yields the following two forms of standardization

$$z = \frac{x}{\max(x) - \min(x)}, \qquad (1.3)$$

and

$$z = \frac{x - \min(x)}{\max(x) - \min(x)}. \qquad (1.4)$$

The standardization in Equation 1.4 is bounded by zero and one, with at least one observed value at each of the end points. The transformed variate given by Equation 1.3 is a linear function of the one determined by Equation 1.4, so data standardized using these transformations will result in identical Euclidean distances.

1.5.3 Sphering the Data

This type of standardization called **sphering** pertains to multivariate data, and it serves a similar purpose as the 1-D standardization methods given above. The transformed variables will have a p-dimensional mean of $\mathbf{0}$ and a covariance matrix given by the identity matrix.

We start off with the p-dimensional sample mean given by

$$\bar{\mathbf{x}} = \frac{1}{n} \sum_{i=1}^{n} \mathbf{x}_i.$$

We then find the sample covariance matrix given by the following

$$\mathbf{S} = \frac{1}{n-1} \sum_{i=1}^{n} (\mathbf{x}_i - \bar{\mathbf{x}})(\mathbf{x}_i - \bar{\mathbf{x}})^T,$$

where we see that the covariance matrix can be written as the sum of n matrices. Each of these rank one matrices is the outer product of the centered observations [Duda and Hart, 1973].

We sphere the data using the following transformation

$$\mathbf{Z}_i = \Lambda^{-1/2} \mathbf{Q}^T (\mathbf{x}_i - \bar{\mathbf{x}}) \qquad i = 1, \ldots, n,$$

where the columns of $\mathbf{Q}$ are the eigenvectors obtained from $\mathbf{S}$, Λ is a diagonal matrix of corresponding eigenvalues, and $\mathbf{x}_i$ is the i-th observation.

Example 1.3

We now provide the MATLAB code to obtain sphered data. First, we generate 2-D multivariate normal random variables that have the following parameters:

$$\mu = \begin{bmatrix} -2 \\ 2 \end{bmatrix},$$

and

$$\Sigma = \begin{bmatrix} 1 & 0.5 \\ 0.5 & 1 \end{bmatrix},$$

where Σ is the covariance matrix. A scatterplot of these data is shown in Figure 1.4 (top).

```
% First generate some 2-D multivariate normal
% random variables, with mean MU and
% covariance SIGMA. This uses a Statistics
% Toolbox function.
n = 100;
mu = [-2, 2];
sigma = [1,.5;.5,1];
X = mvnrnd(mu,sigma,n);
plot(X(:,1),X(:,2),'.')
```

We now apply the steps to sphere the data, and show the transformed data in Figure 1.4 (bottom).

```
% Now sphere the data.
xbar = mean(X);
% Get the eigenvectors and eigenvalues of the
% covariance matrix.
[V,D] = eig(cov(X));
% Center the data.
Xc = X - ones(n,1)*xbar;
% Sphere the data.
Z = ((D)^(-1/2)*V'*Xc')';
plot(Z(:,1),Z(:,2),'.')
```

By comparing these two plots, we see that the transformed data are sphered and are now centered at the origin.
❑

1.6 Further Reading

There are several books that will provide the reader with more information and other perspectives on EDA. Most of these books do not offer software and algorithms, but they are excellent resources for the student or practitioner of exploratory data analysis.

As we stated in the beginning of this chapter, the seminal book on EDA is Tukey [1977], but the text does not include the more up-to-date view based on current computational resources and methodology. Similarly, the short book on EDA by Hartwig and Dearing [1979] is an excellent introduction to the topic and a quick read, but it is somewhat dated. For the graphical approach, the reader is referred to du Toit, Steyn and Stumpf [1986], where the authors use SAS to illustrate the ideas. They include other EDA methods such as multidimensional scaling and cluster analysis. Hoaglin, Mosteller

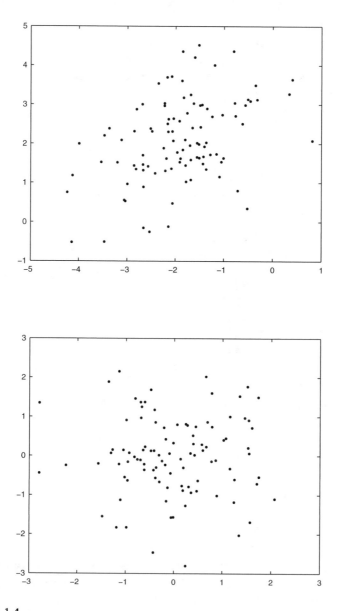

FIGURE 1.4
The top figure shows a scatterplot of the 2-D multivariate normal random variables. Note that these are not centered at the origin, and the cloud is not spherical. The sphered data are shown in the bottom panel. We see that they are now centered at the origin with a spherical spread. This is similar to the z-score standardization in 1-D.

and Tukey [1983] edited an excellent book on robust and exploratory data analysis. It includes several chapters on transforming data, and we recommend the one by Emerson and Stoto [1983]. The chapter includes a discussion of power transformations, as well as plots to assist the data analyst in choosing an appropriate one.

For a more contemporary resource that explains data mining approaches, of which EDA is a part, Hand, Mannila and Smyth [2001] is highly recommended. It does not include computer code, but it is very readable. The authors cover the major parts of data mining: EDA, descriptive modeling, classification and regression, discovering patterns and rules, and retrieval by content. Finally, the reader could also investigate the book by Hastie, Tibshirani and Friedman [2001]. These authors cover a wide variety of topics of interest to exploratory data analysts, such as clustering, nonparametric probability density estimation, multidimensional scaling, and projection pursuit.

As was stated previously, EDA is sometimes defined as an attitude of flexibility and discovery in the data analysis process. There is an excellent article by Good [1982] outlining the philosophy of EDA, where he states that "EDA is more an art, or even a bag of tricks, than a science." While we do not think there is anything "tricky" about the EDA techniques, it is somewhat of an art in that the analyst must try various methods in the discovery process, keeping an open mind and being prepared for surprises! Finally, other summaries of EDA were written by Diaconis [1985] and Weihs [1993]. Weihs describes EDA mostly from a graphical viewpoint and includes descriptions of dimensionality reduction, grand tours, prediction models, and variable selection. Diaconis discusses the difference between exploratory methods and the techniques of classical mathematical statistics. In his discussion of EDA, he considers Monte Carlo techniques such as the bootstrap [Efron and Tibshirani, 1993].

Exercises

1.1 What is exploratory data analysis? What is confirmatory data analysis? How do these analyses fit together?

1.2 Repeat Example 1.1 using the remaining columns (39 – 72) of the **leukemia** data set. Does this follow the same pattern as the others?

1.3 Repeat Example 1.1 using the **lungB** gene expression data set. Is there a pattern?

1.4 Generate some 1-D normally distributed random variables with $\mu = 5$ and $\sigma = 2$ using **normrnd** or **randn** (must transform the results to have the required mean and standard deviation if you use this function). Apply the various standardization procedures described in this chapter and verify the comments regarding the location and spread of the transformed variables.

1.5   Write MATLAB functions that implement the standardizations mentioned in this chapter.

1.6   Using the **mvnrnd** function (see Example 1.3), generate some nonstandard bivariate normal random variables. Sphere the data and verify that the resulting sphered data have mean 0 and identity covariance matrix using the MATLAB functions **mean** and **cov**.

1.7   We will discuss the quartiles and the interquartile range in Chapter 9, but for now look at the MATLAB **help** files on the **iqr** and **median** functions. We can use these robust measures of location and spread to transform our variables. Using Equations 1.1 and 1.2, substitute the median for the sample mean $\bar{x}$ and the interquartile range for the sample standard deviation s. Write a MATLAB function that does this and try it out on the same data you generated in problem 1.4.

1.8   Generate $n = 2$ normally distributed random variables. Find the Euclidean distance between the points after they have been transformed first using Equation 1.1 and then Equation 1.2. Are the distances the same? Hint: Use the **pdist** function from the Statistics Toolbox.

1.9   Repeat problem 1.8 using the standardizations given by Equations 1.3 and 1.4.

1.10 Generate $n = 100$ uniform 1-D random variables using the **rand** function. Construct a histogram using the **hist** function. Now transform the data by taking the logarithm (use the **log** function). Construct a histogram of these transformed values. Did the shape of the distribution change? Comment on the results.

1.11 Try the following transformations on the software data:

$$T(x) = \log(1 + \sqrt{x})$$
$$T(x) = \log(\sqrt{x})$$

Construct a scatterplot and compare with the results in Example 1.2.

Part II

EDA as Pattern Discovery

Chapter 2

Dimensionality Reduction - Linear Methods

In this chapter we describe several linear methods of dimensionality reduction. We first discuss some classical methods such as principal component analysis (PCA), singular value decomposition (SVD), and factor analysis. Finally, we include a brief discussion of methods for determining the intrinsic dimensionality of a data set.

2.1 Introduction

Dimensionality reduction is the process of finding a suitable lower-dimensional space in which to represent the original data. Our hope is that the alternative representation of the data will help us:

- Explore high-dimensional data with the goal of discovering structure or patterns that lead to the formation of statistical hypotheses.
- Visualize the data using scatterplots when dimensionality is reduced to 2-D or 3-D.
- Analyze the data using statistical methods, such as clustering, smoothing, probability density estimation, or classification.

One possible method for dimensionality reduction would be to just select subsets of the variables for processing and analyze them in groups. However, in some cases, that would mean throwing out a lot of useful information. An alternative would be to create new variables that are functions (e.g., linear combinations) of the original variables. The methods we describe in this book are of the second type, where we seek a mapping from the higher-dimensional space to a lower-dimensional one, while keeping information on all of the available variables. This mapping can be linear or nonlinear.

Since the methods in this chapter transform the data using projections, we take a moment to describe this concept before going on to explain PCA and SVD. A projection will be in the form of a matrix that takes the data from the original space to a lower-dimensional one. We illustrate this concept in Example 2.1.

Example 2.1

In this example, we show how projection works when our data set consists of the two bivariate points

$$\mathbf{x}_1 = \begin{bmatrix} 4 \\ 3 \end{bmatrix} \qquad \mathbf{x}_2 = \begin{bmatrix} -4 \\ 5 \end{bmatrix},$$

and we are projecting onto a line that is θ radians from the horizontal or *x*-axis. For this example, the projection matrix is given by

$$\mathbf{P} = \begin{bmatrix} (\cos\theta)^2 & \cos\theta\sin\theta \\ \cos\theta\sin\theta & (\sin\theta)^2 \end{bmatrix}.$$

The following MATLAB code is used to enter this information.

```
% Enter the data as rows of our matrix X.
X = [4 3; -4 5];
% Get theta.
theta = pi/3;
% Now obtain the projection matrix.
c2 = cos(theta)^2;
cs = cos(theta)*sin(theta);
s2 = sin(theta)^2;
P = [c2 cs; cs s2];
```

The coordinates of the projected observations are a weighted sum of the original variables, where the columns of **P** provide the weights. We project a single observation as follows

$$\mathbf{y}_i = \mathbf{P}^T\mathbf{x}_i \qquad i = 1, \dots, n \ .$$

Expanding this out shows the new *y* coordinates as a weighted sum of the *x* variables:

$$\begin{aligned} y_{i1} &= P_{11}x_{i1} + P_{21}x_{i2} \\ y_{i2} &= P_{12}x_{i1} + P_{22}x_{i2} \end{aligned} \qquad i = 1, \dots, n \ .$$

We have to use some linear algebra to project the data matrix **X** because the observations are *rows*. Taking the transpose of both sides of our projection equation above, we have

$$\mathbf{y}_i^T = (\mathbf{P}^T \mathbf{x}_i)^T$$
$$= \mathbf{x}_i^T \mathbf{P} \ .$$

Thus, we can project the data using this MATLAB code:

```
% Now project the data onto the theta-line.
% Since the data are arranged as rows in the
% matrix X, we have to use the following to
% project the data.
Xp = X*P;
plot(Xp(:,1),Xp(:,2),'o') % Plot the data.
```

This projects the data onto a 1-D subspace that is at an angle θ with the original coordinate axes. As an example of a projection onto the horizontal coordinate axis, we can use these commands:

```
% We can project onto the 1-D space given by the
% horizontal axis using the projection:
Px = [1;0];
Xpx = X*Px;
```

These data now have only one coordinate value representing the number of units along the *x*-axis. The projections are shown in Figure 2.1, where the o's represent the projection of the data onto the θ line and the stars represent the projections onto the *x*-axis.
❑

2.2 Principal Component Analysis - PCA

The main purpose of *principal component analysis* (PCA) is to reduce the dimensionality from p to d, where $d < p$, while at the same time accounting for as much of the variation in the original data set as possible. With PCA, we transform the data to a new set of coordinates or variables that are a linear combination of the original variables. In addition, the observations in the new principal component space are uncorrelated. The hope is that we can gain information and understanding of the data by looking at the observations in the new space.

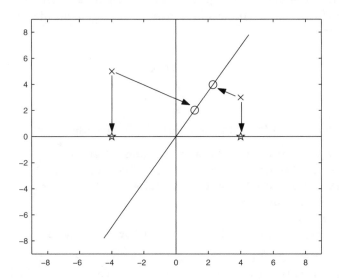

FIGURE 2.1
This figure shows the orthogonal projection of the data points to both the θ line (o's) and the
x-axis (stars).

2.2.1 PCA Using the Sample Covariance Matrix

We start with our centered data matrix $\mathbf{X}_c$ that has dimension $n \times p$. Recall
that this matrix contains observations that are centered about the mean; i.e.,
the sample mean has been subtracted from each row. We then form the
sample covariance matrix $\mathbf{S}$ as

$$\mathbf{S} = \frac{1}{n-1}\mathbf{X}_c^T\mathbf{X}_c,$$

where the superscript T denotes the matrix transpose. The jk-th element of $\mathbf{S}$
is given by

$$s_{jk} = \frac{1}{n-1}\sum_{i=1}^{n}(x_{ij} - \bar{x}_j)(x_{ik} - \bar{x}_k) \qquad j, \; k = 1, \dots, p,$$

with

$$\bar{x}_j = \frac{1}{n}\sum_{i=1}^{n}x_{ij}.$$

The next step is to calculate the eigenvectors and eigenvalues of the matrix **S**. The eigenvalues are found by solving the following equation for each l_j, $j = 1, \ldots, p$

$$|\mathbf{S} - l\mathbf{I}| = 0, \tag{2.1}$$

where **I** is a $p \times p$ identity matrix and $|\bullet|$ denotes the determinant. Equation 2.1 produces a polynomial equation of degree p. The eigenvectors are obtained by solving the following set of equations for $\mathbf{a}_j$

$$(\mathbf{S} - l_j\mathbf{I})\mathbf{a}_j = 0 \qquad j = 1, \ldots, p,$$

subject to the condition that the set of eigenvectors is orthonormal. This means that the magnitude of each eigenvector is one, and they are orthogonal to each other:

$$\mathbf{a}_i\mathbf{a}_i^T = 1$$
$$\mathbf{a}_j\mathbf{a}_i^T = 0,$$

for $i, j = 1, \ldots, p$ and $i \neq j$.

A major result in matrix algebra shows that any square, symmetric, nonsingular matrix can be transformed to a diagonal matrix using

$$\mathbf{L} = \mathbf{A}^T\mathbf{S}\mathbf{A},$$

where the columns of **A** contain the eigenvectors of **S**, and **L** is a diagonal matrix with the eigenvalues along the diagonal. By convention, the eigenvalues are ordered in descending order $l_1 \geq l_2 \geq \ldots \geq l_p$, with the same order imposed on the corresponding eigenvectors.

We use the eigenvectors of **S** to obtain new variables called *principal components* (PCs). The j-th PC is given by

$$z_j = \mathbf{a}_j^T(\mathbf{x} - \bar{\mathbf{x}}) \qquad j = 1, \ldots, p, \tag{2.2}$$

and the elements of **a** provide the weights or coefficients of the old variables in the new PC coordinate space. It can be shown that the PCA procedure defines a principal axis rotation of the original variables about their means [Jackson, 1991; Strang, 1988], and the elements of the eigenvector **a** are the direction cosines that relate the two coordinate systems. Equation 2.2 shows that the PCs are linear combinations of the original variables.

Scaling the eigenvectors to have unit length produces PCs that are uncorrelated and whose variances are equal to the corresponding eigenvalue. Other scalings are possible [Jackson, 1991], such as

$$\mathbf{v}_j = \sqrt{l_j}\,\mathbf{a}_j,$$

and

$$\mathbf{w}_j = \frac{\mathbf{a}_j}{\sqrt{l_j}}.$$

The three eigenvectors $\mathbf{a}_j$, $\mathbf{v}_j$, and $\mathbf{w}_j$ differ only by a scale factor. The $\mathbf{a}_j$ are typically needed for hypothesis testing and other diagnostic methods. Since they are scaled to unity, their components will always be between ± 1. The vectors $\mathbf{v}_j$ are useful at times, because they and their PCs have the same units as the original variables. Using $\mathbf{w}_j$ in the transformation yields PCs that are uncorrelated with unit variance.

We transform the observations to the PC coordinate system via the following

$$\mathbf{Z} = \mathbf{X}_c\mathbf{A}. \tag{2.3}$$

The matrix $\mathbf{Z}$ contains the ***principal component scores***. Note that these PC scores have zero mean (because we are using the data centered about the mean) and are uncorrelated. We could also transform the original observations in $\mathbf{X}$ by a similar transformation, but the PC scores in this case will have mean $\bar{\mathbf{z}}$. We can invert this transformation to get an expression relating the original variables as a function of the PCs, which is given by

$$\mathbf{x} = \bar{\mathbf{x}} + \mathbf{A}\mathbf{z}.$$

To summarize: the *transformed variables* are the PCs and the individual *transformed data values* are the PC scores.

The dimensionality of the principal component scores in Equation 2.3 is still p, so no dimensionality reduction has taken place. We know that the sum of the variances of the original variables is equal to the sum of the eigenvalues. The idea of dimensionality reduction with PCA is that one could include in the analysis only those PCs that have the highest eigenvalues, thus accounting for the highest amount of variation with fewer dimensions or PC variables. We can reduce the dimensionality to d with the following

$$\mathbf{Z}_d = \mathbf{X}_c\mathbf{A}_d, \tag{2.4}$$

where $\mathbf{A}_d$ contains the first d eigenvectors or columns of $\mathbf{A}$. We see that $\mathbf{Z}_d$ is an $n \times d$ matrix (each observation now has only d elements), and $\mathbf{A}_d$ is a $p \times d$ matrix.

2.2.2 PCA Using the Sample Correlation Matrix

We can scale the data first to have standard units, as described in Chapter 1. This means we have the j-th element of $\mathbf{x}^*$ given by

$$x_j^* = \frac{(x_j - \bar{x}_j)}{\sqrt{s_{jj}}}; \qquad j = 1, \ldots, p ,$$

where s_{jj} is the variance of x_j (i.e., the jj-th element of the sample covariance matrix $\mathbf{S}$). The standardized data $\mathbf{x}^*$ are then treated as observations in the PCA process.

The covariance of this standardized data set is the same as the correlation matrix. The ij-th element of the sample correlation matrix $\mathbf{R}$ is given by

$$r_{ij} = \frac{s_{ij}}{\sqrt{s_{ii}}\sqrt{s_{jj}}},$$

where s_{ij} is the ij-th element of $\mathbf{S}$ and s_{ii} is the i-th diagonal element of $\mathbf{S}$. The rest of the results from PCA hold for the $\mathbf{x}^*$. For example, we can transform the standardized data using Equation 2.3 or reduce the dimensionality using Equation 2.4, where now the matrices $\mathbf{A}$ and $\mathbf{A}_d$ contain the eigenvectors of the correlation matrix.

The correlation matrix should be used for PCA when the variances along the original dimensions are very different; i.e., if some variables have variances that are very much greater than the others. In this case, the first few PCs will be dominated by those same variables and will not be very informative. This is often the case when the variables are of different types or units. Another benefit of using the correlation matrix rather than the covariance matrix arises when one wants to compare the results of PCA among different analyses.

PCA based on covariance matrices does have certain advantages, too. Methods for statistical inference based on the sample PCs from covariance matrices are easier and are available in the literature. The PCs obtained from the correlation and covariance matrices do not provide equivalent information. Additionally, the eigenvectors and eigenvalues from one process do not have simple relationships or correspondence with those from the other one [Jolliffe, 1986]. Since this text is primarily concerned with exploratory data analysis, not inferential methods, we do not discuss this further. In any event, in the spirit of EDA, the analyst should take advantage of both methods to describe and to explore the data.

2.2.3 How Many Dimensions Should We Keep?

One of the key questions at this point is how many PCs to keep. We offer the following possible ways to address this question; more details and options (such as hypothesis tests for equality of eigenvalues, cross-validation, and correlation procedures) can be found in Jackson [1991]. All of the following techniques will be explored in Example 2.2.

Cumulative Percentage of Variance Explained

This is a popular method of determining the number of PCs to use in PCA dimensionality reduction and seems to be implemented in many computer packages. The idea is to select those d PCs that contribute a cumulative percentage of total variation in the data, which is calculated using

$$t_d = 100 \sum_{i=1}^{d} l_i \div \sum_{j=1}^{p} l_j.$$

If the correlation matrix is used for PCA, then this is simplified to

$$t_d = \frac{100}{p} \sum_{i=1}^{d} l_i.$$

Choosing a value for t_d can be problematic, but typical values range between 70% and 95%. We note that Jackson [1991] does not recommend using this method.

Scree Plot

A graphical way of determining the number of PCs to retain is called the *scree plot*. The original name and idea is from Cattell [1966], and it is a plot of l_k (the eigenvalue) versus k (the index of the eigenvalue). In some cases, we might plot the log of the eigenvalues when the first eigenvalues are very large. This type of plot is called a *log-eigenvalue* or LEV plot. To use the scree plot, one looks for the 'elbow' in the curve or the place where the curve levels off and becomes almost flat. Another way to look at this is by the slopes of the lines connecting the points. When the slopes start to level off and become less steep, that is the number of PCs one should keep.

The Broken Stick

In this method, we choose the number of PCs based on the size of the eigenvalue or the proportion of the variance explained by the individual PC.

If we take a line segment and randomly divide it into p segments, then the expected length of the k-th longest segment is

$$g_k = \frac{1}{p} \sum_{i=k}^{p} \frac{1}{i}.$$

If the proportion of the variance explained by the k-th PC is greater than g_k, then that PC is kept. We can say that these PCs account for more variance than would be expected by chance alone.

Size of Variance

One rule that can be used for correlation-based PCA is due to Kaiser [1960], but is more commonly used in factor analysis. Using this rule, we would retain PCs whose variances are greater than 1 ($l_k \geq 1$). Some suggest that this is too high [Jolliffe, 1972], so a better rule would be to keep PCs whose variances are greater than 0.7 ($l_k \geq 0.7$). We can use something similar for covariance-based PCA, where we use the average of the eigenvalues rather than 1. In this case, we would keep PCs if

$$l_k \geq 0.7\bar{l} \quad \text{or} \quad l_k \geq \bar{l}.$$

Example 2.2

We show how to perform PCA using the **yeast** cell cycle data set. Recall from Chapter 1 that these contain 384 genes corresponding to five phases, measured at 17 time points. We first load the data and center each row.

```
load yeast
[n,p] = size(data);
% Center the data.
datac = data - repmat(sum(data)/n,n,1);
% Find the covariance matrix.
covm = cov(datac);
```

We are going to use the covariance matrix in PCA since the variables have common units. The reader is asked to explore the correlation matrix approach in the exercises. The **eig** function is used to calculate the eigenvalues and eigenvectors. MATLAB returns the eigenvalues in a diagonal matrix, and they are in ascending order, so they must be flipped to get the scree plot.

```
[eigvec,eigval] = eig(covm);
eigval = diag(eigval); % Extract the diagonal elements
% Order in descending order
eigval = flipud(eigval);
eigvec = eigvec(:,p:-1:1);
```

```
% Do a scree plot.
figure, plot(1:length(eigval),eigval,'ko-')
title('Scree Plot')
xlabel('Eigenvalue Index - k')
ylabel('Eigenvalue')
```

We see from the scree plot in Figure 2.2 that keeping four PCs seems reasonable. Next we calculate the cumulative percentage of variance explained.

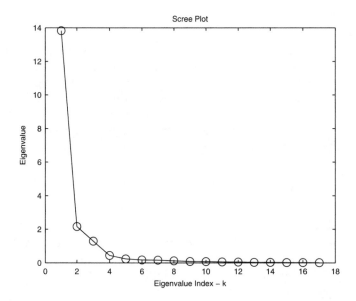

FIGURE 2.2
This is the scree plot for the **yeast** data. The elbow in the curve seems to occur at $k = 4$.

```
% Now for the percentage of variance explained.
pervar = 100*cumsum(eigval)/sum(eigval);
```

The first several values are:

    73.5923   85.0875   91.9656   94.3217   95.5616

Depending on the cutoff value, we would keep four to five PCs (if we are using the higher end of the range of t_d). Now we show how to do the broken stick test.

```
% First get the expected sizes of the eigenvalues.
g = zeros(1,p);
for k = 1:p
    for i = k:p
```

```
        g(k) = g(k) + 1/i;
    end
    end
    g = g/p;
```

The next step is to find the proportion of the variance explained.

```
    propvar = eigval/sum(eigval);
```

Looking only at the first several values, we get the following for g_k and the proportion of variance explained by each PC:

```
    g(1:4) =            0.2023     0.1435     0.1141     0.0945
    propvar(1:4) = 0.7359         0.1150     0.0688     0.0236
```

Thus, we see that only the first PC would be retained using this method. Finally, we look at the size of the variance.

```
    % Now for the size of the variance.
    avgeig = mean(eigval);
    % Find the length of ind:
    ind = find(eigval > avgeig);
    length(ind)
```

According to this test, the first three PCs would be retained. So, we see that different values of d are obtained using the various procedures. Because of visualization issues, we will use the first three PCs to reduce the dimensionality of the data, as follows.

```
    % So, using 3, we will reduce the dimensionality.
    P = eigvec(:,1:3);
    Xp = datac*P;
    figure,plot3(Xp(:,1),Xp(:,2),Xp(:,3),'k*')
    xlabel('PC 1'),ylabel('PC 2'),zlabel('PC 3')
    grid on, axis tight
```

These results are shown in Figure 2.3.
❑

We illustrated the use of the **eig** function that comes in the main MATLAB package. It contains another useful function called **eigs** that can be used to find the PCs and eigenvalues of sparse matrices. For those who have the Statistics Toolbox, there is a function called **princomp**. It returns the PCs, the PC scores and other useful information for making inferences regarding the eigenvalues. It centers the observations at the means, but does not rescale (Statistics Toolbox, Version 5). For PCA using the covariance matrix, the **pcacov** function is provided.

Before moving on to the next topic, we recall that the PCA described in this book is based on the sample covariance or sample correlation matrix. The procedure is similar if the population version of these matrices is used. Many interesting and useful properties are known about principal components and

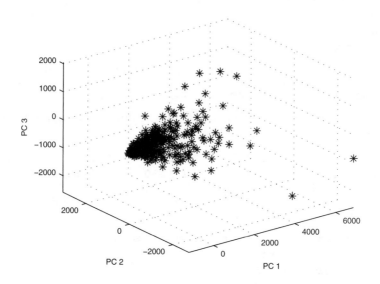

FIGURE 2.3
This shows the results of projecting the **yeast** data onto the first three PCs.

the associated data transformations, but are beyond the scope and purpose of this book. We provide references in the last section.

2.3 Singular Value Decomposition - SVD

Singular value decomposition (SVD) is an important method from matrix algebra and is related to PCA. In fact, it provides a way to find the PCs without explicitly calculating the covariance matrix [Gentle, 2002]. It also enjoys widespread use in the analysis of gene expression data [Alter, Brown and Botstein, 2000; Wall, Dyck and Brettin, 2001; Wall, Rechtsteiner and Rocha, 2003] and in information retrieval applications [Deerwester, et al., 1990; Berry, Dumais and O'Brien, 1995; Berry, Drmac and Jessup, 1999], so it is an important technique its own right.

As before, we start with our data matrix $\mathbf{X}$, where in some cases, we will center the data about their mean to get $\mathbf{X}_c$. We use the noncentered form in the explanation that follows, but the technique is valid for an arbitrary matrix; i.e., the matrix does not have to be square. The SVD of $\mathbf{X}$ is given by

$$\mathbf{X} = \mathbf{UDV}^T, \tag{2.5}$$

where $\mathbf{U}$ is an $n \times n$ matrix, $\mathbf{D}$ is a diagonal matrix with n rows and p columns, and $\mathbf{V}$ has dimensions $p \times p$. The columns of $\mathbf{U}$ and $\mathbf{V}$ are orthonormal. $\mathbf{D}$ is a matrix containing the *singular values* along its diagonal, which are the square roots of the eigenvalues of $\mathbf{X}^T\mathbf{X}$, and zeros everywhere else.

The columns of $\mathbf{U}$ are called the *left singular vectors* and are calculated as the eigenvectors of $\mathbf{X}\mathbf{X}^T$ (this is an $n \times n$ matrix). Similarly, the columns of $\mathbf{V}$ are called the *right singular vectors*, and these are the eigenvectors of $\mathbf{X}^T\mathbf{X}$ (this is a $p \times p$ matrix). An additional point of interest is that the singular values are the square roots of the eigenvalues of $\mathbf{X}^T\mathbf{X}$ and $\mathbf{X}\mathbf{X}^T$.

Let the rank of the matrix $\mathbf{X}$ be given by r, where

$$r \leq \min(n, \ p).$$

Then the first r columns of $\mathbf{V}$ form an orthonormal basis for the column space of $\mathbf{X}$, and the first r columns of $\mathbf{U}$ form a basis for the row space of $\mathbf{X}$ [Strang, 1993]. As with PCA, we order the singular values largest to smallest and impose the same order on the columns of $\mathbf{U}$ and $\mathbf{V}$. A lower rank approximation to the original matrix $\mathbf{X}$ is obtained via

$$\mathbf{X}_k = \mathbf{U}_k\mathbf{D}_k\mathbf{V}_k^T, \tag{2.6}$$

where $\mathbf{U}_k$ is an $n \times k$ matrix containing the first k columns of $\mathbf{U}$, $\mathbf{V}_k$ is the $p \times k$ matrix whose columns are the first k columns of $\mathbf{V}$, and $\mathbf{D}_k$ is a $k \times k$ diagonal matrix whose diagonal elements are the k largest singular values of $\mathbf{X}$. It can be shown that the approximation given in Equation 2.6 is the best one in a least squares sense.

To illustrate the SVD approach, we look at an example from information retrieval called *latent semantic indexing* or LSI [Deerwester, et al., 1990]. Many applications of information retrieval (IR) rely on lexical matching, where words are matched between a user's query and those words assigned to documents in a corpus or database. However, the words people use to describe documents can be diverse and imprecise, so the results of the queries are often less than perfect. LSI uses SVD to derive vectors that are more robust at representing the meaning of words and documents.

Example 2.3

This illustrative example of SVD applied to information retrieval (IR) is taken from Berry, Drmac and Jessup [1999]. The documents in the corpus comprise a small set of book titles, and a subset of words have been used in the analysis, where some have been replaced by their root words (e.g., *bake* and *baking* are both *bake*). The documents and terms are shown in Table 2.1. We start with a data matrix, where each row corresponds to a term, and each column corresponds to a document in the corpus. The elements of the term-document matrix $\mathbf{X}$ denote the number of times the word appears in the

TABLE 2.1

Document Information for Example 2.3

Number	Title
Doc 1	How to *Bake Bread* Without *Recipes*
Doc 2	The Classic Art of Viennese *Pastry*
Doc 3	Numerical *Recipes*: The Art of Scientific Computing
Doc 4	*Breads, Pastries, Pies* and *Cakes*: Quantity *Baking Recipes*
Doc 5	*Pastry*: A Book of Best French *Recipes*
Term 1	bak (e, ing)
Term 2	recipes
Term 3	bread
Term 4	cake
Term 5	pastr (y, ies)
Term 6	pie

document. In this application, we are not going to center the observations, but we do pre-process the matrix by normalizing each column such that the magnitude of the column is 1. This is done to ensure that relevance between the document and the query is not measured by the absolute count of the terms.[1] The following MATLAB code starts the process:

```
load lsiex
% Loads up variables: X, termdoc, docs and words.
% Convert the matrix to one that has columns
% with a magnitude of 1.
[n,p] = size(termdoc);
for i = 1:p
    termdoc(:,i) = X(:,i)/norm(X(:,i));
end
```

Say we want to find books about *baking bread,* then the vector that represents this query is given by a column vector with a 1 in the first and fourth positions:

```
q1 = [1 0 1 0 0 0]';
```

If we are seeking books that pertain only to *baking,* then the query vector is:

```
q2 = [1 0 0 0 0 0]';
```

We can find the most relevant documents using the original term-document matrix by finding the cosine of the angle between the query vectors and the columns (i.e., the vectors representing documents or books); the higher cosine values indicate greater similarity between the query and the document. This would be a straightforward application of lexical matching. Recall from matrix algebra that the cosine of the angle between two vectors $\mathbf{x}$ and $\mathbf{y}$ is given by

---

[1] Other term weights can be found in the literature [Berry and Browne, 1999].

$$\cos\theta_{x,y} = \frac{\mathbf{x}^T\mathbf{y}}{\sqrt{\mathbf{x}^T\mathbf{x}}\sqrt{\mathbf{y}^T\mathbf{y}}}.$$

The MATLAB code to find the cosines for the query vectors in our example is:

```
% Find the cosine of the angle between
% columns of termdoc and query.
% Note that the magnitude of q1 is not 1.
m1 = norm(q1);
cosq1a = q1'*termdoc/m1;
% Note that the magnitude of q2 is 1.
cosq2a = q2'*termdoc;
```

The resulting cosine values are:

```
cosq1a = 0.8165, 0, 0, 0.5774, 0
cosq2a = 0.5774, 0, 0, 0.4082, 0
```

If we use a cutoff value of 0.5, then the relevant books for our first query are the first and the fourth ones, which are those that describe *baking bread*. On the other hand, the second query matches with the first book, but misses the fourth one, which would be relevant. Researchers in the area of IR have applied several techniques to alleviate this problem, one of which is LSI. One of the powerful uses of LSI is in matching a user's query with existing documents in the corpus whose representations have been reduced to lower rank approximations via the SVD. The idea being that some of the dimensions represented by the full term-document matrix are noise and that documents will have closer semantic structure after dimensionality reduction using SVD. So, we now find the singular value decomposition using the function **svd**.

```
% Find the singular value decomposition.
[u,d,v] = svd(termdoc);
```

We then find the representation of the query vector in the reduced space given by the first k columns of $\mathbf{U}$ in the following manner

$$\mathbf{q}_k = \mathbf{U}_k^T\mathbf{q},$$

which is a vector with k elements. We note that in some applications of LSI, the following is used as the reduced query

$$\mathbf{q}_k = \mathbf{D}_k^{-1}\mathbf{U}_k^T\mathbf{q}.$$

This is simply a scaling by the singular values, since $\mathbf{D}$ is diagonal. The following code projects the query into the reduced space and also finds the cosine of the angle between the query vector and the columns. Berry, Drmac and Jessup show that we do not have to form the full reduced matrix $\mathbf{X}_k$. Instead, we can use the columns of $\mathbf{V}_k$, saving storage space.

```
% Project the query vectors.
q1t = u(:,1:3)'*q1;
q2t = u(:,1:3)'*q2;
% Now find the cosine of the angle between the query
% vector and the columns of the reduced rank matrix,
% scaled by D.
for i = 1:5
    sj = d(1:3,1:3)*v(i,1:3)';
    m3 = norm(sj);
    cosq1b(i) = sj'*q1t/(m3*m1);
    cosq2b(i) = sj'*q2t/(m3);
end
```

From this we have

```
cosq1b = 0.7327, -0.0469, 0.0330, 0.7161, -0.0097
cosq2b = 0.5181, -0.0332, 0.0233, 0.5064, -0.0069
```

Using a cutoff value of 0.5, we now correctly have documents 1 and 4 as being relevant to our queries on *baking bread* and *baking*.
❑

Note that in the above loop, we are using the magnitudes of the original query vectors as in Berry, Drmac and Jessup [Equation 6.1, 1999]. This saves on computations and also improves precision (disregarding irrelevant information). We could divide by the magnitudes of the reduced query vectors (**q1t** and **q2t**) in the loop to improve recall (retrieving relevant information) at the expense of precision.

Before going on to the next topic, we point out that the literature describing SVD applied to LSI and gene expression data defines the matrices of Equation 2.5 in a different way. The decomposition is the same, but the difference is in the size of the matrices **U** and **D**. Some definitions have the dimensions of **U** as $n \times p$ and **D** with dimensions $p \times p$ [Golub & Van Loan, 1996]. We follow the definition in Strang [1988, 1993], which is also used in MATLAB.

2.4 Factor Analysis

There is much confusion in the literature as to the exact definition of the technique called *factor analysis* [Jackson, 1981], but we follow the commonly used definition given in Jolliffe [1986]. In the past, this method has also been confused with PCA, mostly because PCA is sometimes provided in software packages as a special case of factor analysis. Both of these techniques attempt to reduce the dimensionality of a data set, but they are different from one another in many ways. We describe these differences at the end of the discussion of factor analysis.

The idea underlying factor analysis is that the p observed random variables can be written as linear functions of $d < p$ unobserved *latent variables* or *common factors* f_j, as follows:

$$x_1 = \lambda_{11} f_1 + \ldots + \lambda_{1d} f_d + \varepsilon_1$$
$$\ldots \tag{2.7}$$
$$x_p = \lambda_{p1} f_1 + \ldots + \lambda_{pd} f_d + \varepsilon_p.$$

The λ_{ij} ($i = 1, \ldots , p$ and $j = 1, \ldots , d$) in the above model are called the *factor loadings*, and the error terms ε_i are called the *specific factors*. Note that the error terms ε_i are specific to *each* of the original variables, while the f_j are common to *all* of the variables. The sum of the squared factor loadings for the i-th variable

$$\lambda_{i1}^2 + \ldots + \lambda_{id}^2$$

is called the *communality* of x_i.

We see from the model in Equation 2.7 that the original variables are written as a linear function of a smaller number of variables or factors, thus reducing the dimensionality. It is hoped that the factors provide a summary or clustering of the original variables (not the observations), such that insights are provided about the underlying structure of the data. While this model is fairly standard, it can be extended to include more error terms, such as measurement error. It can also be made nonlinear [Jolliffe, 1986].

The matrix form of the factor analysis model is

$$\mathbf{x} = \Lambda \mathbf{f} + \mathbf{e}. \tag{2.8}$$

Some assumptions are made regarding this model, which are

$$E[\mathbf{e}] = 0 \qquad E[\mathbf{f}] = 0 \qquad E[\mathbf{x}] = 0,$$

where $E[\bullet]$ denotes the expected value. If the last of these assumptions is violated, the model can be adjusted to accommodate this, yielding

$$\mathbf{x} = \Lambda \mathbf{f} + \mathbf{e} + \mu, \tag{2.9}$$

where $E[\mathbf{x}] = \mu$. We also assume that the error terms ε_i are uncorrelated with each other, and that the common factors are uncorrelated with the specific factors f_j. Given these assumptions, the sample covariance (or correlation) matrix is of the form

$$\mathbf{S} = \Lambda^T \Lambda + \Psi,$$

where Ψ is a diagonal matrix representing $E[\mathbf{ee}^T]$. The variance of ε_i is called the *specificity* of x_i, so the matrix Ψ is also called the *specificity matrix*. Factor analysis obtains a reduction in dimensionality by postulating a model that relates the original variables x_i to the d hypothetical variables or factors.

The matrix form of the factor analysis model is reminiscent of a regression problem, but here both Λ and $\mathbf{f}$ are unknown and must be estimated. This leads to one problem with factor analysis: the estimates are not unique. Estimation of the parameters in the factor analysis model is usually accomplished via the matrices Λ and Ψ. The estimation proceeds in stages, where an initial estimate is found by placing conditions on Λ. Once this initial estimate is obtained, other solutions can be found by rotating Λ. The goal of some rotations is to make the structure of Λ more interpretable, by making the λ_{ij} close to one or zero. Several methods that find rotations such that a desired criterion is optimized are described in the literature. Some of the commonly used methods are varimax, equimax, orthomax, quartimax, promax, and procrustes.

These factor rotation methods can either be **orthogonal** or **oblique**. In the case of orthogonal rotations, the axes are kept at 90 degrees. If this constraint is relaxed, then we have oblique rotations. The orthogonal rotation methods include quartimax, varimax, orthomax, and equimax. The promax and procrustes rotations are oblique.

The goal of the quartimax rotation is to simplify the rows of the factor matrix by getting a variable with a high loading on one factor and small loadings on all other factors. The varimax rotation focuses on simplifying the columns of the factor matrix. From the varimax approach, perfect simplification is obtained if there are only ones and zeros in a single column. The output from this method tends to have high loadings close to ± 1 and some near zero in each column. The equimax rotation is a compromise between these two methods, where both the rows and the columns of the factor matrix are simplified as much as possible.

Just as we did in PCA, we might want to transform the observations using the estimated factor analysis model either for plotting purposes or for further analysis methods, such as clustering or classification. We could think of these observations as being transformed to the 'factor space.' These are called *factor scores*, similarly to PCA. However, unlike PCA, there is no single method for finding the factor scores, and the analyst must keep in mind that the factor scores are really *estimates* and depend on the method that is used.

An in-depth discussion of the many methods for estimating the factor loadings, the variances Ψ, and the factor scores, as well as the rotations, is beyond the scope of this book. For more information on the details of the methods for estimation and rotation in factor analysis, see Jackson [1991], Lawley and Maxwell [1971], or Cattell [1978]. Before we go on to an example of factor analysis, we note that the MATLAB Statistics Toolbox uses the maximum likelihood method to obtain the factor loadings, and it also implements some of the various rotation methods mentioned earlier.

Example 2.4

In this example, we examine some data provided with the Statistics Toolbox, called **stockreturns**. An alternative analysis of these data is provided in the Statistics Toolbox User's Guide. The data set consists of 100 observations, representing the percent change in stock prices for 10 companies. Thus, the data set has $n = 100$ observations and $p = 10$ variables. It turns out that the first four companies can be classified as technology, the next three as financial, and the last three as retail. We can use factor analysis to see if there is any structure in the data that supports this grouping. We first load up the data set and perform factor analysis using the function **factoran**.

```
load stockreturns
% Loads up a variable called stocks.
% Perform factor analysis:3 factors,default rotation.
[LamVrot,PsiVrot] = factoran(stocks,3);
```

This is the basic syntax for **factoran**, where the user must specify the number of factors (3 in this case), and the default is to use the **varimax** rotation, which optimizes a criterion based on the variance of the loadings. See the MATLAB **help** on **factoran** for more details on the rotations. Next, we specify no rotation, and we plot the matrix **Lam** (the factor loadings) in Figure 2.4.

```
[Lam,Psi] = factoran(stocks,3,'rotate','none');
```

These plots show the pairwise factor loadings, and we can see that the factor loadings are not close to one of the factor axes, making it more difficult to interpret the factors. We can try rotating the matrix next using one of the oblique (nonorthogonal) rotations called **promax**, and we plot these results in Figure 2.5.

```
% Now try the promax rotation.
[LProt,PProt]=factoran(stocks,3,'rotate','promax');
```

Note that we now have a more interpretable structure with the factor loadings, and we are able to group the stocks. We might also be interested in estimating the factor scores. The user is asked to explore this aspect of it in the exercises.
❑

It can be very confusing trying to decide whether to use PCA or factor analysis. Since the objective of this book is to describe exploratory data analysis techniques, we suggest that both methods be used to explore the

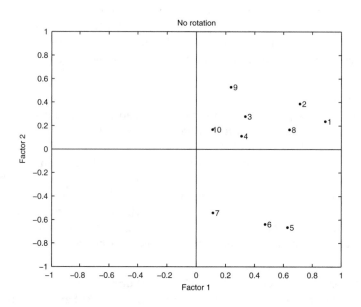

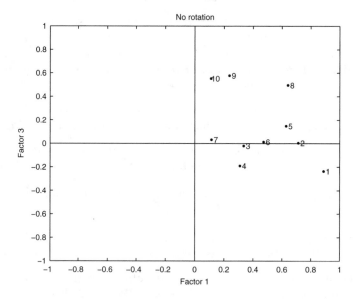

FIGURE 2.4
These show the factor loadings in their unrotated form. We see that the loadings are not grouped around the factor axes, although it is interesting to note that we have the three financial companies (points 5, 6, & 7) grouped together in the upper plot (factors 1 and 2), while the three retail companies (points 8, 9, & 10) are grouped together in the lower plot (factors 1 and 3).

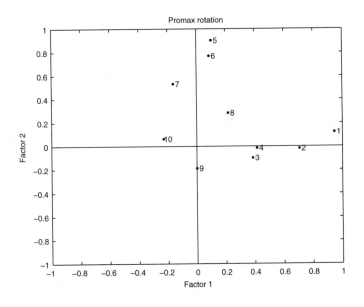

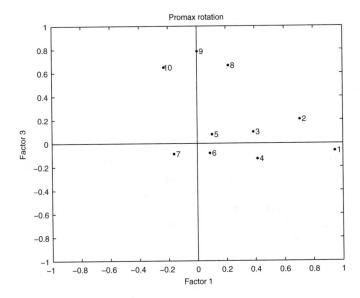

FIGURE 2.5
These plots show the factor loadings after the promax rotation. We see that the stocks can be grouped as technology companies {1, 2, 3, 4}, financial {5, 6, 7}, and retail {8, 9, 10}. The rotation makes the factors somewhat easier to interpret.

data, because they take different approaches to the problem and can uncover different facets of the data. We now outline the differences between PCA and factor analysis; a discussion of which method to use can be found in Velicer and Jackson [1990].

- Both factor analysis and PCA try to represent the structure of the data set based on the covariance or correlation matrix. Factor analysis tries to explain the off-diagonal elements, while PCA explains the variance or diagonal elements of the matrix.

- Factor analysis is typically performed using the correlation matrix, and PCA can be used with either the correlation or the covariance matrix.

- Factor analysis has a model, as given in Equation 2.8 and 2.9, but PCA does not have an explicit model associated with it (unless one is interested in inferential methods associated with the eigenvalues and PCs, in which case, distributional assumptions are made).

- If one changes the number of PCs to keep, then the existing PCs do not change. If we change the number of factors, then the entire solution changes; i.e., existing factors must be re-estimated.

- PCA has a unique solution, but factor analysis does not.

- The PC scores are found in an exact manner, but the factor scores are estimates.

2.5 Intrinsic Dimensionality

Knowing the *intrinsic dimensionality* (sometimes called *effective dimensionality*) of a data set is useful information when exploring a data set. This is defined as the smallest number of dimensions or variables needed to model the data without loss [Kirby, 2001]. Fukunaga [1990] provides a similar definition of intrinsic dimensionality as "the minimum number of parameters needed to account for the observed properties of the data."

Several approaches to estimating the intrinsic dimensionality of a data set have been proposed in the literature. Trunk [1968, 1976] describes a statistical approach using hypothesis testing regarding the most likely local dimensionality. Fukunaga and Olsen [1971] present an algorithm where the data are divided into small subregions, and the eigenvalues of the local covariance matrix are computed for each region. The intrinsic dimensionality is then defined based on the size of the eigenvalues. Pettis, et al. [1979] develop an algorithm for estimating intrinsic dimensionality that is based on nearest neighbor information and a density estimator. We choose to implement the original Pettis algorithm as outlined below; details of the derivation can be found in the original paper.

We first set some notation, using 1-D representation for the observations as in the original paper. However, this method works with the *distance* between observations, so it easily extends to the multi-dimensional case. Let $r_{k,x}$ represent the distance from x to the k-th nearest neighbor of x. The average k-th nearest neighbor distance is given by

$$\bar{r}_k = \frac{1}{n} \sum_{i=1}^{n} r_{k,x_i} . \tag{2.10}$$

Pettis, et al. [1979] show that the expected value of the average distance in Equation 2.10 is

$$E(\bar{r}_k) = \frac{1}{G_{k,d}} k^{1/d} C_n , \tag{2.11}$$

where

$$G_{k,d} = \frac{k^{1/d} \Gamma(k)}{\Gamma\left(k + \frac{1}{d}\right)} ,$$

and C_n is independent of k. If we take logarithms of Equation 2.11 and do some rearranging, then we obtain the following

$$\log(G_{k,d}) + \log E(\bar{r}_k) = (1/d)\log(k) + \log(C_n). \tag{2.12}$$

We can get an estimate for $E(\bar{r}_k)$ by taking the observed value of $\bar{r}_k$ based on the sample, yielding the following estimator $\hat{d}$ for the intrinsic dimensionality d

$$\log(G_{k,\hat{d}}) + \log(\bar{r}_k) = (1/\hat{d})\log(k) + \log(C_n). \tag{2.13}$$

This is similar to a regression problem, where the slope is given by $(1/\hat{d})$. The term $\log(C_n)$ affects the intercept, not the slope, so we can disregard this in the estimation process.

The estimation procedure must be iterative since $\hat{d}$ also appears on the response side of Equation 2.13. To get started, we set the term $\log(G_{k,\hat{d}})$ equal to zero and find the slope using least squares, where the predictor values are given by $\log(k)$, and the responses are $\log(\bar{r}_k)$, for $k = 1,...,K$. Once we have this initial value for $\hat{d}$, we then find $\log(G_{k,\hat{d}})$ using Equation 2.13. Using this value, we find the slope where the responses are now $\log(G_{k,\,\hat{d}}) + \log(\bar{r}_k)$. The algorithm continues in this manner until the estimates of intrinsic dimensionality converge.

We outline the algorithm shortly, but we need to discuss the effect of outliers first. Pettis, et al. [1979] found that their algorithm works better if potential outliers are removed before estimating the intrinsic dimensionality. To define outliers in this context, we first define a measure of the maximum average distance given by

$$m_{\max} = \frac{1}{n}\sum_{i=1}^{n} r_{K,\,x_i} \, ,$$

where r_{K,x_i} represents the i-th K nearest neighbor distance (i.e., the maximum distance). A measure of the spread is found by

$$s_{\max}^2 = \frac{1}{n-1}\sum_{i=1}^{n}(r_{K,x_i} - m_{\max})^2.$$

The data points x_i for which

$$r_{K,\,x_i} \le m_{\max} + s_{\max} \tag{2.14}$$

are used in the nearest neighbor estimate of intrinsic dimensionality. We are now ready to describe the algorithm.

Procedure - Intrinsic Dimensionality

1. Set a value for the maximum number of nearest neighbors K.
2. Determine all of the distances r_{k,x_i}.
3. Remove outliers; i.e., keep only those points that satisfy Equation 2.14.
4. Calculate $\log(\bar{r}_k)$.
5. Get the initial estimate $\hat{d}_0$ by fitting a line to

$$\log(\bar{r}_k) = (1/\hat{d})\log(k),$$

and taking the inverse of the slope.
6. Calculate $\log(G_{k,\hat{d}_j})$ using Equation 2.13.
7. Update the estimate of intrinsic dimensionality by fitting a line to

$$\log(G_{k,\hat{d}_j}) + \log(\overline{r}_k) = (1/\hat{d})\log(k).$$

8. Repeat steps 6 and 7 until the estimates converge.

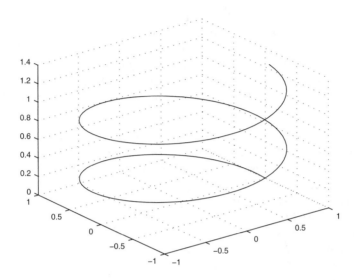

FIGURE 2.6
This shows the helix used in Example 2.5. This is a 1-D structure embedded in 3-D.

Example 2.5

The following MATLAB code implements the Pettis algorithm for estimating intrinsic dimensionality, which is also contained in the EDA Toolbox function called **idpettis**. We first generate some data to illustrate the functionality of the algorithm. The helix is described by the following equations, and points are randomly chosen along this path:

$$x = \cos\theta$$
$$y = \sin\theta$$
$$z = 0.1\theta$$

for $0 \leq \theta \leq 4\pi$. For this data set, the dimensionality is 3, but the intrinsic dimensionality is 1. We show a picture of the helix in Figure 2.6. We obtain the data by generating uniform random numbers in the interval $0 \leq \theta \leq 4\pi$.

```
% Generate the random numbers
% unifrnd is from the Statistics Toolbox.
n = 500;
```

```
theta = unifrnd(0,4*pi,1,n);
% Use in the equations for a helix.
x = cos(theta);y=sin(theta);z = 0.1*(theta);
% Put into a data matrix.
X = [x(:),y(:),z(:)];
```

We note that the function **unifrnd** is from the Statistics Toolbox, but users can also employ the **rand** function and scale to the correct interval [Martinez and Martinez, 2002]. In the interest of space and clarity, we show only the code for the algorithm after the outliers have been removed. Thus, you cannot run the code given below, as it stands. We refer curious readers to the MATLAB function **idpettis** for the implementation of the rest of the procedure.

```
% Get initial value for d. Values n, k, & logrk
% are defined in the idpettis function. This is
% an extract from that function.
logk = log(k);
[p,s] = polyfit(logk,logrk,1);
dhat = 1/p(1);
dhatold = realmax;
maxiter = 100;
epstol = 0.01;
i = 0;
while abs(dhatold - dhat) >= epstol & i < maxiter
    % Adjust the y values by adding logGRk
    logGRk = (1/dhat)*log(k)+...
            gammaln(k)-gammaln(k+1/dhat);
    [p,s] = polyfit(logk,logrk + logGRk,1);
    dhatold = dhat;
    dhat = 1/p(1);
    i = i+1;
end
idhat = dhat;
```

As an example of using the **idpettis** function, we offer the following:

```
% Get the distances using the pdist function.
% This returns the interpoint distances.
ydist = pdist(X);
idhat = idpettis(ydist,n);
```

where the input variable **X** is the helix data. The resulting estimate of the intrinsic dimensionality is 1.14. We see from this result, that in most cases the estimate will need to be rounded to the nearest integer. Thus, the estimate in this case is the correct one: 1-D.

❑

2.6 Summary and Further Reading

In this chapter, we introduced the concept of dimensionality reduction by presenting several methods that find a linear mapping from a high-dimensional space to a lower one. These methods include principal component analysis, singular value decomposition, and factor analysis. We also addressed the issue of determining the intrinsic dimensionality of a data set.

For a general treatment of linear and matrix algebra, we recommend Strang [1985, 1993] or Golub and van Loan [1996]. Discussions of PCA and factor analysis can be found in most multivariate analysis books; some examples are Manly [1994] or Seber [1984]. Readers who would like a more detailed treatment of these subjects are referred to Jackson [1991] and Jolliffe [1986]. These texts provide extensive discussions (including many extensions and applications) of PCA, SVD, and factor analysis. There are many recent applications of PCA and SVD to microarray analysis, some examples include gene shaving [Hastie, et al., 2001], predicting the clinical status of human breast cancer [West, et al., 2001], obtaining a global picture of the dynamics of gene expression [Alter, Brown and Botstein, 2000], and clustering [Yeung and Ruzzo, 2001].

The Fukunaga-Olsen method of estimating intrinsic dimensionality is further explored in Verveer and Duin [1995] and Bruske and Sommer [1998]. The Pettis algorithm is also described in Fukunaga [1990] and is expanded in Verveer and Duin [1995]. Verveer and Duin [1995] revise both the Pettis and the Fukunaga-Olsen algorithms and provide an evaluation of their performance. Levina and Bickel [2004] describe a new method for estimating the intrinsic dimensionality of a data set by applying maximum likelihood to the distances between close neighbors. They derive the estimator by a Poisson process approximation. Costa and Hero [2004] propose a geometric approach based on entropic graph methods. Their method is called the geodesic minimal spanning tree (GMST), and it yields estimates of the manifold dimension and the α-entropy of the sample density on the manifold.

Exercises

2.1 Generate $n = 50$, $p = 3$ normally distributed random variables that have high variance in one dimension. For example, you might use the following MATLAB code to do this:

```
x1 = randn(50,1)*100;
```

```
x2 = randn(50,2);
X = [x1,x2];
```

Try PCA using both correlation and covariance matrices. Is the one with the covariance matrix very informative? Which one would be better to use in this case?

2.2   Do a scree plot of the data set used in Example 2.2. Generate multivariate, uncorrelated normal random variables for the same n and p. (Hint: use **randn(n,p)**.) Perform PCA for this data set, and construct its scree plot. It should look approximately like a straight line. Plot both scree plots together; where they intersect is an estimate of the number of PCs to keep. How does this compare with previous results? [Jackson, 1991, p. 46].

2.3   Generate a set of $n = 30$ trivariate normal random variables using **randn(30,3)**.

a. Subtract the mean from the observations and find the covariance matrix, using **cov**. Get the eigenvectors and eigenvalues based on the covariance matrix of the centered observations. Is the total variance of the original data equal to the sum of the eigenvalues? Verify that the eigenvectors are orthonormal. Find the PC scores and their mean.

b. Impose a mean of $[2, 2, 2]^T$ on the original variables. Perform PCA using the covariance of these noncentered data. Find the PC scores and their mean.

c. Center and scale the original variables, so that each one has zero mean and a standard deviation of one. Find the covariance of these transformed data. Find the correlation matrix of the original, non-transformed data. Are these two matrices the same?

d. Verify that the PC scores produce data that are uncorrelated by finding the correlation matrix of the PC scores.

2.4   Repeat Example 2.2 using the correlation matrix. Compare with the previous results.

2.5   Generate multivariate normal random variables, centered at the origin (i.e., centered about 0). Form the matrix X^TX, find the eigenvectors. Form the matrix XX^T, find the eigenvectors. Now use the SVD on X. Compare the columns of U and V with the eigenvectors. Are they the same? Note that the columns of U and V are unique up to a sign change. What can you say about the eigenvalues and singular values?

2.6   Generate a set of multivariate normal random variables, centered at the origin. Obtain the eigenvalue decomposition of the covariance matrix. Multiply them to get the original matrix back, and also perform the multiplication to get the diagonal matrix L.

2.7   Construct a plot based on the slopes of the lines connecting the points in the scree plot. Apply to the **yeast** data. Does this help in the analysis? Hint: use the **diff** function.

2.8 Apply PCA to the following data sets. What value of d would you get using the various methods for choosing the number of dimensions?

a. Other gene expression data sets.

b. **oronsay** data.

c. **sparrow** data.

2.9 Repeat Example 2.3 for the SVD - LSI using $k = 2$. Comment on the results and compare to the document retrieval using $k = 3$.

2.10 Using the term-document matrix in Example 2.3, find the cosine of the angle between the reduced query vectors and the reduced document vectors. However, this time use the magnitude of the reduced query vectors in the loop, not the magnitudes of the original queries. Discuss how this affects your results in terms of precision and recall.

2.11 Generate a bivariate data set using either **rand** or **randn**. Verify that the singular values are the square roots of the eigenvalues of $\mathbf{X}^T\mathbf{X}$ and $\mathbf{X}\mathbf{X}^T$.

2.12 Repeat Example 2.4, using other rotations and methods implemented in MATLAB for estimating the factor loadings. Do they change the results?

2.13 Plot factors 2 and 3 in Example 2.4 for no rotation and promax rotation. Discuss any groupings that might be evident.

2.14 Try Example 2.4 with different values for the number of factors - say two and four. How do the results change?

2.15 The MATLAB function **factoran** includes the option of obtaining estimates of the factor scores. Using the data in Example 2.4, try the following code

```
[lam,psi,T,stats,F]=...
    factoran(stocks,3,'rotate','promax');
```

The factor scores are contained in the variable **F**. These can be viewed using **plotmatrix**.

2.16 Try factor analysis on some of the gene expression data sets to cluster the patients or experiments.

2.17 Generate 2-D standard bivariate random variables using **randn** for increasing values of n. Use the **idpettis** function to estimate the intrinsic dimensionality, repeating for several Monte Carlo trials for each value of n. The true intrinsic dimensionality of these data is 2. How does the algorithm perform on these data sets?

2.18 This data set is taken from Fukunaga [1990], where he describes a Gaussian pulse characterized by three parameters **a**, **m**, and σ. The waveform is given by

$$\mathbf{x}(t) = \mathbf{a}\exp\left(-\frac{(t-\mathbf{m})^2}{2\sigma^2}\right).$$

These parameters are generated randomly in the following ranges:

$$0.7 \le \mathbf{a} \le 1.3$$
$$0.3 \le \mathbf{m} \le 0.7$$
$$0.2 \le \sigma \le 0.4.$$

The signals are generated at 8 time steps in the range $0 \le t \le 1.05$, so each of the signals is an 8-D random vector. The intrinsic dimensionality of these data is 3. Generate random vectors for various values of n, and apply **idpettis** to estimate the intrinsic dimensionality and assess the results.

2.19   Estimate the intrinsic dimensionality of the **yeast** data. Does this agree with the results in Example 2.2?

2.20   Estimate the intrinsic dimensionality of the following data sets. Where possible, compare with the results from PCA.

a. All BPM data sets.

b. **oronsay** data.

c. **sparrow** data

d. Other gene expression data.

2.21   The Statistics Toolbox, Version 5, has a new function called **rotatefactors** that can be used with either factor analysis or principal component analysis loadings. Do a **help** on this function. Apply it to the factor loadings from the **stockreturns** data. Apply it to the PCA results of the **yeast** data in Example 2.2.

Chapter 3

Dimensionality Reduction - Nonlinear Methods

This chapter covers various methods for nonlinear dimensionality reduction, where the nonlinear aspect refers to the mapping between the high-dimensional space and the low-dimensional space. We start off by discussing a method that has been around for many years called multidimensional scaling. We follow this with several recently developed nonlinear dimensionality reduction techniques called locally linear embedding, isometric feature mapping, and Hessian eigenmaps. We conclude by discussing two related methods from the machine learning community called self-organizing maps and generative topographic maps.

3.1 Multidimensional Scaling - MDS

In general, *multidimensional scaling* (MDS) is a set of techniques for the analysis of proximity data measured on a set of objects in order to reveal hidden structure. The purpose of MDS is to find a configuration of the data points in a low-dimensional space such that the proximity between objects in the full-dimensional space is represented with some degree of fidelity by the distances between points in the low-dimensional space. This means that observations that are close together in a high-dimensional space should be close in the low-dimensional space. Many aspects of MDS were originally developed by researchers in the social science community, and the method is now widely available in most statistical packages, including the MATLAB Statistics Toolbox.

We first provide some definitions and notation before we go on to describe the various categories of MDS [Cox and Cox, 2001]. As before, we assume that we have a data set with n observations. In general, MDS starts with measures of proximity that quantify how close objects are to one another or how similar they are. They can be of two types: those that indicate similarity or dissimilarity. A measure of dissimilarity between objects r and s is denoted by δ_{rs}, and the similarity is denoted by s_{rs}. For most definitions of these proximity measures, we have

$$\delta_{rs} \geq 0 \qquad \delta_{rr} = 0$$

and

$$0 \leq s_{rs} \leq 1 \qquad s_{rr} = 1.$$

Thus, we see that for a dissimilarity measure δ_{rs}, small values correspond to observations that are close together, while the opposite is true for similarity measures s_{rs}. These two main types of proximity measures are easily converted into the other one when necessary (see Appendix A for more information on this issue). Thus, for the rest of this chapter, we will assume that *proximity* measures *dissimilarity* between objects. We also assume that the dissimilarities are arranged in matrix form, which will be denoted by Δ. In many cases, this will be a symmetric $n \times n$ matrix (sometimes given in either lower or upper triangular form).

In the lower-dimensional space, we denote the *distance* between observation r and s by d_{rs}. It should also be noted that in the MDS literature the matrix of coordinate values in the *lower-dimensional space* is denoted by $\mathbf{X}$. We follow that convention here, knowing that it might be rather confusing with our prior use of $\mathbf{X}$ as representing the original set of n p-dimensional observations.

In MDS, one often starts with or might only have the dissimilarities Δ, not the original observations. In fact, in the initial formulations of MDS, the experiments typically involved qualitative judgements about differences between subjects, and p-dimensional observations do not make sense in that context. So, to summarize, with MDS we start with dissimilarities Δ and end up with d-dimensional[1] transformed observations $\mathbf{X}$. Usually, $d = 2$ or $d = 3$ is chosen to allow the analyst to view and explore the results for interesting structure, but any $d < p$ is also appropriate.

There are many different techniques and flavors of MDS, most of which fall into two main categories: metric MDS and nonmetric MDS. The main characteristic that divides them arises from the different assumptions of how the dissimilarities δ_{rs} are transformed into the configuration of distances d_{rs} [Cox and Cox, 2001]. **Metric MDS** assumes that the dissimilarities δ_{rs} calculated from the p-dimensional data and distances d_{rs} in a lower-dimensional space are related as follows

$$d_{rs} \approx f(\delta_{rs}), \tag{3.1}$$

---

[1] We realize that the use of the notation d as both the lower dimensionality of the data ($d < p$) and the distance d_{rs} between points in the configuration space might be confusing. However, the meaning should be clear from the context.

where f is a continuous monotonic function. The form of $f(\bullet)$ specifies the MDS model. For example, we might use the formula

$$f(\delta_{rs}) = b\delta_{rs} \ . \tag{3.2}$$

Mappings that follow Equation 3.2 are called *ratio* MDS [Borg and Groenen, 1997]. Another common choice is *interval* MDS, with $f(\bullet)$ given by

$$f(\delta_{rs}) = a + b\delta_{rs} \ ,$$

where a and b are free parameters. Other possibilities include higher degree polynomials, logarithmic, and exponential functions.

   Nonmetric MDS relaxes the metric properties of $f(\bullet)$ and stipulates only that the rank order of the dissimilarities be preserved. The transformation or scaling function must obey the monotonicity constraint:

$$\delta_{rs} < \delta_{ab} \Rightarrow f(\delta_{rs}) \leq f(\delta_{ab}) \ ,$$

for all objects. Because of this, nonmetric MDS is also known as *ordinal MDS*.

3.1.1 Metric MDS

Most of the methods in metric MDS start with the fact that we seek a transformation that satisfies Equation 3.1. We can tackle this problem by defining an objective function and then using a method that will optimize it. One way to define the objective function is to use the squared discrepancies between d_{rs} and $f(\delta_{rs})$ as follows

$$\sqrt{\frac{\left(\sum_r \sum_s (f(\delta_{rs}) - d_{rs})^2\right)}{\text{scale factor}}} \ . \tag{3.3}$$

In general, Equation 3.3 is called the *stress*; different forms for the scale factor give rise to different forms of stress and types of MDS. The scale factor used most often is

$$\sum_r \sum_s d_{rs}^2 \ ,$$

in which case, we have an expression called *Stress-1* [Kruskal, 1964a]. The summation is taken over all dissimilarities, skipping those that are missing.

As we stated previously, sometimes the dissimilarities are symmetric, in which case, we only need to sum over $1 = r < s = n$.

Thus, in MDS, we would scale the dissimilarities using $f(\bullet)$ and then find a configuration of points in a d-dimensional space such that when we calculate the distances between them the stress is minimized. This can now be solved using numerical methods and operations research techniques (e.g., gradient or steepest descent). These methods are usually iterative and are not guaranteed to find a global solution to the optimization problem. We will expand on some of the details later, but first we describe a case where a closed form solution is possible.

The phrase *metric multidimensional scaling* is often used in the literature to refer to a specific technique called *classical* MDS. However, metric MDS includes more than this one technique, such as least squares scaling and others [Cox and Cox, 2001]. For metric MDS approaches, we first describe classical MDS followed by a method that optimizes a loss function using a majorization technique.

Classical MDS

If the proximity measure in the original space and the distance are taken to be Euclidean, then a closed form solution exists to find the configuration of points in a d-dimensional space. This is the classical MDS approach. The function $f(\bullet)$ relating dissimilarities and distances is the identity function, so we seek a mapping such that

$$d_{rs} = \delta_{rs} .$$

This technique originated with Young and Householder [1938], Torgerson [1952], and Gower [1966]. Gower was the one that showed the importance of classical scaling, and he gave it the name *principal coordinates analysis,* because it uses ideas similar to those in PCA. Principal coordinate analysis and classical scaling are the same thing, and they have become synonymous with metric scaling.

We now describe the steps of the method only, without going into the derivation. Please see any of the following for the derivation of classical MDS: Cox and Cox [2001], Borg and Groenen [1997], or Seber [1984].

Procedure - Classical MDS

1. Using the matrix of dissimilarities, Δ, find matrix $\mathbf{Q}$, where each element of $\mathbf{Q}$ is given by

$$q_{rs} = -\frac{1}{2}\delta_{rs}^2 .$$

2. Find the centering matrix **H** using

$$\mathbf{H} = \mathbf{I} - n^{-1}\mathbf{1}\mathbf{1}^T,$$

where **I** is the $n \times n$ identity matrix, and **1** is a vector of n ones.

3. Find the matrix **B**, as follows

$$\mathbf{B} = \mathbf{HQH}.$$

4. Determine the eigenvectors and eigenvalues of **B**:

$$\mathbf{B} = \mathbf{ALA}^T.$$

5. The coordinates in the lower-dimensional space are given by

$$\mathbf{X} = \mathbf{A}_d \mathbf{L}_d^{1/2},$$

where $\mathbf{A}_d$ contains the eigenvectors corresponding to the d largest eigenvalues, and $\mathbf{L}_d^{1/2}$ contains the square root of the d largest eigenvalues along the diagonal.

We use similar ideas from PCA to determine the dimensionality d to use in Step 5 of the algorithm. Although, $d = 2$ is often used in MDS, since the data can then be represented in a scatterplot.

For some data sets, the matrix **B** might not be positive semi-definite, in which case some of the eigenvalues will be negative. One could ignore the negative eigenvalues and proceed to step 5 or add an appropriate constant to the dissimilarities to make **B** positive semi-definite. We do not address this second option here, but the reader is referred to Cox and Cox [2001] for more details. If the dissimilarities are in fact Euclidean distances, then this problem does not arise.

Since this uses the decomposition of a square matrix, some of the properties and issues discussed about PCA are applicable here. For example, the lower-dimensional representations are nested. The first two dimensions of the 3-D coordinate representation are the same as the 2-D solution. It is also interesting to note that PCA and classical MDS provide equivalent results when the dissimilarities are Euclidean distances [Cox and Cox, 2001].

Example 3.1

For this example, we use the BPM data described in Chapter 1. Recall that one of the applications for these data is to discover different topics or sub-topics. For ease of discussion and presentation, we will look at only two of the topics in this example: the comet falling into Jupiter (topic 6) and DNA in the O. J.

Simpson trial (topic 9). First, using the match distances, we extract the required observations from the full interpoint distance matrix.

```
% First load the data - use the 'match' interpoint
% distance matrix.
load matchbpm
% Now get the data for topics 9 and 6.
% Find the indices where they are not equal to 6 or 9.
indlab = find(classlab ~= 6 & classlab ~=9);
% Now get rid of these from the distance matrix.
matchbpm(indlab,:) = [];
matchbpm(:,indlab) = [];
classlab(indlab) = [];
```

The following piece of code shows how to implement the steps for classical MDS in MATLAB.

```
% Now implement the steps for classical MDS
% Find the matrix Q:
Q = -0.5*matchbpm.^2;
% Find the centering matrix H:
n = 73;
H = eye(n) - ones(n)/n;
% Find the matrix B:
B = H*Q*H;
% Get the eigenvalues of B.
[A,L] = eig(B);
% Find the ordering largest to smallest.
[vals, inds] = sort(diag(L));
inds = flipud(inds);
vals = flipud(vals);
% Re-sort based on these.
A = A(:,inds);
L = diag(vals);
```

We are going to plot these results using $d = 2$ for ease of visualization, but we can also construct a scree-type plot to look for the 'elbow' in the curve. As in PCA, this can help us determine a good value to use. The code for constructing this plot and finding the coordinates in a 2-D space is given next.

```
% First plot a scree-type plot to look for the elbow.
% The following uses a log scale on the y-axis.
semilogy(vals(1:10),'o')
% Using 2-D for visualization purposes,
% find the coordinates in the lower-dimensional space.
X = A(:,1:2)*diag(sqrt(vals(1:2)));
% Now plot in a 2-D scatterplot
ind6 = find(classlab == 6);
```

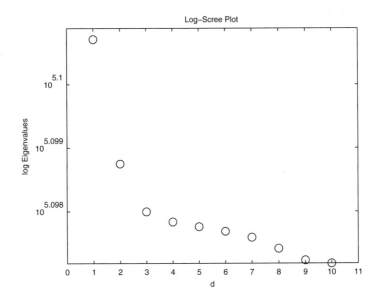

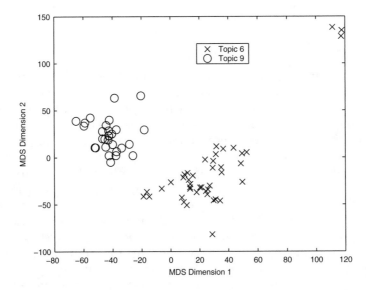

FIGURE 3.1

The top plot shows the logarithm of the eigenvalues. The bottom plot is a scatterplot of the 2-D coordinates after classical MDS. Note the good separation between the two topics, as well as the appearance of sub-topics for topic 6.

```
ind9 = find(classlab == 9);
plot(X(ind6,1),X(ind6,2),'x',X(ind9,1),X(ind9,2),'o')
legend({'Topic 6';'Topic 9'})
```

The scree plot is shown in Figure 3.1 (top), where we see that the elbow looks to be around $d = 3$. The scatterplot of the 2-D coordinates is given in Figure 3.1 (bottom). We see clear separation between topics 6 and 9. However, it is also interesting to note that there seems to be several sub-topics in topic 6.
❑

MATLAB provides a function called **cmdscale** for constructing lower-dimensional coordinates using classical MDS. This is available in the Statistics Toolbox.

Metric MDS - SMACOF

The general idea of the *majorization method* for optimizing a function is as follows. Looking only at the one-dimensional case, we want to minimize a complicated function $f(x)$ by using a function $g(x, y)$ that is easily minimized. The function g has to satisfy

$$f(x) \leq g(x, y),$$

for a given y such that

$$f(y) = g(y, y).$$

Looking at graphs of these function one would see that the function g is always above f, and they coincide at the point $x = y$. The method of minimizing f is iterative. We start with an initial value x_0, and we minimize $g(x, x_0)$ to get x_1. We then minimize $g(x, x_1)$ to get the next value, continuing in this manner until convergence. Upon convergence of the algorithm, we will also have minimized $f(x)$.

The SMACOF (Scaling by Majorizing a Complicated Function) method goes back to de Leeuw [1977], and others have refined it since then, including Groenen [1993]. The method is simple, and it can be used for both metric and nonmetric applications. We now follow the notation of Borg and Groenen [1997] to describe the algorithm for the metric case. They show that the SMACOF algorithm satisfies the requirements for minimizing a function using majorization, as described above. We leave out the derivation, but those who are interested in this and in the nonmetric case can refer to the above references as well as Cox and Cox [2001].

The *raw stress* is written as

$$\sigma(\mathbf{X}) = \sum_{r<s} w_{rs} (d_{rs}(\mathbf{X}) - \delta_{rs})^2$$

$$= \sum_{r<s} w_{rs} \delta_{rs}^2 + \sum_{r<s} w_{rs} d_{rs}^2(\mathbf{X}) - 2 \sum_{r<s} w_{rs} \delta_{rs} d_{rs}(\mathbf{X}),$$

for all available dissimilarities δ_{rs}. The inequality $r < s$ in the summation means that we only sum over half of the data, since we are assuming that the dissimilarities and distances are symmetric. We might have some missing values, so we define a weight w_{rs} with a value of 1 if the dissimilarity is present and a value of 0 if it is missing. The notation $d_{rs}(\mathbf{X})$ makes it explicit that the distances are a function of $\mathbf{X}$ (d-dimensional transformed observations), and that we are looking for a configuration $\mathbf{X}$ that minimizes stress.

Before we describe the algorithm, we need to present some relationships and notation. Let $\mathbf{Z}$ be a possible configuration of points. The matrix $\mathbf{V}$ has elements given by the following

$$v_{ij} = -w_{ij}, \qquad i \neq j$$

and

$$v_{ii} = \sum_{j=1, \, i \neq j}^{n} w_{ij}.$$

This matrix is not of full rank, and we will need the inverse in one of the update steps. So we turn to the **Moore-Penrose inverse**, which will be denoted by $\mathbf{V}^+$. [2]

We next define matrix $\mathbf{B}(\mathbf{Z})$ with elements

$$b_{ij} = \begin{cases} -\dfrac{w_{ij} \delta_{ij}}{d_{ij}(\mathbf{Z})} & d_{ij}(\mathbf{Z}) \neq 0 \\ 0 & d_{ij}(\mathbf{Z}) = 0 \end{cases}, \quad i \neq j$$

and

$$b_{ii} = -\sum_{j=1, \, j \neq i}^{n} b_{ij}.$$

---

[2] The Moore-Penrose inverse is also called the **pseudoinverse** and can be computed using the singular value decomposition. MATLAB has a function called **pinv** that provides this inverse.

We are now ready to define the **Guttman transform**. The general form of the transform is given by

$$\mathbf{X}^k = \mathbf{V}^+\mathbf{B}(\mathbf{Z})\mathbf{Z},$$

where the k represents the iteration number in the algorithm. If all of the weights are one (none of the dissimilarities are missing), then the transform is much simpler:

$$\mathbf{X}^k = n^{-1}\mathbf{B}(\mathbf{Z})\mathbf{Z}.$$

The SMACOF algorithm is outlined below.

SMACOF Algorithm

1. Find an initial configuration of points in R^d. This can either be random or nonrandom (i.e., some regular grid). Call this $\mathbf{X}^0$.
2. Set $\mathbf{Z} = \mathbf{X}^0$ and the counter to $k = 0$.
3. Compute the raw stress $\sigma(\mathbf{X}^0)$.
4. Increase the counter by 1: $k = k + 1$.
5. Obtain the Guttman transform $\mathbf{X}^k$.
6. Compute the stress for this iteration, $\sigma(\mathbf{X}^k)$.
7. Find the difference in the stress values between the two iterations. If this is less than some pre-specified tolerance or if the maximum number of iterations has been reached, then stop.
8. Set $\mathbf{Z} = \mathbf{X}^k$, and go to step 4.

We illustrate this algorithm in the next example.

Example 3.2

We turn to a different data set for this example and look at the **leukemia** data. Recall that we can look at either genes or patients as our observations; in this example we will be looking at the genes. To make things easier, we only implement the case where all the weights are 1, but the more general situation is easily implemented using the above description. First we load the data and get the distances.

```
% Use the Leukemia data, using the genes (columns)
% as the observations.
load leukemia
y = leukemia';
% Get the interpoint distance matrix.
% pdist gets the interpoint distances.
```

```
% squareform converts them to a square matrix.
D = squareform(pdist(y,'seuclidean'));
[n,p] = size(D);
% Turn off this warning... :
warning off MATLAB:divideByZero
```

Next we get the required initial configuration and the stress associated with it.

```
% Get the first term of stress.
% This is fixed - does not depend on the configuration.
stress1 = sum(sum(D.^2))/2;
% Now find an initial random configuration.
d = 2;
% Part of Statistics Toolbox
Z = unifrnd(-2,2,n,d);
% Find the stress for this.
DZ = squareform(pdist(Z));
stress2 = sum(sum(DZ.^2))/2;
stress3 = sum(sum(D.*DZ));
oldstress = stress1 + stress2 - stress3;
```

Now we iteratively adjust the configuration until the stress converges.

```
% Iterate until stress converges.
tol = 10^(-6);
dstress = realmax;
numiter = 1;
dstress = oldstress;
while dstress > tol & numiter <= 100000
    numiter = numiter + 1;
    % Now get the update.
    BZ = -D./DZ;
    for i = 1:n
        BZ(i,i) = 0;
        BZ(i,i) = -sum(BZ(:,i));
    end
    X = n^(-1)*BZ*Z;
    Z = X;
    % Now get the distances.
    % Find the stress.
    DZ = squareform(pdist(Z));
    stress2 = sum(sum(DZ.^2))/2;
    stress3 = sum(sum(D.*DZ));
    newstress = stress1 + stress2 - stress3;
    dstress = oldstress - newstress;
    oldstress = newstress;
end
```

A scatterplot of the resulting configuration is shown in Figure 3.2, with two different ways to classify the data. We see that some clustering is visible.
❑

3.1.2 Nonmetric MDS

An algorithm for solving the nonmetric MDS problem was first discussed by Shepard [1962a, 1962b]. However, he did not introduce the idea of using a loss function. That came with Kruskal [1964a, 1964b] who expanded the ideas of Shepard and gave us the concept of minimizing a loss function called *stress*.

Not surprisingly, we first introduce some more notation and terminology for nonmetric MDS. The **disparity** is a measure of how well the distance d_{rs} matches the dissimilarity δ_{rs}. We represent the disparity as $\hat{d}_{rs}$. The r-th point in our configuration $\mathbf{X}$ will have coordinates

$$\mathbf{x}_r = (x_{r1}, \ldots, x_{rd})^T.$$

We will use the Minkowski dissimilarity to measure the distance between points in our d-dimensional space. It is defined as

$$d_{rs} = \left[\sum_{i=1}^{d} |x_{ri} - x_{si}|^{\lambda} \right]^{1/\lambda},$$

where $\lambda > 0$. See Appendix A for more information on this distance and the parameter λ.

We can view the disparities as a function of the distance, such as

$$\hat{d}_{rs} = f(d_{rs}),$$

where

$$\delta_{rs} < \delta_{ab} \Rightarrow \hat{d}_{rs} \leq \hat{d}_{ab}.$$

Thus, the order of the original dissimilarities is preserved by the disparities. Note that this condition allows for possible ties in the disparities.

We define a loss function L, which is really stress, as follows

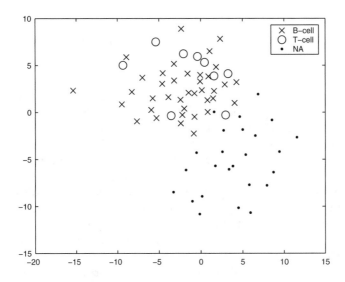

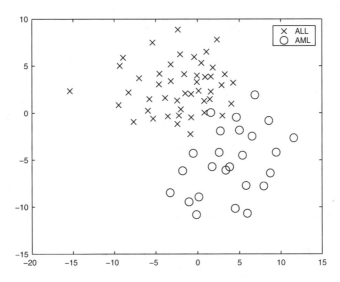

FIGURE 3.2

Here we show the results of using the SMACOF algorithm on the **leukemia** data set. The top panel shows the data transformed to 2-D and labeled by B-cell and T-cell. The lower panel illustrates the same data using different symbols based on the ALL or AML labels. There is some indication of being able to group the genes in a reasonable fashion.

$$L = S = \sqrt{\frac{\sum_{r<s}(d_{rs} - \hat{d}_{rs})^2}{\sum_{r<s}d_{rs}^2}} = \sqrt{\frac{S^*}{T^*}} .$$

It could be the case that we have missing dissimilarities or the values might be meaningless for some reason. If the analyst is faced with this situation, then the summations in the definition of stress are restricted to those pairs (r,s) for which the dissimilarity is available.

As with other forms of MDS, we seek a configuration of points X, such that the stress is minimized. Note that the coordinates of the configuration enter into the loss function through the distances d_{rs}. The original dissimilarities enter into the stress by imposing an ordering on the disparities. Thus, the stress is minimized subject to the constraint on the disparities. This constraint is satisfied by using *isotonic regression* (also known as *monotone regression*)[3] to obtain the disparities.

We now pause to describe the isotonic regression procedure. This was first described in Kruskal [1964b], where he developed an up-and-down-blocks algorithm. A nice explanation of this is given in Borg and Groenen [1997], as well as in the original paper by Kruskal. We choose to describe and implement the method for isotonic regression outlined in Cox and Cox [2001]. Isotonic regression of the d_{rs} on the δ_{rs} partitions the dissimilarities into blocks over which the $\hat{d}_{rs}$ are constant. The estimated disparities $\hat{d}_{rs}$ are given by the mean of the d_{rs} values within the block.

Example 3.3
The easiest way to explain (and hopefully to understand) isotonic regression is through an example. We use a simple one given in Cox and Cox [2001]. There are four objects with the following dissimilarities

$$\delta_{12} = 2.1, \ \delta_{13} = 3.0, \ \delta_{14} = 2.4, \ \delta_{23} = 1.7, \ \delta_{24} = 3.9, \ \delta_{34} = 3.2 .$$

A configuration of points yields the following distances

$$d_{12} = 1.6, \ d_{13} = 4.5, \ d_{14} = 5.7, \ d_{23} = 3.3, \ d_{24} = 4.3, \ d_{34} = 1.3 .$$

We now order the dissimilarities from smallest to largest and use a single subscript to denote the rank. We also impose the same order (and subscript) on the corresponding distances. This yields

---

[3] This is also known as *monotonic least squares regression*.

$$\delta_1 = 1.7, \; \delta_2 = 2.1, \; \delta_3 = 2.4, \; \delta_4 = 3.0, \; \delta_5 = 3.2, \; \delta_6 = 3.9$$
$$d_1 = 3.3, \; d_2 = 1.6, \; d_3 = 5.7, \; d_4 = 4.5, \; d_5 = 1.3, \; d_6 = 4.3.$$

The constraint on the disparities requires these distances to be ordered such that $d_i < d_{i+1}$. If this is what we have from the start, then we need not adjust things further. Since that is not true here, we must use isotonic regression to get $\hat{d}_{rs}$. To do this, we first get the cumulative sums of the distances d_i defined as

$$D_i = \sum_{j=1}^{i} d_j, \qquad i = 1, \; \dots, \; N,$$

where N is the total number of dissimilarities available. In essence, the algorithm finds the greatest convex minorant of the graph of D_i, going through the origin. See Figure 3.3 for an illustration of this. We can think of this procedure as taking a string, attaching it to the origin on one end and the last point on the other end. The points on the greatest convex minorant are those where the string touches the points D_i. These points partition the distances d_i into blocks, over which we will have disparities of constant value. These disparities are the average of the distances that fall into the block. Cox and Cox [2001] give a proof that this method yields the required isotonic regression. We now provide some MATLAB code that illustrates these ideas.

```
% Enter the original data.
dissim = [2.1 3 2.4 1.7 3.9 3.2];
dists = [1.6 4.5 5.7 3.3 4.3 1.3];
N = length(dissim);
% Now re-order the dissimilarities.
[dissim,ind] = sort(dissim);
% Now impose the same order on the distances.
dists = dists(ind);
% Now find the cumulative sums of the distances.
D = cumsum(dists);
% Add the origin as the first point.
D = [0 D];
```

It turns out that we can find the greatest convex minorant by finding the slope of each D_i with respect to the origin. We first find the smallest slope, which defines the first partition (i.e., it is on the greatest convex minorant). We then find the next smallest slope, after removing the first partition from further consideration. We continue in this manner until we reach the end of the points. The following code implements this process.

```
% Now find the slope of these.
slope = D(2:end)./(1:N);
```

```
% Find the points on the convex minorant by looking
% for smallest slopes.
i = 1;
k = 1;
while i <= N
    val = min(slope(i:N));
    minpt(k) = find(slope == val);
    i = minpt(k) + 1;
    k = k + 1;
end
```

It turns out that this procedure yields extra points, ones that are *not* on the convex minorant. MATLAB has a function called **convhull** that finds all of the points that are on the convex hull[4] of a set of points. This also yields extra points because we only want those points that are on the 'bottom' of the convex hull. To get the desired points, we take the intersection of the two sets.[5]

```
K = convhull(D, 0:N);
minpt = intersect(minpt+1,K) - 1;
```

Now that we have the points that divide the distances into blocks, we find the disparities, which are given by the average distance in that block.

```
% Now that we have all of the minorant points
% that divide into blocks, the disparities are
% the averages of the distances over those blocks.
j = 1;
for i = 1:length(minpt)
    dispars(j:minpt(i)) = mean(dists(j:minpt(i)));
    j = minpt(i) + 1;
end
```

The disparities are given by

$$\hat{d}_1 = 2.45, \ \hat{d}_2 = 2.45, \ \hat{d}_3 = 3.83, \ \hat{d}_4 = 3.83, \ \hat{d}_5 = 3.83, \ \hat{d}_6 = 4.3 \ .$$

The graphs that illustrate these concepts are shown in Figure 3.3, where we see that the disparities fall into three groups, as given above.
❑

Now that we know how to do isotonic regression, we can describe Kruskal's algorithm for nonmetric MDS. This is outlined below.

---

[4] The convex hull of a data set is the smallest convex region that contains the data set.
[5] Please see the M-file for an alternative approach, courtesy of Tom Lane of The MathWorks.

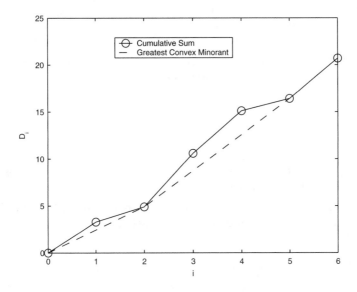

FIGURE 3.3

This shows the idea behind isotonic regression using the greatest convex minorant. The greatest convex minorant is given by the dashed line. The places where it touches the graph of the cumulative sums of the distances partition the distances into blocks. Each of the disparities in the blocks are given by the average of the distances falling in that partition.

Procedure - Kruskal's Algorithm

1. Choose an initial configuration X_0 for a given dimensionality d. Set the iteration counter k equal to 0. Set the step size α to some desired starting value.

2. Normalize the configuration to have mean of zero (i.e., centroid is at the origin) and a mean square distance from the origin equal to 1.

3. Compute the interpoint distances for this configuration.

4. Check for equal dissimilarities. If some are equal, then order those such that the corresponding distances form an increasing sequence within that block.

5. Find the disparities $\hat{d}_{rs}$ using the current configuration of points and isotonic regression.

6. Append all of the coordinate points into one vector such that

$$\mathbf{x} = (x_{11}, \ \ldots, \ x_{1d}, \ \ldots, \ x_{n1}, \ \ldots, \ x_{nd})^T.$$

7. Compute the gradient for the k-th iteration given by

$$\frac{\partial S}{\partial x_{ui}} = g_{ui} = S \sum_{r<s} (\delta^{ur} - \delta^{us}) \left[\frac{d_{rs} - \hat{d}_{rs}}{S^*} - \frac{d_{rs}}{T^*} \right] \frac{|x_{ri} - x_{si}|^{\lambda - 1}}{d_{rs}^{\lambda - 1}} \, \text{sgn}(x_{ri} - x_{si}),$$

where δ^{ur} represents the Kronecker delta function, *not* dissimilarities. This function is defined as follows

$$\delta^{ur} = 1 \qquad \text{if } u = r$$

$$\delta^{ur} = 0 \qquad \text{if } u \neq r$$

The value of the sgn($\bullet$) function is +1, if the argument is positive and −1 if it is negative.

8. Check for convergence. Determine the magnitude of the gradient vector for the current configuration using

$$\text{mag}(\mathbf{g}_k) = \sqrt{\frac{1}{n} \sum_{u, i} g_{k_{ui}}^2}.$$

If this magnitude is less than some small value ε or if some maximum allowed number of iterations has been reached, then the process stops.

9. Find the step size α

$$\alpha_k = \alpha_{k-1} \times \text{angle factor} \times \text{relaxation factor} \times \text{good luck factor},$$

where k represents the iteration number and

$$\text{angle factor} = 4^{(\cos\theta)^3} \text{ with } \left(\cos\theta = \frac{\sum\limits_{r<s} g_{k_{rs}} g_{k-1_{rs}}}{\sqrt{\sum\limits_{r<s} g_{k_{rs}}^2} \sqrt{\sum\limits_{r<s} g_{k-1_{rs}}^2}} \right)$$

$$\text{relaxation factor} = \frac{1.3}{1 + \beta^5} \text{ with } \left(\beta = \min\left[1, \frac{S_k}{S_{k-5}}\right] \right)$$

$$\text{good luck factor} = \min\left[1, \frac{S_k}{S_{k-1}}\right].$$

10. The new configuration is given by

$$\mathbf{x}_{k+1} = \mathbf{x}_k - \alpha_k \frac{\mathbf{g}_k}{\text{mag}(\mathbf{g}_k)}.$$

11. Increment the counter: $k = k + 1$. Go to step 2.

A couple of programming issues should be noted. When we are in the beginning of the iterative algorithm, we will not have values of the stress for $k - 5$ or $k - 1$ (needed in Step 9). In these cases, we will simply use the value at the first iteration until we have enough. We can use a similar idea for the gradient. Kruskal [1964b] states that this step size provides large steps during the initial iterations and small steps at the end. There is no claim of optimality with this procedure, and the reader should be aware that the final configuration is not guaranteed to be a global solution. It is likely that the configuration will be a local minimum for stress, which is typically the case for greedy iterative algorithms. It is recommended that several different starting configurations be used, and the one with the minimum stress be accepted as the final answer.

This leads to another programming point. How do we find an initial configuration to start the process? We could start with a grid of points that are evenly laid out over a d-dimensional space. Another possibility is to start from the configuration given by classical MDS. Finally, one could also generate random numbers according to a Poisson process in R^d.

Another issue that must be addressed with nonmetric MDS is how to handle ties in the dissimilarities. There are two approaches to this. The primary approach states that if $\delta_{rs} = \delta_{tu}$, then $\hat{d}_{rs}$ is not required to be equal to $\hat{d}_{tu}$. The secondary and more restrictive approach requires the disparities to be equal when dissimilarities are tied. We used the primary approach in the above procedure and in our MATLAB implementation.

Example 3.4

The **nmmds** function (included in the EDA Toolbox) that implements Kruskal's nonmetric MDS is quite long, and it involves several helper functions. So, we just show how it would be used rather than repeating all of the code here. We return to our BPM data, but this time we will look at two different topics: topic 8 (the death of the North Korean leader) and topic 11 (Hall's helicopter crash in North Korea). Since these are both about North Korea, we would expect some similarity between them. Previous experiments showed that the documents from these two topics were always grouped together, but in several sub-groups [Martinez, 2002]. We apply the nonmetric MDS method to these data using the Ochiai measure of semantic dissimilarity.

```
load ochiaibpm
% First get out the data for topics 8 and 11.
% Find the indices where they are not equal to 8 or 11.
indlab = find(classlab ~= 8 & classlab ~= 11);
% Now get rid of these from the distance matrix.
ochiaibpm(indlab,:) = [];
ochiaibpm(:,indlab)=[];
```

```
classlab(indlab) = [];
% We only need the upper part.
n = length(classlab);
dissim = [];
for i = 1:n
    dissim = [dissim, ochiaibpm(i,(i+1):n)];
end
% Find configuration for R^2.
d = 2;
r = 1;
% The nmds function is in the EDA Toolbox.
[Xd,stress,dhats] = nmmds(dissim,d,r);
ind8 = find(classlab == 8);
ind11 = find(classlab == 11);
% Plot with symbols mapped to class (topic).
plot(Xd(ind8,1),Xd(ind8,2),'.',...
    Xd(ind11,1),Xd(ind11,2),'o')
legend({'Class 8';'Class 11'})
```

The resulting plot is shown in Figure 3.4. We see that there is no clear separation between the two topics, but we do have some interesting structure in this scatterplot indicating the possible presence of sub-topics.
❑

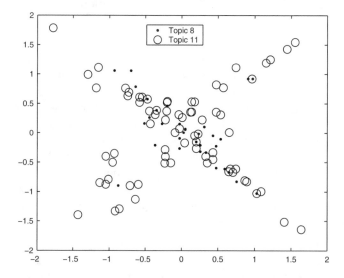

FIGURE 3.4
This shows the results of applying Kruskal's nonmetric MDS to topics 8 and 11, both of which concern North Korea. We see some interesting structure in this configuration and the possible presence of sub-topics.

What should the dimensionality be when we are applying MDS to a set of data? This is a rather difficult question to answer for most MDS methods. In the spirit of EDA, one could apply the procedure for several values of d and record the value for stress. We could then construct a plot similar to the scree plot in PCA, where we have the dimension d on the horizontal axis and the stress on the vertical axis. Stress always decreases with the addition of more dimensions, so we would be looking for the point at which the payoff from adding more dimensions decreases (i.e., the value of d where we see an elbow in the curve).

The Statistics Toolbox, Version 5 includes a new function for multi-dimensional scaling called **mdscale**. It does both metric and nonmetric MDS, and it includes several choices for the stress. With this function, one can use weights, specify the type of starting configuration, and request replicates for different random initial configurations.

3.2 Manifold Learning

Some recently developed approaches tackle the problem of nonlinear dimensionality reduction by assuming that the data lie on a submanifold of Euclidean space M. The common goal of these methods is to produce coordinates in a lower dimensional space, such that the neighborhood structure of the submanifold is preserved. In other words, points that are neighbors along the submanifold are also neighbors in the reduced parameterization of the data. (See Figure 3.5 and Example 3.5 for an illustration of these concepts.)

These new methods are discussed in this section. We will first present locally linear embedding, which is an unsupervised learning algorithm that exploits the local symmetries of linear reconstructions. Next, we cover isometric feature mapping, which is an extension to classical MDS. Finally, we present a new method called Hessian eigenmaps, which addresses one of the limitations of isometric feature mapping.

All of these methods are implemented in MATLAB, and the code is freely available for download. (We provide the URLs in Appendix B.) Because of this, we will not be including all of the code, but will only show how to use the existing implementations of the techniques.

3.2.1 Locally Linear Embedding

Locally linear embedding (LLE) was developed by Roweis and Saul [2000]. The method is an eigenvector-based method, and its optimizations do not involve local minima or iterative algorithms. The technique is based on simple geometric concepts. First, we assume that the data are sampled in

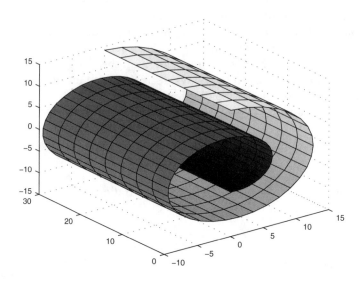

FIGURE 3.5
This shows the submanifold for the Swiss roll data set. We see that this is really a 2-D manifold (or surface) embedded in a 3-D space.

sufficient quantity from a smooth submanifold. We also assume that each data point and its neighbors lie on or close to a locally linear patch of the manifold M.

The LLE algorithm starts by characterizing the local geometry of the patches by finding linear coefficients that reconstruct each data point by using only its k nearest neighbors, with respect to Euclidean distance. There will be errors in the reconstruction, and these are measured by

$$\varepsilon(W) = \sum_i \left| \mathbf{x}_i - \sum_j W_{ij}\mathbf{x}_j \right|^2, \tag{3.4}$$

where the subscript j ranges over those points that are in the neighborhood of $\mathbf{x}_i$. The weights are found by optimizing Equation 3.4 subject to the following constraint:

$$\sum_j W_{ij} = 1.$$

The optimal weights are found using least squares, the details of which are omitted here.

Once the weights W_{ij} are found, we fix these and find a representation $\mathbf{y}_i$ of the original points in a low-dimensional space. This is also done by optimizing a cost function, which in this case is given by

$$\Phi(\mathbf{y}) = \sum_i \left| \mathbf{y}_i - \sum_j W_{ij} \mathbf{y}_j \right|^2 . \tag{3.5}$$

This defines a quadratic form in $\mathbf{y}_i$, and it can be minimized by solving a sparse eigenvector problem. The d eigenvectors corresponding to the smallest nonzero eigenvalues provide a set of orthogonal coordinates centered at the origin. The method is summarized below.

Locally Linear Embedding

1. Determine a value for k and d.
2. For each data point $\mathbf{x}_i$, find the k nearest neighbors.
3. Compute the weights W_{ij} that optimally reconstruct each data point $\mathbf{x}_i$ from its neighbors (Equation 3.4).
4. Find the d-dimensional points $\mathbf{y}_i$ that are optimally reconstructed using the same weights found in step 3 (Equation 3.5).

Note that the algorithm requires a value for k, which governs the size of the neighborhood, and a value for d. Of course, different results can be obtained when we vary these values. We illustrate the use of LLE in Example 3.5.

3.2.2 Isometric Feature Mapping - ISOMAP

Isometric feature mapping (ISOMAP) was developed by Tenenbaum, de Silva and Langford [2000] as a way of enhancing classical MDS. The basic idea is to use distances along a geodesic path (presumably measured along the manifold M) as measures of dissimilarity. As with LLE, ISOMAP assumes that the data lie on an unknown submanifold M that is embedded in a p-dimensional space. It seeks a mapping $f: X \rightarrow Y$ that preserves the intrinsic metric structure of the observations. That is, the mapping preserves the distances between observations, where the distance is measured along the geodesic path of M. It also assumes that the manifold M is globally isometric to a convex subset of a low-dimensional Euclidean space.

In Figure 3.6, we show an example that illustrates this idea. The Euclidean distance between two points on the manifold is shown by the straight line connecting them. If our goal is to recover the manifold, then a truer indication of the distance between these two points is given by their distance on the manifold, i.e., the geodesic distance along the manifold between the points.

The ISOMAP algorithm has three main steps. The first step is to find the neighboring points based on the interpoint distances d_{ij}. This can be done by either specifying a value of k to define the number of nearest neighbors or a radius ε. The distances are typically taken to be Euclidean, but they can be any valid metric. The neighborhood relations are then represented as a

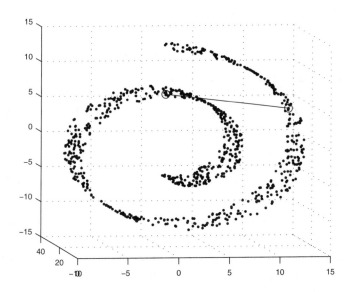

FIGURE 3.6

This is a data set that was randomly generated according to the Swiss roll parametrization [Tenenbaum, de Silva and Langford, 2000]. The Euclidean distance between two points indicated by the circles is given by the straight line shown here. If we are seeking the neighborhood structure as given by the submanifold *M*, then it would be better to use the geodesic distance (the distance along the manifold or the roll) between the points. One can think of this as the distance a bug would travel if it took the shortest path between these two points, while walking on the manifold.

weighted graph, where the edges of the graph are given weights equal to the distance d_{ij}. The second step of ISOMAP provides estimates of the geodesic distances between all pairs of points *i* and *j* by computing their shortest path distance using the graph obtained in step one. In the third step, classical MDS is applied to the geodesic distances and an embedding is found in *d*-dimensional space as described in the first section of this chapter.

Procedure - ISOMAP

1. Construct the neighborhood graph over all observations by con-necting the *ij*-th point if point *i* is one of the *k* nearest neighbors of *j* (or if the distance between them is less than ε). Set the lengths of the edges equal to d_{ij}.

2. Calculate the shortest paths between points in the graph.

3. Obtain the *d*-dimensional embedding by applying classical MDS to the geodesic paths found in step 2.

The input to this algorithm is a value for k or ε and the interpoint distance matrix. It should be noted that different results are obtained when the value for k (or ε) is varied. In fact, it could be the case that ISOMAP will return fewer than n lower-dimensional points. If this happens, then the value of k (or ε) should be increased.

3.2.3 Hessian Eigenmaps

We stated in the description of ISOMAP that the manifold M is assumed to be convex. Donoho and Grimes [2003] developed a method called **Hessian eigenmaps** that will recover a low-dimensional parametrization for data lying on a manifold that is locally isometric to an open, connected subset of Euclidean space. This significantly expands the class of data sets where manifold learning can be accomplished using isometry principles. This method can be viewed as a modification of the LLE method. Thus, it is also called **Hessian locally linear embedding** (HLLE).

We start with a parameter space $\Theta \subset R^d$ and a smooth mapping $\psi: \Theta \to R^p$, where R^p is the embedding space and $d < p$. Further, we assume that Θ is an open, connected subset of R^d, and ψ is a locally isometric embedding of Θ into R^p. The manifold can be written as a function of the parameter space, as follows

$$M = \psi(\Theta).$$

One can think of the manifold as the enumeration $m = \psi(\theta)$ of all possible measurements as one varies the parameters θ for a given process.

We assume that we have some observations m_i that represent measurements over many different choices of control parameters θ_i $(i = 1,...,n)$. These measurements are the same as our observations x_i.[6] The HLLE description given in Donoho and Grimes [2003] considers the case where all data points lie exactly on the manifold M. The goal is to recover the underlying parameterization ψ and the parameter settings θ_i, up to a rigid motion.

We now describe the main steps of the HLLE algorithm. We leave out the derivation and proofs because we just want to provide the general ideas underlying the algorithm. The reader is referred to the original paper and the MATLAB code for more information and implementation details.

The two main assumptions of the HLLE algorithm are:

1. In a small enough neighborhood of each point m, geodesic distances to nearby points m' (both on the manifold M) are identical to Euclidean distances between associated parameter points θ and θ'.

---

[6] We use different notation here to be consistent with the original paper.

This is called *local isometry*. (ISOMAP deals with the case where *M* assumes a globally isometric parameterization.)

2. The parameter space Θ is an open, connected subset of $\mathbf{R}^d$, which is a weaker condition than the convexity assumption of ISOMAP.

In general, the idea behind HLLE is to define a neighborhood around some *m* in *M* and obtain local tangent coordinates. These local coordinates are used to define the Hessian of a smooth function $f: M \to R$. The function *f* is differentiated in the tangent coordinate system to produce the tangent Hessian. A quadratic form $\mathscr{H}(f)$ is obtained using the tangent Hessian, and the isometric coordinates θ can be recovered by identifying a suitable basis for the null space of $\mathscr{H}(f)$.

The inputs required for the algorithm are a set of *n* *p*-dimensional data points, a value for *d*, and the number of nearest neighbors *k* to determine the neighborhood. The only constraint on these values is that $\min(k,p) > d$. The algorithm estimates the tangent coordinates by using the SVD on the neighborhood of each point, and it uses the empirical version of the operator $\mathscr{H}(f)$. The output of HLLE is a set of *n* *d*-dimensional embedding coordinates.

Example 3.5

We generated some data from an S-curve manifold to be used with LLE and ISOMAP and saved it in a file called **scurve.mat**.[7] We show both the true manifold and the data randomly generated from this manifold in Figure 3.7.

```
load scurve
% The scurve file contains our data matrix X.
% First set up some parameters for LLE.
K = 12;
d = 2;
% Run LLE - note that LLE takes the input data
% as rows corresponding to dimensions and
% columns corresponding to observations. This is
% the transpose of our usual data matrix.
Y = lle(X,K,d);
% Plot results in scatter plot.
scatter(Y(1,:),Y(2,:),12,[angle angle],'+','filled');
```

We show the results in Figure 3.8. Note by the colors that neighboring points on the manifold are mapped into neighboring points in the 2-D embedding. Now we use ISOMAP on the same data.

```
% Now run the ISOMAP - we need the distances for input.
% We need the data matrix as n x p.
X = X';
dists = squareform(pdist(X));
```

---

[7] Look at the file called **example35.m** for more details on how to generate the data.

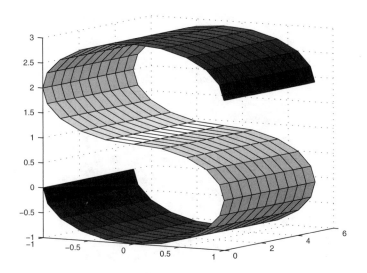

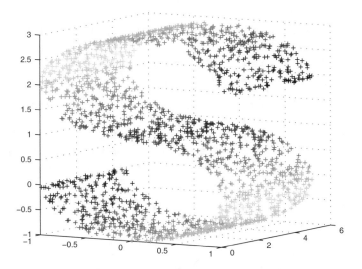

FIGURE 3.7

The top panel shows the true S-curve manifold, which is a 2-D manifold embedded in 3-D. The bottom panel is the data set randomly generated from the manifold. The gray scale values are an indication of the neighborhood. See the associated color figure following page 144.

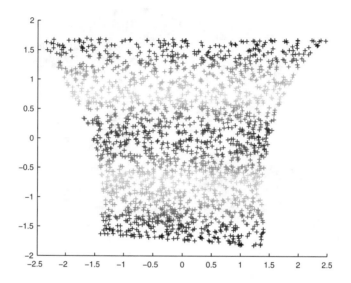

FIGURE 3.8
This is the embedding recovered from LLE. Note that the neighborhood structure is preserved. See the associated color figure following page 144.

```
options.dims = 1:10;      % These are for ISOMAP.
options.display = 0;
[Yiso, Riso, Eiso] = isomap(dists, 'k', 7, options);
```

Constructing a scatter plot of this embedding to compare with LLE is left as an exercise to the reader. As stated in the text, LLE and ISOMAP have some problems with data sets that are not convex. We show an example here using both ISOMAP and HLLE to discover such an embedding. These data were generated according to the code provided by Donoho and Grimes [2003]. Essentially, we have the Swiss roll manifold with observations removed that fall within a rectangular area along the surface.

```
% Now run the example from Grimes and Donoho.
load swissroll
options.dims = 1:10;
options.display = 0;
dists = squareform(pdist(X'));
[Yiso, Riso, Eiso] = isomap(dists, 'k', 7, options);
% Now for the Hessian LLE.
Y2 = hlle(X,K,d);
scatter(Y2(1,:),Y2(2,:),12,tt,'+');
```

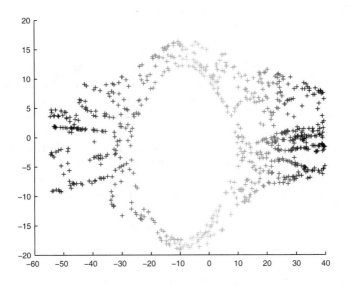

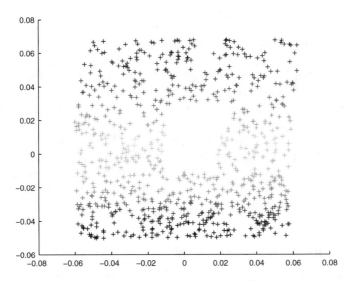

FIGURE 3.9
The top panel contains the 2-D coordinates from ISOMAP. We do see the hole in the embedding, but it looks like an oval rather than a rectangle. The bottom panel shows the 2-D coordinates from HLLE. HLLE was able to recover the correct embedding. See the associated color figure following page 144.

We see in Figure 3.9 that ISOMAP was unable to recover the correct embedding. We did find the hole, but it is distorted. HLLE was able to find the correct embedding with no distortion.
❑

We note a few algorithmic complexity issues. The two locally linear embedding methods, LLE and HLLE, can be used on problems with many data points (large n), because the initial computations are performed on small neighborhoods and they can exploit sparse matrix techniques. On the other hand, ISOMAP requires the calculation of a full matrix of geodesic distances for its initial step. However, LLE and HLLE are affected by the dimensionality p, because they must estimate a local tangent space at each data point. Also, in the HLLE algorithm, we must estimate second derivatives, which can be difficult with high-dimensional data.

3.3 Artificial Neural Network Approaches

We now discuss two other methods that we categorize as ones based on artificial neural network (ANN) ideas: the self-organizing map and the generative topographic mapping. These ANN approaches also look for intrinsically low-dimensional structures that are embedded nonlinearly in a high-dimensional space. As in MDS and the manifold learning approaches, these seek a single global low-dimensional nonlinear model of the observations. Both of these algorithms try to fit a grid or predefined topology (usually 2-D) to the data, using greedy algorithms that first fit the large-scale linear structure of the data and then make small-scale nonlinear refinements.

There are MATLAB toolboxes available for both self-organizing maps and generative topographic maps. They are free, and they come with extensive documentation and examples. Because the code for these methods can be very extensive and would not add to the understanding of the reader, we will not be showing the code in the book. Instead, we will show how to use some of the functions in the examples.

3.3.1 Self-Organizing Maps - SOM

The *self-organizing map* or SOM is a tool for the exploration and visualization of high-dimensional data [Kohonen, 1998]. It derives an orderly mapping of data onto a regular, low-dimensional grid. The dimensionality of the grid is usually $d = 2$ for ease of visualization. It converts complex, nonlinear relationships in the high-dimensional space into simpler geometric relationships such that the important topological and metric relationships are conveyed. The data are organized on the grid in such a way that observations that are close together in the high-dimensional space are also closer to each other on the grid. Thus, this is very similar to the ideas of MDS, except that the positions in the low-dimensional space are restricted to the grid. The grid locations are denoted by $\mathbf{r}_i$.

There are two methods used in SOM: incremental learning and batch mapping. We will describe both of them, and then illustrate some of the functions in the SOM Toolbox[8] in Example 3.6. We will also present a brief discussion of the various methods for visualizing and exploring the results of SOM.

The *incremental* or *sequential learning method* for SOM is an iterative process. We start with the set of observations $\mathbf{x}_i$ and a set of p-dimensional model vectors $\mathbf{m}_j$. These *model vectors* are also called *neurons*, *prototypes*, or *codebooks*. Each model vector is associated with a vertex ($\mathbf{r}_i$) on a 2-D lattice that can be either hexagonal or rectangular and starts out with some initial value ($\mathbf{m}_i(t=0)$). This could be done by setting them to random values, by setting them equal to a random subset of the original data points or by using PCA to provide an initial ordering.

At each training step, a vector $\mathbf{x}_i$ is selected and the distance between it and all the model vectors is calculated, where the distance is typically Euclidean. The SOM Toolbox handles missing values by excluding them from the calculation, and variable weights can also be used. The best-matching unit (BMU) or model vector is found and is denoted by $\mathbf{m}_c$. Once the closest model vector $\mathbf{m}_c$ is found, the model vectors are updated so that $\mathbf{m}_c$ is moved closer to the data vector $\mathbf{x}_i$. The neighbors of the BMU $\mathbf{m}_c$ are also updated, in a weighted fashion. The update rule for the model vector $\mathbf{m}_i$ is

$$\mathbf{m}_i(t+1) = \mathbf{m}_i(t) + \alpha(t)h_{ci}(t)[\mathbf{x}(t) - \mathbf{m}_i(t)],$$

where t denotes time or iteration number. The learning rate is given by $\alpha(t)$ and $0 < \alpha(t) < 1$, which decreases monotonically as the iteration proceeds.

The neighborhood is governed by the function $h_{ci}(t)$, and a Gaussian centered at the best-matching unit is often used. The SOM Toolbox has other neighborhood functions available, as well as different learning rates $\alpha(t)$. If the Gaussian is used, then we have

$$h_{ci}(t) = \exp\left(-\frac{\|\mathbf{r}_i - \mathbf{r}_c\|^2}{2\sigma^2(t)}\right),$$

where the symbol $\|\cdot\|$ denotes the distance, and the $\mathbf{r}_i$ are the coordinates on the grid. The size or width of the neighborhood is governed by $\sigma(t)$, and this value decreases monotonically.

The training is usually done in two phases. The first phase corresponds to a large learning rate and neighborhood radius $\sigma(t)$, which provides a large-scale approximation to the data. The map is fine tuned in the second phase,

---

[8]See Appendix B for the website where one can download the SOM Toolbox.

where the learning rate and neighborhood radius are small. Once the process finishes, we have the set of prototype or model vectors over the 2-D coordinates on the map grid.

The *batch training method* or Batch Map is also iterative, but it uses the whole data set before adjustments are made rather than a single vector. At each step of the algorithm, the data set is partitioned such that each observation is associated with its nearest model vector. The updated model vectors are found as a weighted average of the data, where the weight of each observation is the value of the neighborhood function $h_{ic}(t)$ at its BMU c.

Several methods exist for visualizing the resulting map and prototype vectors. These methods can have any of the following goals. The first is to get an idea of the overall shape of the data and whether clusters are present. The second goal is to analyze the prototype vectors for characteristics of the clusters and correlations among components or variables. The third task is to see how new observations fit with the map or to discover anomalies. We will focus on one visualization method called the U-matrix that is often used to locate clusters in the data [Ultsch and Siemon, 1990].

The U-matrix is based on distances. First, the distance of each model vector to each of its immediate neighbors is calculated. This distance is then visualized using a color scale. Clusters are seen as those map units that have smaller distances, surrounded by borders indicating larger distances. Another approach is to use the size of the symbol to represent the average distance to its neighbors, so cluster borders would be shown as larger symbols. We illustrate the SOM and the U-matrix visualization method in Example 3.6.

Example 3.6

We turn to the **oronsay** data set to illustrate some of the basic commands in the SOM Toolbox. First we have to load the data and put it into a MATLAB data structure that is recognized by the functions. This toolbox comes with several normalization methods, and it is recommended that they be used before building a map.

```
% Use the oronsay data set.
load oronsay
% Convert labels to cell array of strings.
for i = 1:length(beachdune)
    mid{i} = int2str(midden(i));
    % Use next one in exercises.
    bd{i} = int2str(beachdune(i));
end
% Normalize each variable to have unit variance.
D = som_normalize(oronsay,'var');
% Convert to a data structure.
sD = som_data_struct(D);
% Add the labels - must be transposed.
```

```
sD = som_set(sD,'labels',mid');
```

We can visualize the results in many ways, most of which will be left as an exercise. We show the U-matrix in Figure 3.10 and include labels for the codebook vectors.

```
% Make the SOM
sM = som_make(sD);
sM = som_autolabel(sM,sD,'vote');
% Plot U matrix.
som_show(sM,'umat','all');
% Add labels to an existing plot.
som_show_add('label',sM,'subplot',1);
```

Note that the larger values indicate cluster borders, and low values indicate clusters. By looking at the colors, we see a couple of clusters - one in the lower left corner and one in the top. The labels indicate some separation into groups.

❑

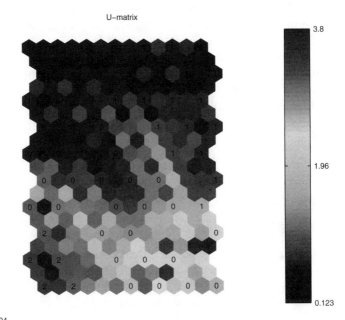

SOM 26–Jun–2004

FIGURE 3.10
This is the SOM for the **oronsay** data set. We can see some cluster structure here by the colors. One is in the upper part of the map, and the other is in the lower left corner. The labels on the map elements also indicate some clustering. Recall that the classes are midden (0), *Cnoc Coig* (1), and *Caisteal nan Gillean* (2). See the associated color figure following page 144.

3.3.2 Generative Topographic Maps - GTM

The SOM has several limitations [Bishop, Svensén and Williams, 1996]. First, the SOM is based on heuristics, and the algorithm is not derived from the optimization of an objective function. Preservation of the neighborhood structure is not guaranteed by the SOM method, and there could be problems with convergence of the prototype vectors. The SOM does not define a density model, and the use of the codebook vectors as a model of the distribution of the original data is limited. Finally, the choice of how the neighborhood function should shrink during training is somewhat of an art, so it is difficult to compare different runs of the SOM procedure on the same data set. The *generative topographic mapping* (GTM) was inspired by the SOM and attempts to overcome its limitations.

The GTM is described in terms of a latent variable model (or space) with dimensionality d [Bishop, Svensén and Williams, 1996, 1998]. The goal is to find a representation for the distribution $p(\mathbf{x})$ of p-dimensional data, in terms of a smaller number of d latent variables $\mathbf{m} = (m_1, \dots , m_d)$. As usual, we want $d < p$, and often take $d = 2$ for ease of visualization. We can achieve this goal by finding a mapping $\mathbf{y}(\mathbf{m};\mathbf{W})$, where $\mathbf{W}$ is a matrix containing weights, that takes a point $\mathbf{m}$ in the latent space and maps it to a point $\mathbf{x}$ in the data space. This might seem somewhat backward from what we considered previously, but we will see later how we can use the inverse of this mapping to view summaries of the observations in the reduced latent space.

We start with a probability distribution $p(\mathbf{m})$ defined on the latent variable space (with d dimensions), which in turn induces a distribution $p(\mathbf{y} \mid \mathbf{W})$ in the data space (with p dimensions). For a given $\mathbf{m}$ and $\mathbf{W}$, we choose a Gaussian centered at $\mathbf{y}(\mathbf{m};\mathbf{W})$ as follows:

$$p(\mathbf{x} \mid \mathbf{m}, \mathbf{W}, \beta) = \left(\frac{\beta}{2\pi}\right)^{p/2} \exp\left\{\frac{-\beta}{2}\|\mathbf{y}(\mathbf{m};\mathbf{W}) - \mathbf{x}\|^2\right\},$$

where the variance is β^{-1},[9] and $\|\bullet\|$ denotes the inner product. Other models can be used, as discussed in Bishop, Svensén and Williams [1998]. The matrix $\mathbf{W}$ contains the parameters or weights that govern the mapping. To derive the desired mapping, we must estimate β and the matrix $\mathbf{W}$.

The next step is to assume some form for $p(\mathbf{m})$ defined in the latent space. To make this similar to the SOM, the distribution is taken as a sum of delta functions centered on the nodes of a regular grid in latent space:

$$p(\mathbf{m}) = \frac{1}{K} \sum_{k=1}^{K} \delta(\mathbf{m} - \mathbf{m}_k),$$

---

[9] We note that this is different than the usual notation, familiar to statisticians. However, we are keeping it here to be consistent with the original derivation.

where K is the total number of grid points or delta functions. Each point $\mathbf{m}_k$ in the latent space is mapped to a corresponding point $\mathbf{y}(\mathbf{m}_k;\mathbf{W})$ in the data space, where it becomes the center of a Gaussian density function. This is illustrated in Figure 3.11.

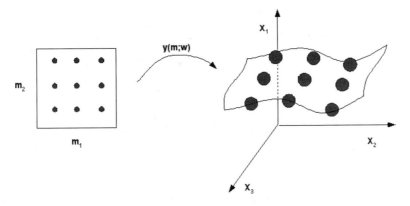

FIGURE 3.11
This illustrates the mapping used in GTM. Our latent space is on the left, and the data space is on the right. A Gaussian is centered at each of the data points, represented by the spheres.

We can use maximum likelihood and the Expectation-Maximization (EM) algorithm (see Chapter 6) to get estimates for β and $\mathbf{W}$. Bishop, Svensén and Williams [1998] show that the log-likelihood function for the distribution described here is given by

$$L(\mathbf{W}, \beta) = \sum_{i=1}^{n} \ln \left[\frac{1}{K} \sum_{k=1}^{K} p(\mathbf{x}_i | \mathbf{m}_k, \mathbf{W}, \beta) \right]. \qquad (3.6)$$

They next choose a model for $\mathbf{y}(\mathbf{m};\mathbf{W})$, which is given by

$$\mathbf{y}(\mathbf{x};\mathbf{W}) = \mathbf{W}\phi(\mathbf{x}),$$

where the elements of $\phi(\mathbf{m})$ have M fixed basis functions $\phi_j(\mathbf{m})$, and $\mathbf{W}$ is of size $p \times M$. For basis functions, they choose Gaussians whose centers are distributed on a uniform grid in latent space. Note that the centers for these basis functions are *not* the same as the grid points $\mathbf{m}_i$. Each of these Gaussian basis functions ϕ has a common width parameter σ. The smoothness of the manifold is determined by the value of σ, the number of basis functions M, and their spacing.

Looking at Equation 3.6, we can view this as a missing-data problem, where we do not know which component k generated each data point $\mathbf{x}_i$. The

EM algorithm for estimating β and **W** consists of two steps that are done iteratively until the value of the log-likelihood function converges. In the E-step, we use the current values of the parameters to evaluate the posterior probabilities of each component k for every data point. This is calculated according to the following

$$\tau_{ki}(\mathbf{W}_{\text{old}}, \beta_{\text{old}}) = \frac{p(\mathbf{x}_i | \mathbf{m}_k, \mathbf{W}_{\text{old}}, \beta_{\text{old}})}{\displaystyle\sum_{c=1}^{K} p(\mathbf{x}_i | \mathbf{m}_c, \mathbf{W}_{\text{old}}, \beta_{\text{old}})}, \tag{3.7}$$

where the subscript 'old' indicates the current values.

In the M-step, we use the posterior probabilities to find weighted updates for the parameters. First, we calculate a new version of the weight matrix from the following equation

$$\Phi^T \mathbf{G}_{\text{old}} \Phi \mathbf{W}_{\text{new}}^T = \Phi^T \mathbf{T}_{\text{old}} \mathbf{X}, \tag{3.8}$$

where Φ is a $K \times M$ matrix with elements $\Phi_{kj} = \phi_j(\mathbf{m}_k)$, **X** is the data matrix, **T** is a $K \times n$ matrix with elements τ_{ki}, and **G** is a $K \times K$ diagonal matrix where the elements are given by

$$G_{kk} = \sum_{i=1}^{n} \tau_{ki}(\mathbf{W}, \beta).$$

Equation 3.8 is solved for $\mathbf{W}_{\text{new}}$ using standard linear algebra techniques. A time-saving issue related to this update equation is that Φ is constant, so it only needs to be evaluated once.

We now need to determine an update for β that maximizes the log-likelihood. This is given by

$$\frac{1}{\beta_{\text{new}}} = \frac{1}{np} \sum_{i=1}^{n} \sum_{k=1}^{K} \tau_{kn}(\mathbf{W}_{\text{old}}, \beta_{\text{old}}) \| \mathbf{W}_{\text{old}} \phi(\mathbf{m}_k - \mathbf{x}_i) \|^2. \tag{3.9}$$

To summarize, the algorithm requires starting points for the matrix **W** and the inverse variance β. We must also specify a set of points $\mathbf{m}_i$, as well as a set of basis functions $\phi_j(\mathbf{m})$. The parameters **W** and β define a mixture of Gaussians with centers $\mathbf{W}\phi(\mathbf{m}_k)$ and equal covariance matrices given by $\beta^{-1}\mathbf{I}$. Given initial values, the EM algorithm is used to estimate these parameters. The E-step finds the posterior probabilities using Equation 3.7, and the M-step updates the estimates for **W** and β using Equations 3.8 and 3.9. These steps are repeated until the log-likelihood (Equation 3.6) converges.

So, the GTM gives us a mapping from this latent space to the original p-dimensional data space. For EDA purposes, we are really interested in going the other way: mapping our p-dimensional data into some lower-dimensional space. As we see from the development of the GTM, each datum x_i provides a posterior distribution in our latent space. Thus, this posterior *distribution* in latent space provides information about a *single* observation. Visualizing all observations in this manner would be too difficult, so each distribution should be summarized in some way. Two summaries that come to mind are the mean and the mode, which are then visualized as individual points in our latent space. The mean for observation x_i is calculated from

$$\overline{\mathbf{m}}_i = \sum_{k=1}^{K} \tau_{ki} \mathbf{m}_k .$$

The mode for the i-th observation (or posterior distribution) is given by the maximum value τ_{ki} over all values of k. The values $\overline{\mathbf{m}}_i$ or modes are shown as symbols in a scatterplot or some other visualization scheme.

Example 3.7

We again turn to the **oronsay** data to show the basic functionality of the GTM Toolbox.[10] The parameters were set according to the example in their documentation.

```
load oronsay
% Initialize parameters for GTM.
noLatPts = 400;
noBasisFn = 81;
sigma = 1.5;
% Initialize required variables for GTM.
[X,MU,FI,W,beta] = gtm_stp2(oronsay,noLatPts,...
    noBasisFn,sigma);
lambda = 0.001;
cycles = 40;
[trndW,trndBeta,llhLog] = gtm_trn(oronsay,FI,W,...
    lambda,cycles,beta,'quiet');
```

The function **gtm_stp2** initializes the required variables, and **gtm_trn** does the training. Each observation gets mapped into a probability distribution in the 2-D map, so we need to find either the mean or mode of each one to show as a point. We can do this as follows:

```
% Get the means in latent space.
mus = gtm_pmn(oronsay,X,FI,trndW,trndBeta);
% Get the modes in latent space.
```

---

[10] See Appendix B for information on where to download the GTM Toolbox.

```
modes = gtm_pmd(oronsay,X,FI,trndW);
```

We now plot the values in the lower-dimensional space using symbols
corresponding to their class.

```
ind0 = find(midden == 0);
ind1 = find(midden == 1);
ind2 = find(midden == 2);
plot(mus(ind0,1),mus(ind0,2),'k.',mus(ind1,1),...
     mus(ind1,2),'kx',mus(ind2,1),mus(ind2,2),'ko')
```

The resulting plot is shown in Figure 3.12, where we can see some separation
into the three groups.
❑

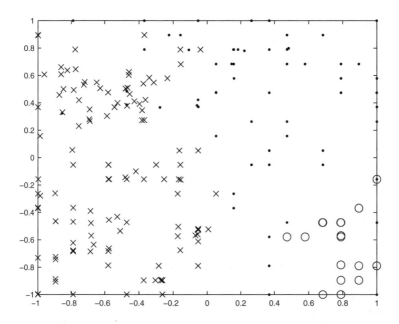

FIGURE 3.12

This shows the map obtained from GTM, where the distribution for each point is summarized by the mean. Each mean is displayed using a different symbol: Class 0 is ' . '; Class 1
is 'x'; and Class 2 is 'o'. We can see some separation into groups from this plot.

3.4 Summary and Further Reading

In this chapter, we discussed several methods for finding a nonlinear
mapping from a high-dimensional space to one with lower dimensionality.

The first set of methods was grouped under the name multidimensional scaling, and we presented both metric and nonmetric MDS. We note that in some cases, depending on how it is set up, MDS finds a linear mapping. We also presented several methods for learning manifolds, where the emphasis is on nonlinear manifolds. These techniques are locally linear embedding, ISOMAP, and Hessian locally linear embedding. ISOMAP is really an enhanced version of classical MDS, where geodesic distances are used as input to classical MDS. HLLE is similar in spirit to LLE, and its main advantage is that it can handle data sets that are not convex. Finally, we presented two artificial neural network approaches called self-organizing maps and generative topographic maps.

We have already mentioned some of the major MDS references, but we also include them here with more information on their content. Our primary reference for terminology and methods was Cox and Cox [2001]. This is a highly readable book, suitable for students and practitioners with a background in basic statistics. The methods are clearly stated for both classical MDS and nonmetric MDS. Also, the authors provide a CD-ROM with programs (running under DOS) and data sets so the reader can use these tools.

Another book by Borg and Groenen [1997] brings many of the algorithms and techniques of MDS together in a way that can be understood by those with a two-semester course in statistics for social sciences or business. The authors provide the derivation of the methods, along with ways to interpret the results. They do not provide computer software or show how this can be done using existing software packages, but their algorithms are very clear and understandable.

A brief introduction to MDS can be obtained from Kruskal and Wish [1978]. This book is primarily focused on applications in the social sciences, but it would be useful for those who need to come to a quick understanding of the basics of MDS. Computational and algorithmic considerations are not stressed in this book. There are many overview papers on MDS in the various journals and encyclopedias; these include Mead [1992], Siedlecki, Siedlecka and Sklansky [1988], Steyvers [2002], and Young [1985].

Some of the initial work in MDS was done by Shepard in 1962. The first paper in the series [Shepard, 1962a] describes a computer program to reconstruct a configuration of points in Euclidean space, when the only information that is known about the distance between the points is some unknown monotonic function of the distance. The second paper in the series [Shepard, 1962b] presents two applications of this method to artificial data. Kruskal continued the development of MDS by introducing an objective function and developing a method for optimizing it [Kruskal, 1964a, 1964b]. We highly recommend reading the original Kruskal papers; they are easily understood and should provide the reader with a nice understanding of the origins of MDS.

Because they are relatively recent innovations, there are not a lot of references for the manifold learning methods (ISOMAP, LLE, and HLLE).

However, each of the websites has links to technical reports and papers that describe the work in more depth (see Appendix B). A nice overview of manifold learning is given in Saul and Roweis [2002], with an emphasis on LLE. Further issues with ISOMAP are explored in Balasubramanian and Schwartz [2002]. More detailed information regarding HLLE can be found in a technical report written by Donoho and Grimes [2002].

There is one book dedicated to SOM written by Kohonen [2001]. Many papers on SOM have appeared in the literature, and a 1998 technical report lists 3,043 works that are based on the SOM [Kangas and Kaski, 1998]. A nice, short overview of SOM can be found in Kohonen [1998]. Some recent applications of SOM have been in the area of document clustering [Kaski, et al., 1998; Kohonen, et al., 2000] and the analysis of gene microarrays [Tamayo, et al., 1999]. Theoretical aspects of the SOM are discussed in Cottrell, Fort and Pages [1998]. The use of the SOM for clustering is described in Kiang [2001] and Vesanto and Alhoniemi [2000]. Visualization and EDA methods for SOM are discussed in Mao and Jain [1995], Ultsch and Siemon [1990], Deboeck and Kohonen [1998], and Vesanto [1997; 1999]. GTM is a recent addition to this area, so there are fewer papers, but for those who want further information, we recommend Bishop, Svensén and Williams [1996, 1997a, 1997b, and 1998], Bishop, Hinton and Strachan [1997], and Bishop and Tipping [1998].

Exercises

3.1   Try the classical MDS approach using the **skull** data set. Plot the results in a scatterplot using the text labels as plotting symbols (see the **text** function). Do you see any separation between the categories of gender? [Cox and Cox, 2001]. Try PCA on the **skull** data. How does this compare with the classical MDS results?

3.2   Apply the SMACOF and nonmetric MDS methods to the **skull** data set. Compare your results with the configuration obtained through classical MDS.

3.3   Use **plot3** (similar to **plot**, but in three dimensions) to construct a 3-D scatterplot of the data in Example 3.1. Describe your results.

3.4   Apply the SMACOF method to the **oronsay** data set and comment on the results.

3.5   Repeat the Examples 3.2, 3.4 and problem 3.4 for several values of d. See the **help** on **gplotmatrix** and use it to display the results for $d > 2$. Do a scree-like plot of stress versus d to see what d is best.

3.6   The Shepard diagram for nonmetric MDS is a plot where the ordered dissimilarities are on the horizontal axis. The distances (shown as points) and the disparities (shown as a line) are on the vertical. With large data sets, this is not too useful. However, with small data sets, it can be used to see the shape of the regression curve. Implement this in MATLAB and test it on one of the smaller data sets.

3.7 Try using **som_show(sM)** in Example 3.6. This shows a U-matrix for each variable. Look at each one individually and add the labels to the elements: **som_show(sM,'comp',1)**, etc. See the SOM Toolbox documentation for more information on how these functions work.

3.8 Repeat Example 3.6 and problem 3.7 using the other labels for the **oronsay** data set. Discuss your results.

3.9 Repeat the plot in Example 3.7 (GTM) using the modes instead of the means. Do you see any difference between them?

3.10 Do a help on the Statistics Toolbox (version 5) function **mdscale**. Apply the methods (metric and nonmetric MDS) to the **skulls** and **oronsay** data sets.

3.11 Apply the ISOMAP method to the **scurve** data from Example 3.5. Construct a scatterplot of the data and compare to the results from LLE.

3.12 Apply the LLE method to the **swissroll** data of Example 3.5. Construct a scatterplot and compare with HLLE and ISOMAP.

3.13 What is the intrinsic dimensionality of the **swissroll** and **scurve** data sets?

3.14 Where possible, apply MDS, ISOMAP, LLE, HLLE, SOM, and GTM to the following data sets. Discuss and compare the results.

   a. BPM data sets

   b. gene expression data sets

   c. **iris**

   d. **pollen**

   e. **posse**

   f. **oronsay**

   g. **skulls**

3.15 Repeat Example 3.2 with different starting values to search for different structures. Analyze your results.

Chapter 4

Data Tours

In previous chapters we searched for lower-dimensional representations of our data that might show interesting structure. However, there are an infinite number of possibilities, so we might try touring through space, looking at many of these representations. This chapter includes a description of several tour methods that can be roughly categorized into the following groups:

1. **Grand Tours**: If the goal is to look at the data from all possible viewpoints to get an idea of the overall distribution of the p-dimensional data, then we might want to look at a random sequence of lower-dimensional projections. Typically, there is little user interaction with these methods other than to set the step size for the sequence and maybe to stop the tour when interesting structure is found. This was the idea behind the original torus grand tour of Asimov [1985].

2. **Interpolated Tours**: In this type of tour, the user chooses a starting plane and an ending plane. The data are projected onto the first plane. The tour then proceeds from one plane to the other by interpolation. At each step of the sequence, the user is presented with a different view of the data, usually in the form of a scatter plot [Hurley and Buja, 1990].

3. **Guided Tours**: These tours can be either partly or completely guided by the data. An example of this type of tour is the EDA projection pursuit method. While this is not usually an interactive or visual tour in the sense of the others, it does look at many projections of the data searching for interesting structure such as holes, clusters, etc.

Each of these methods is described in more detail in the following sections.

4.1 Grand Tour

In the grand tour methods, we want to view the data from 'all' possible perspectives. This is done by projecting the data onto a 2-D subspace and then viewing it as a scatterplot. We do this repeatedly and rapidly, so the user ends up seeing an animated sequence (or movie) of scatterplots. We could also project to spaces with dimensionality greater than 2, but the visualization would have to be done differently (see Chapter 10 for more on this subject). Projections to 1-D (a line) are also possible, where we could show the individual points on the line or as a distribution (e.g., histogram or other estimate of data density). We describe two grand tour methods in this section: the torus winding method and the pseudo grand tour.

In general, grand tours should have the following desirable characteristics:

1. The sequence of planes (or projections) should be dense in the space of all planes, so the tour eventually comes close to any given 2-D projection.

2. The sequence should become dense rapidly, so we need an efficient algorithm to compute the sequence, project the data, and present it to the user.

3. We want our sequence of planes to be uniformly distributed, because we do not want to spend a lot of time in one area.

4. The sequence of planes should be 'continuous' to aid user understanding and to be visually appealing. However, a trade-off between continuity and speed of the tour must be made.

5. The user should be able to reconstruct the sequence of planes after the tour is over. If the user stops the tour at a point where interesting structure is found, then that projection should be recovered easily.

To achieve these, the grand tour algorithm requires a continuous, space-filling path through the set of 2-D subspaces in p-dimensional space.

To summarize, the grand tour provides an overview or tour of a high-dimensional space by visualizing a sequence of 2-D scatterplots, in such a way that the tour is representative of all projections of the data. The tour continues until the analyst sees some interesting structure, at which time it is halted. The output of the grand tour method is a movie or animation with information encoded in the smooth motion of the 2-D scatterplots. A benefit from looking at a moving sequence of scatterplots is that two additional dimensions of information are available in the speed vectors of the data points [Buja & Asimov, 1986]. For example, the further away a point is from the computer screen, the faster the point rotates.

4.1.1 Torus Winding Method

The *torus winding method* was originally proposed by Asimov [1985] and Buja and Asimov [1986] as a way of implementing the grand tour. We let $\{\lambda_1,...,\lambda_N\}$ be a set of real numbers that are linearly independent over the integers. We also define a function a(t) as

$$a(t) = (\lambda_1 t, \ ... \ , \lambda_N t), \tag{4.1}$$

where the coordinates $\lambda_i t$ are interpreted modulo 2π. It is known that the mapping in Equation 4.1 defines a space-filling path [Asimov, 1985; Wegman and Solka, 2002] that winds around a torus.

We let the vector e_i represent the canonical basis vector. It contains zeros everywhere except in the i-th position where it has a one. Next we let $R_{ij}(\theta)$ denote a $p \times p$ matrix that rotates the $e_i e_j$ plane through an angle of size θ. This is given by the identity matrix, but with the following changes:

$$r_{ii} = r_{jj} = \cos(\theta); \ r_{ij} = -\sin(\theta); \ r_{ji} = \sin(\theta).$$

We then define a function f as follows

$$f(\theta_{1,2}, ..., \ \theta_{p-1,p}) = Q = R_{12}(\theta_{12}) \times R_{13}(\theta_{13}) \times ... \times R_{p-1,p}(\theta_{p-1,p}). \tag{4.2}$$

Note that we have N arguments (or angles) in the function f, and it is subject to the restrictions that $0 \le \theta_{ij} \le 2\pi$ and $1 \le i < j \le p$. We use a reduced form of this function with fewer terms [Asimov, 1985] in our procedure outlined below.

Procedure - Torus Method

1. The number of factors in Equation 4.2 is given by $N = 2p - 3$.[1] We use only $R_{ij}(\theta_{ij})$ with $i = 1$ or $i = 2$. If $i = 1$, then $2 \le j \le p$. If $i = 2$, then $3 \le j \le p$.
2. Choose real numbers $\{\lambda_1, \ ... \ , \lambda_N\}$ and a stepsize t such that the numbers $\{2\pi, \ \lambda_1 t, \ ... \ , \lambda_N t\}$ are linearly independent. The stepsize t should be chosen to yield a continuous sequence of planes.
3. The values $\lambda_1 Kt, \ ... \ , \lambda_N Kt, K = 1, 2, ...$ are used as the arguments to the function $f(\bullet)$. K is the iteration number.
4. Form the product Q_K of all the rotations (Equation 4.2).
5. Rotate the first two basis vectors using

---

[1] We are using the reduced form described in the appendix of Asimov [1985]. The complete form of this rotation consists of $(p^2 - p)/2$ possible plane rotations, corresponding to the distinct 2–planes formed by the canonical basis vectors.

$$\mathbf{A}_K = \mathbf{Q}_K \mathbf{E}_{12},$$

where the columns of $\mathbf{E}_{12}$ contain the first two basis vectors, $\mathbf{e}_1$ and $\mathbf{e}_2$.

6. Project the data onto the rotated coordinate system for the K-th step:

$$\mathbf{X}_K = \mathbf{X}\mathbf{A}_K.$$

7. Display the points as a scatterplot.

8. Repeat from step 3 for the next value of K.

We need to choose λ_i and λ_j such that the ratio λ_i/λ_j is irrational for every i and j. Additionally, we must choose these such that no λ_i/λ_j is a rational multiple of any other ratio. It is also recommended that the time step t be a small positive irrational number. Two possible ways to obtain irrational values are the following [Asimov, 1985]:

1. Let $\lambda_i = \sqrt{P_i}$, where P_i is the i-th prime number.
2. Let $\lambda_i = e^i \mod 1$.

We show how to implement the torus grand tour in Example 4.1.

Example 4.1

We use the **yeast** data in this example of a torus grand tour. First we load the data and set some constants.

```
load yeast
[n,p] = size(data);
% Set up vector of frequencies.
N = 2*p - 3;
% Use second option from above.
lam = mod(exp(1:N),1);
% This is a small irrational number:
delt = exp(-5);
% Get the indices to build the rotations.
% As in step 1 of the torus method.
J = 2:p;
I = ones(1,length(J));
I = [I, 2*ones(1,length(J)-1)];
J = [J, 3:p];
E = eye(p,2);      % Basis vectors
% Just do the tour for some number of iterations.
maxit = 2150;
```

Next we implement the tour itself.

```
% Get an initial plot.
z = zeros(n,2);
ph = plot(z(:,1),z(:,2),'o','erasemode','normal');
axis equal, axis off
% Use some Handle Graphics to remove flicker.
set(gcf,'backingstore','off','renderer',...
    'painters','DoubleBuffer','on')
% Start the tour.
for k = 1:maxit
    % Find the rotation matrix.
    Q = eye(p);
    for j = 1:N
        dum = eye(p);
        dum([I(j),J(j)],[I(j),J(j)]) = ...
         cos(lam(j)*k*delt);
        dum(I(j),J(j)) = -sin(lam(j)*k*delt);
        dum(J(j),I(j)) = sin(lam(j)*k*delt);
        Q = Q*dum;
    end
    % Rotate basis vectors.
    A = Q*E;
    % Project onto the new basis vectors.
    z = data*A;
    % Plot the transformed data.
    set(ph,'xdata',z(:,1),'ydata',z(:,2))
    % Forces Matlab to plot the data.
    pause(0.02)
end
```

We provide a function called **torustour** that implements this code. The configuration obtained at the end of this tour is shown in Figure 4.1.
❑

4.1.2 Pseudo Grand Tour

Asimov [1985] and Buja and Asimov [1986] described other ways of implementing a grand tour called the *at-random method* and the *random-walk method*. These methods, along with the torus grand tour, have some limitations. With the torus method we may end up spending too much time in certain regions, and it can be computationally intensive. Other techniques are better computationally, but cannot be reversed easily (to recover the projection) unless the set of random numbers used to generate the tour is retained.

We now discuss the *pseudo grand tour* first described in Wegman and Shen [1993] and later implemented in MATLAB by Martinez and Martinez [2002]. One of the important aspects of the torus grand tour is that it provides a

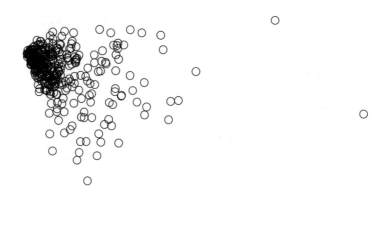

FIGURE 4.1
This shows the end of the torus grand tour using the **yeast** data.

continuous space-filling path through the manifold of planes. The following method does not employ a space-filling curve; thus it is called a pseudo grand tour. Another limitation is that the pseudo grand tour does not generalize to higher dimensional tours like the torus method. In spite of this, the pseudo grand tour has many benefits, such as speed, ease of calculation, uniformity of the tour, and ease of recovering the projection.

For the tour we need unit vectors that comprise the desired projection. Our first unit vector is denoted as $\alpha(t)$, such that

$$\alpha^T(t)\alpha(t) = \sum_{i=1}^{p}\alpha_i^2(t) = 1,$$

for every t, where t represents the stepsize as before. We need a second unit vector $\beta(t)$ that is orthonormal to $\alpha(t)$, so

$$\beta^T(t)\beta(t) = \sum_{i=1}^{p}\beta_i^2(t) = 1, \qquad \alpha^T(t)\beta(t) = 0.$$

For the pseudo grand tour, $\alpha(t)$ and $\beta(t)$ must be continuous functions of t and should produce 'all' possible orientations of a unit vector.

Before we continue in our development, we consider an observation **x**. If p is odd, then we augment each data point with a zero, to get an even number of elements. In this case,

$$\mathbf{x} = [x_1, \ldots, x_p, 0]^T, \qquad \text{for } p \text{ odd.}$$

This will not affect the projection. So, without loss of generality, we present the method with the understanding that p is even. We take the vector $\alpha(t)$ to be

$$\alpha(Kt) = \sqrt{\frac{2}{p}}[\sin\omega_1 Kt, \ \cos\omega_1 Kt, \ \ldots, \ \sin\omega_{p/2} Kt, \ \cos\omega_{p/2} Kt]^T, \quad (4.3)$$

for $K = 1, 2, \ldots$ and the vector $\beta(t)$ as

$$\beta K(t) = \sqrt{\frac{2}{p}}[\cos\omega_1 Kt, \ -\sin\omega_1 Kt, \ \ldots, \ \cos\omega_{p/2} Kt, \ -\sin\omega_{p/2} Kt]^T. \quad (4.4)$$

We choose ω_i and ω_j in a similar manner to the λ_i and λ_j in the torus grand tour. The steps for implementing the 2-D pseudo grand tour are given here, and the details on how to implement this in MATLAB are given in Example 4.2.

Procedure- Pseudo Grand Tour

1. Set each ω_i to an irrational number. Determine a small positive irrational number for the stepsize t.
2. Find vectors $\alpha(Kt)$ and $\beta(Kt)$ using Equations 4.3 and 4.4.
3. Project the data onto the plane spanned by these vectors.
4. Display the projected points in a 2-D scatterplot.
5. Repeat from step 2 for the next value of K.

Example 4.2

We will use the **oronsay** data set for this example that illustrates the pseudo grand tour. We provide a function called **pseudotour** that implements the pseudo grand tour, but details are given below. Since the **oronsay** data set has an even number of variables, we do not have to augment the observations with a zero.

```
load oronsay
x = oronsay;
maxit = 10000;
[n,p] = size(x);
```

```
% Set up vector of frequencies as in grand tour.
th = mod(exp(1:p),1);
% This is a small irrational number:
delt = exp(-5);
cof = sqrt(2/p);
% Set up storage space for projection vectors.
a = zeros(p,1); b = zeros(p,1);
z = zeros(n,2);
% Get an initial plot.
ph = plot(z(:,1),z(:,2),'o','erasemode','normal');
axis equal, axis off
set(gcf,'backingstore','off','renderer',...
  'painters','DoubleBuffer','on')
for t = 0:delt:(delt*maxit)
 % Find the transformation vectors.
 for j = 1:p/2
    a(2*(j-1)+1) = cof*sin(th(j)*t);
    a(2*j) = cof*cos(th(j)*t);
    b(2*(j-1)+1) = cof*cos(th(j)*t);
    b(2*j) = cof*(-sin(th(j)*t));
 end
 % Project onto the vectors.
 z(:,1) = x*a;
 z(:,2) = x*b;
 set(ph,'xdata',z(:,1),'ydata',z(:,2))
 drawnow
end
```

A scatterplot showing an interesting configuration of points is shown in Figure 4.2. The reader is encouraged to view this tour as it shows some interesting structure along the way.
❑

4.2 Interpolation Tours

We present a version of the *interpolation tour* described in Young and Rheingans [1991] and Young, Faldowski and McFarlane [1993]. The mathematics underlying this type of tour were presented in Hurley and Buja [1990] and Asimov and Buja [1994]. The idea behind interpolation tours is that it starts with two subspaces: an initial one and a target subspace. The tour proceeds by traveling from one to the other via geodesic interpolation paths between the two spaces. Of course, we also display the projected data

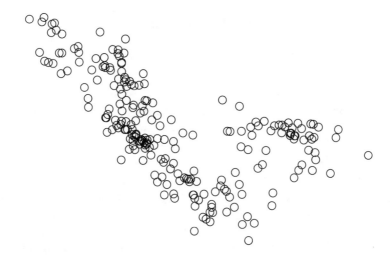

Iteration: 1822

FIGURE 4.2
This shows the scatterplot of points for an interesting projection of the **oronsay** data found during the pseudo grand tour in Example 4.2.

in a scatterplot at each step in the path for a movie view, and one can continue to tour the data by going from one target space to another.

We assume that the data matrix **X** is column centered; i.e., the centroid of the data space is at the origin. As with the other tours, we must have a visual space to present the data to the user. The visual space will be denoted by $\mathbf{V}_t$, which is an $n \times 2$ matrix of coordinates in 2-D.

One of the difficulties of the interpolation tour is getting the target spaces. Some suggested spaces include those spanned by subsets of the eigenvectors in PCA, which is what we choose to implement here. So, assuming that we have the principal component scores (see chapter 2), the initial visible space and target space will be $n \times 2$ matrices, whose columns contain different principal components.

The interpolation path is obtained through the following rotation:

$$\mathbf{V}_t = \mathbf{T}_k[\cos \mathbf{U}_t] + \mathbf{T}_{k+1}[\sin \mathbf{U}_t], \qquad (4.5)$$

where $\mathbf{V}_t$ is the visible space at the t-th step in the path, $\mathbf{T}_k$ indicates the k-th target in the sequence, and $\mathbf{U}_t$ is a diagonal 2×2 matrix with values θ_k between 0 and $\pi/2$. At each value of t, we increment the value of θ_k for some small stepsize. Note that the subscript k indicates the k-th plane in the target sequence, since we can go from one target plane to another.

Example 4.3

We use the **oronsay** data set to illustrate our function that implements the interpolation tour. The function takes the data matrix as its first argument. The next two inputs to the function contain column indices to the matrix of principal components. The first vector designates the starting plane and the second one corresponds to the target plane.

```
load oronsay
% Set up the vector of indices to the columns spanning
% the starting and target planes.
T1 = [3 4];
T2 = [5 6];
intour(oronsay, T1, T2);
```

We show the scatterplots corresponding to the starting plane and the target plane in Figure 4.3. This function actually completes a full rotation back to the starting plane after pausing at the target. Those readers who are interested in the details of the tour can refer to the M-file **intour** for more information.
❑

4.3 Projection Pursuit

In contrast to the grand tour, the *projection pursuit* method performs a directed search based on some index that indicates a type of structure one is looking for. In this sense, the tour is guided by the data, because it keeps touring until possible structure is found. Like the grand tour method, projection pursuit seeks to find projections of the data that are interesting; i.e., show departures from normality, such as clusters, linear structures, holes, outliers, etc. The objective is to find a projection plane that provides a 2-D view of our data such that the structure (or departure from normality) is maximized over all possible 2-D projections.

Friedman and Tukey [1974] describe projection pursuit as a way of searching for and exploring nonlinear structure in multi-dimensional data by examining many 2-D projections. The idea is that 2-D orthogonal projections of the data should reveal structure in the original data. The projection pursuit technique can also be used to obtain 1-D projections, but we look only at the 2-D case. Extensions to this method are also described in the literature by Friedman [1987], Posse [1995a, 1995b], Huber [1985], and Jones and Sibson [1987]. In our presentation of projection pursuit exploratory data analysis, we follow the method of Posse [1995a, 1995b].

Projection pursuit exploratory data analysis (PPEDA) is accomplished by visiting many projections in search of something interesting, where *interesting* is measured by an index. In most cases, the projection pursuit index measures the departure from normality. We use two indexes in our

Start Axes: 3 4; Target Axes: 5 6

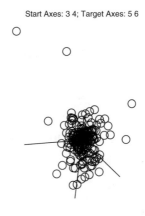

PAUSE: Hit Any
Key to Continue

Start Axes: 3 4; Target Axes: 5 6

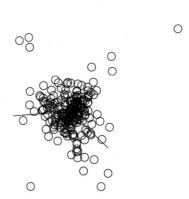

FIGURE 4.3

This shows the start plane (top) and the target plane (bottom) for an interpolation tour using the **oronsay** data set.

implementation. One is the ***chi-square index*** developed in Posse [1995a, 1995b], and the other is the ***moment index*** of Jones and Sibson [1987].

PPEDA consists of two parts:

1) a projection pursuit index that measures the degree of departure from normality, and

2) a method for finding the projection that yields the highest value for the index.

Posse [1995a, 1995b] uses a random search to locate a plane with an optimal value of the projection index and combines it with the structure removal of Friedman [1987] to get a sequence of interesting 2-D projections. Each projection found in this manner shows a structure that is less important (in terms of the projection index) than the previous one. Before we describe this method for PPEDA, we give a summary of the notation that we use to present the method.

Notation

Z is the matrix of sphered data.

α, β are orthonormal p-dimensional vectors that span the projection plane.

(α, β) is the projection plane spanned by α and β.

z_i^α, z_i^β are the sphered observations projected onto the vectors α and β.

(α^*, β^*) denotes the plane where the index is at a current maximum.

$PI_{\chi^2}(\alpha, \beta)$ denotes the chi-square projection index evaluated using the data projected onto the plane spanned by α and β.

$PI_M(\alpha, \beta)$ denotes the moment projection index.

c is a scalar that determines the size of the neighborhood around (α^*, β^*) that is visited in the search for planes that provide better values for the projection pursuit index.

v is a vector uniformly distributed on the unit p-dimensional sphere.

half specifies the number of steps without an increase in the projection index, at which time the value of the neighborhood is halved.

m represents the number of searches or random starts to find the best plane.

Finding the Structure

How we calculate the projection pursuit indexes $PI_{\chi^2}(\alpha, \beta)$ and $PI_M(\alpha, \beta)$ for each candidate plane is discussed at the end of this chapter. So, we first turn our attention to the second part of PPEDA, where we must optimize the projection index over all possible projections onto 2-D planes. Posse [1995a] shows that his random search optimization method performs better than the steepest-ascent techniques [Friedman and Tukey, 1974] typically used in optimization problems of this type.

The Posse algorithm starts by randomly selecting a starting plane, which becomes the current best plane (α^*, β^*). The method seeks to improve the current best solution by considering two candidate solutions within its neighborhood. These candidate planes are given by

$$a_1 = \frac{\alpha^* + cv_1}{\|\alpha^* + cv_1\|} \qquad b_1 = \frac{\beta^* - (a_1^T \beta^*)a_1}{\|\beta^* - (a_1^T \beta^*)a_1\|}$$

$$a_2 = \frac{\alpha^* - cv_2}{\|\alpha^* - cv_2\|} \qquad b_2 = \frac{\beta^* - (a_2^T \beta^*)a_2}{\|\beta^* - (a_2^T \beta^*)a_2\|}. \tag{4.6}$$

We start a global search by looking in large neighborhoods of the current best solution plane (α^*, β^*). We gradually focus in on a plane that yields a maximum index value by decreasing the neighborhood after a specified number of steps with no improvement in the value of the projection pursuit index. The optimization process is terminated when the neighborhood becomes small.

Because this method is a random search, the result could be a locally optimal solution. So, one typically goes through this procedure several times for different starting planes, choosing the final configuration as the one corresponding to the largest value of the projection pursuit index.

A summary of the steps for the exploratory projection pursuit procedure is given here. The complete search for the best plane involves repeating steps 2 through 9 of the procedure *m* times, using different random starting planes. The 'best' plane (α^*, β^*) chosen is the plane where the projected data exhibit the greatest departure from normality as measured by the projection pursuit index.

Procedure - PPEDA

1. Sphere the data to obtain **Z**. See Chapter 1 for details on sphering data.
2. Generate a random starting plane, (α_0, β_0). This is the current best plane, (α^*, β^*).
3. Evaluate the projection index $PI_{\chi^2}(\alpha_0, \beta_0)$ or $PI_M(\alpha_0, \beta_0)$ for the starting plane.
4. Generate two candidate planes (a_1, b_1) and (a_2, b_2) according to Equation 4.6.
5. Calculate the projection index for these candidate planes.

6. Choose the candidate plane with a higher value of the projection pursuit index as the current best plane (α^*, β^*).

7. Repeat steps 4 through 6 while there are improvements in the projection pursuit index.

8. If the index does not improve for *half* times, then decrease the value of c by half.

9. Repeat steps 4 through 8 until c is some small number.

Structure Removal

We have no reason to assume that there is only one interesting projection, and there might be other views that reveal insights about our data. To locate other views, Friedman [1987] devised a method called **structure removal**. The overall procedure is to perform projection pursuit as outlined above, remove the structure found at that projection, and repeat the projection pursuit process to find a projection that yields another maximum value of the projection pursuit index. Proceeding in this manner will provide a sequence of projections providing informative views of the data.

Structure removal in two dimensions is an iterative process. The procedure repeatedly transforms the projected data to standard normal until they stop becoming more normal as measured by the projection pursuit index. We start with a $p \times p$ matrix $\mathbf{U}^*$, where the first two rows of the matrix are the vectors of the projection obtained from PPEDA. The rest of the rows of $\mathbf{U}^*$ have ones on the diagonal and zero elsewhere. For example, if $p = 4$, then

$$\mathbf{U}^* = \begin{bmatrix} \alpha_1^* & \alpha_2^* & \alpha_3^* & \alpha_4^* \\ \beta_1^* & \beta_2^* & \beta_3^* & \beta_4^* \\ 0 & 0 & 1 & 0 \\ 0 & 0 & 0 & 1 \end{bmatrix}.$$

We use the Gram-Schmidt process [Strang, 1988] to make the rows of $\mathbf{U}^*$ orthonormal. We denote the orthonormal version as $\mathbf{U}$. The next step in the structure removal process is to transform the $\mathbf{Z}$ matrix using the following

$$\mathbf{T} = \mathbf{U}\mathbf{Z}^T. \tag{4.7}$$

In Equation 4.7, $\mathbf{T}$ is $p \times n$, so each column of the matrix corresponds to a p-dimensional observation. With this transformation, the first two dimensions (the first two rows of $\mathbf{T}$) of every transformed observation are the projection onto the plane given by (α^*, β^*).

We now remove the structure that is represented by the first two dimensions. We let Θ be a transformation that transforms the first two rows of $\mathbf{T}$ to a standard normal and the rest remain unchanged. This is where we actually remove the structure, making the data normal in that projection (the first two rows). Letting $\mathbf{T}_1$ and $\mathbf{T}_2$ represent the first two rows of $\mathbf{T}$, we define the transformation as follows

$$\Theta(\mathbf{T}_1) = \Phi^{-1}[F(\mathbf{T}_1)]$$

$$\Theta(\mathbf{T}_2) = \Phi^{-1}[F(\mathbf{T}_2)] \tag{4.8}$$

$$\Theta(\mathbf{T}_i) = \mathbf{T}_i; \qquad i = 3, \dots, p \ ,$$

where Φ^{-1} is the inverse of the standard normal cumulative distribution function and F is a function defined below (see Equations 4.9 and 4.10). We see from Equation 4.8 that we will be changing only the first two rows of $\mathbf{T}$.

We now describe the transformation of Equation 4.8 in more detail, working only with $\mathbf{T}_1$ and $\mathbf{T}_2$. First, we note that $\mathbf{T}_1$ can be written as

$$\mathbf{T}_1 = (z_1^{\alpha^*}, \ \dots, \ z_j^{\alpha^*}, \ \dots, \ z_n^{\alpha^*}),$$

and $\mathbf{T}_2$ as

$$\mathbf{T}_2 = (z_1^{\beta^*}, \ \dots, \ z_j^{\beta^*}, \ \dots, \ z_n^{\beta^*}).$$

Recall that $z_j^{\alpha^*}$ and $z_j^{\beta^*}$ would be coordinates of the j-th observation projected onto the plane spanned by (α^*, β^*).

Next, we define a rotation about the origin through the angle γ as follows

$$\tilde{z}_j^{1(t)} = z_j^{1(t)} \cos\gamma + z_j^{2(t)} \sin\gamma$$

$$\tilde{z}_j^{2(t)} = z_j^{2(t)} \cos\gamma - z_j^{1(t)} \sin\gamma, \tag{4.9}$$

where $\gamma = 0, \ \pi \div 4, \ \pi \div 8, \ 3\pi \div 8$ and $z_j^{1(t)}$ represents the j-th element of $\mathbf{T}_1$ at the t-th iteration of the process. We now apply the following transformation to the rotated points,

$$z_j^{1(t+1)} = \Phi^{-1}\left\{\frac{r(\tilde{z}_j^{1(t)}) - 0.5}{n}\right\} \qquad z_j^{2(t+1)} = \Phi^{-1}\left\{\frac{r(\tilde{z}_j^{2(t)}) - 0.5}{n}\right\}, \tag{4.10}$$

where $r(\tilde{z}_j^{1(t)})$ represents the rank (position in the ordered list) of $\tilde{z}_j^{1(t)}$.

This transformation replaces each rotated observation by its normal score in the projection. With this procedure, we are deflating the projection index by making the data more normal. It is evident in the procedure given below, that this is an iterative process. Friedman [1987] states that during the first few iterations, the projection index should decrease rapidly. After approximate normality is obtained, the index might oscillate with small changes. Usually, the process takes between 5 to 15 complete iterations to remove the structure.

Once the structure is removed using this process, we must transform the data back using

$$\mathbf{Z}' = \mathbf{U}^T \Theta (\mathbf{U} \mathbf{Z}^T). \tag{4.11}$$

From matrix theory [Strang, 1988], we know that all directions orthogonal to the structure (i.e., all rows of $\mathbf{T}$ other than the first two) have not been changed, whereas the structure has been Gaussianized and then transformed back.

Procedure - Structure Removal

1. Create the orthonormal matrix $\mathbf{U}$, where the first two rows of $\mathbf{U}$ contain the vectors α^*, β^*.
2. Transform the data $\mathbf{Z}$ using Equation 4.7 to get $\mathbf{T}$.
3. Using only the first two rows of $\mathbf{T}$, rotate the observations using Equation 4.9.
4. Normalize each rotated point according to Equation 4.10.
5. For angles of rotation $\gamma = 0$, $\pi \div 4$, $\pi \div 8$, $3\pi \div 8$, repeat steps 3 through 4.
6. Evaluate the projection index using $z_j^{1(t+1)}$ and $z_j^{2(t+1)}$, after going through an entire cycle of rotation (Equation 4.9) and normalization (Equation 4.10).
7. Repeat steps 3 through 6 until the projection pursuit index stops changing.
8. Transform the data back using Equation 4.11.

Example 4.4

We use the **oronsay** data to illustrate the projection pursuit procedure, which is implemented in the **ppeda** function provided with this text. First we do some preliminaries, such as loading the data and setting the parameter values.

```
load oronsay
X = oronsay;
[n,p] = size(X);
% For m = 5 random starts, find the N = 2
% best projection planes.
N = 2;
m = 5;
```

```
% Set values for other constants.
c = tan(80*pi/180);
half = 30;
% These will store the results for the
% 2 structures.
astar = zeros(p,N);
bstar = zeros(p,N);
ppmax = zeros(1,N);
```

Next we sphere the data to obtain matrix **Z**.

```
% Sphere the data.
muhat = mean(X);
[V,D] = eig(cov(X));
Xc = X - ones(n,1)*muhat;
Z = ((D)^(-1/2)*V'*Xc')';
```

Now we find each of the desired number of structures using the **ppeda** function with the index argument set to the Posse chi-square index.

```
% Now do the PPEDA: Find a structure, remove it,
% and look for another one.
Zt = Z;
for i = 1:N
    % Find one structure
    [astar(:,i),bstar(:,i),ppmax(i)] = ...
        ppeda(Zt,c,half,m,'chi');
    % Now remove the structure.
    % Function comes with text.
    Zt = csppstrtrem(Zt,astar(:,i),bstar(:,i));
end
```

The following MATLAB code shows how to project the data to each of these projection planes and then plot them. The plots are shown in Figure 4.4. The first projection has an index of 9.97, and the second has an index of 5.54.

```
% Now project and see the structure.
proj1 = [astar(:,1), bstar(:,1)];
proj2 = [astar(:,2), bstar(:,2)];
Zp1 = Z*proj1;
Zp2 = Z*proj2;
figure
plot(Zp1(:,1),Zp1(:,2),'k.'),title('Structure 1')
xlabel('\alpha*'),ylabel('\beta*')
figure
plot(Zp2(:,1),Zp2(:,2),'k.'),title('Structure 2')
xlabel('\alpha*'),ylabel('\beta*')
```

We repeat this **for** loop, but this time use the moment index to the **ppeda** function by replacing **chi** with **mom**. The first projection from this procedure has a moment index of 425.71, and the second one yields an index of 424.51. Scatterplots of the projected data onto these two planes are given in Figure 4.5. We see from these plots that the moment index tends to locate projections with outliers.
❑

4.4 Projection Pursuit Indexes

We briefly describe the two projection pursuit indexes (PPIs) that are implemented in the accompanying MATLAB code. Other projection indexes for PPEDA are given in the literature (see some of the articles mentioned in the last section). A summary of these indexes, along with a simulation analysis of their performance, can be found in Posse [1995b].

4.4.1 Posse Chi-Square Index

Posse [1995a, 1995b] developed an index for projection pursuit that is based on the chi-square. We present only the empirical version here, but we first provide some notation.

Notation

ϕ_2 is the standard bivariate normal density.

c_k is the probability evaluated over the k-th region using the standard bivariate normal,

$$c_k = \iint\limits_{B_k} \phi_2 dz_1 dz_2 \ .$$

B_k is a box in the projection plane.

I_{B_k} is the indicator function for region B_k.

$\lambda_j = \pi j/36$, $j = 0, \ldots, 8$ is the angle by which the data are rotated in the plane before being assigned to regions B_k.

$\alpha(\lambda_j)$ and $\beta(\lambda_j)$ are given by

$$\alpha(\lambda_j) = \alpha \cos \lambda_j - \beta \sin \lambda_j$$
$$\beta(\lambda_j) = \alpha \sin \lambda_j + \beta \cos \lambda_j$$

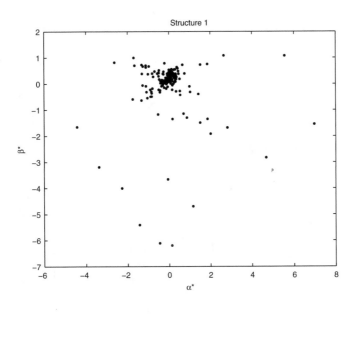

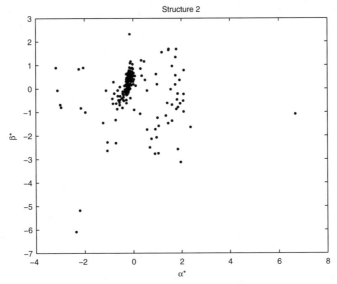

FIGURE 4.4
Here we show the results from applying PPEDA to the **oronsay** data set. The top configuration has a chi-square index of 9.97, and the second one has a chi-square index of 5.54.

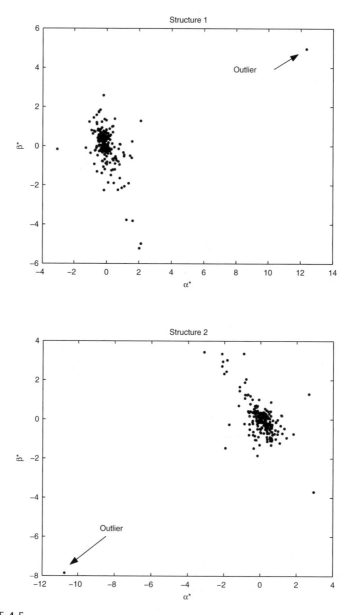

FIGURE 4.5
Here we see scatterplots from two planes found using the moment projection pursuit index. This index tends to locate projections with outliers. In the first structure, there is an outlying point in the upper right corner. In the second one, there is an outlying point in the lower left corner.

The plane is first divided into 48 regions or boxes B_k that are distributed in rings. See Figure 4.6 for an illustration of how the plane is partitioned. All regions have the same angular width of 45 degrees and the inner regions have the same radial width of $(2\log 6)^{1/2}/5$. This choice for the radial width provides regions with approximately the same probability for the standard bivariate normal distribution. The regions in the outer ring have probability $1/48$. The regions are constructed in this way to account for the radial symmetry of the bivariate normal distribution. The projection index is given by

$$PI_{\chi^2}(\alpha, \beta) = \frac{1}{9} \sum_{j=0}^{8} \sum_{k=1}^{48} \frac{1}{c_k} \left[\frac{1}{n} \sum_{i=1}^{n} I_{B_k}(z_i^{\alpha(\lambda_j)}, z_i^{\beta(\lambda_j)}) - c_k \right]^2.$$

The chi-square projection index is not affected by the presence of outliers. It is sensitive to distributions that have a hole in the core, and it will also yield projections that contain clusters. The chi-square projection pursuit index is fast and easy to compute, making it appropriate for large sample sizes. Posse [1995a] provides a formula to approximate the percentiles of the chi-square index so the analyst can assess the significance of the observed value of the projection index.

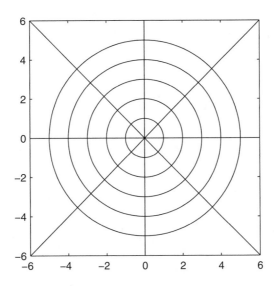

FIGURE 4.6
This shows the layout of the regions B_k for the chi-square projection index. [Posse, 1995a]

4.4.2 Moment Index

This index was developed in Jones and Sibson [1987] and is based on bivariate third and fourth moments. This is very fast to compute, so it is useful for large data sets. However, a problem with this index is that it tends to locate structure in the tails of the distribution. It is given by

$$PI_M(\alpha, \beta) = \frac{1}{12}\left\{\kappa_{30}^2 + 3\kappa_{21}^2 + 3\kappa_{12}^2 + \kappa_{03}^2 + \frac{1}{4}(\kappa_{40}^2 + 4\kappa_{31}^2 + 6\kappa_{22}^2 + 4\kappa_{13}^2 + \kappa_{04}^2)\right\},$$

where

$$\kappa_{21} = \frac{n}{(n-1)(n-2)}\sum_{i=1}^{n}(z_i^\alpha)^2 z_i^\beta \qquad \kappa_{12} = \frac{n}{(n-1)(n-2)}\sum_{i=1}^{n}(z_i^\beta)^2 z_i^\alpha$$

$$\kappa_{22} = \frac{n(n+1)}{(n-1)(n-2)(n-3)}\left\{\sum_{i=1}^{n}(z_i^\alpha)^2 (z_i^\beta)^2 - \frac{(n-1)^3}{n(n+1)}\right\}$$

$$\kappa_{30} = \frac{n}{(n-1)(n-2)}\sum_{i=1}^{n}(z_i^\alpha)^3 \qquad \kappa_{03} = \frac{n}{(n-1)(n-2)}\sum_{i=1}^{n}(z_i^\beta)^3$$

$$\kappa_{31} = \frac{n(n+1)}{(n-1)(n-2)(n-3)}\sum_{i=1}^{n}(z_i^\alpha)^3 z_i^\beta$$

$$\kappa_{13} = \frac{n(n+1)}{(n-1)(n-2)(n-3)}\sum_{i=1}^{n}(z_i^\beta)^3 z_i^\alpha$$

$$\kappa_{04} = \frac{n(n+1)}{(n-1)(n-2)(n-3)}\left\{\sum_{i=1}^{n}(z_i^\beta)^4 - \frac{3(n-1)^3}{n(n+1)}\right\}$$

$$\kappa_{40} = \frac{n(n+1)}{(n-1)(n-2)(n-3)}\left\{\sum_{i=1}^{n}(z_i^\alpha)^4 - \frac{3(n-1)^3}{n(n+1)}\right\}$$

4.5 Summary and Further Reading

In this chapter, we discussed several methods for data tours that can be used to search for interesting features or structure in high-dimensional data. These include the torus winding method for the grand tour, the pseudo grand tour, the interpolation tour, and projection pursuit for EDA. The grand tour methods are dynamic, but are not typically interactive. The interpolation tour is interactive in the sense that the user can specify starting and target planes to guide the tour. Finally, projection pursuit is not a visual tour (although it could be implemented that way); this tour seeks planes that have maximal structure as defined by some measure.

Some excellent papers that describe the underlying mathematical foundation for tour methods and motion graphics include Wegman and Solka [2002], Hurley and Buja [1990], Buja and Asimov [1986], and Asimov and Buja [1994]. A new method for implementing a tour based on a fractal space-filling curve is described in Wegman and Solka [2002]. The grand tour combined with projection pursuit is described in Cook, et al. [1995].

Many articles have been written on projection pursuit and its use in EDA and other applications. Jones and Sibson [1987] describe a steepest-ascent algorithm that starts from either principal components or random starts. Friedman [1987] combines steepest-ascent with a stepping search to look for a region of interest. Crawford [1991] uses genetic algorithms to optimize the projection index. An approach for projection pursuit in three dimensions is developed by Nason [1995]. A description of other projection pursuit indexes can also be found in Cook, Buja and Cabrera [1993].

Other uses for projection pursuit have been proposed. These include projection pursuit probability density estimation [Friedman, Stuetzle and Schroeder, 1984], projection pursuit regression [Friedman and Stuetzle, 1981], robust estimation [Li and Chen, 1985], and projection pursuit for pattern recognition [Flick, et al., 1990]. For a theoretical and comprehensive description of projection pursuit, the reader is directed to Huber [1985], where he discusses the important matter of sphering the data before exploring the data with projection pursuit. This invited paper with discussion also presents applications of projection pursuit to computer tomography and to the deconvolution of time series. Another paper that provides applications of projection pursuit is Jones and Sibson [1987]. Montanari and Lizzani [2001] apply projection pursuit to the variable selection problem. Bolton and Krzanowski [1999] describe the connection between projection pursuit and principal component analysis.

Exercises

4.1   Run the tour as in Example 4.1 and vary the number of iterations. Do
        you see any interesting structure along the way?

4.2   Apply the torus grand tour to the following data sets and comment on
        the results.

        a. **environmental**

        b. **oronsay**

        c. **iris**

        d. **posse** data sets

        e. **skulls**

        f. **spam**

        g. **pollen**

        h. gene expression data sets.

4.3   Run the pseudo grand tour in Example 4.2. Comment on the struc-
        tures that are found.

4.4   Apply the pseudo grand tour to the following data sets and comment
        on the results. Compare with the results you got with the grand tour.

        a. **environmental**

        b. **yeast**

        c. **iris**

        d. **posse** data sets

        e. **skulls**

        f. **spam**

        g. **pollen**

        h. gene expression data sets

4.5   Apply the interpolation tour to the data sets in problem 4.4.

4.6   Repeat the interpolation tour in Example 4.3 using other target planes
        (Hint: 9 and 10 makes an interesting one). Do you see any structure?

4.7   Apply projection pursuit EDA to the data sets in problem 4.4. Search
        for several structures and use both projection pursuit indexes. Show
        your results in a scatterplot and discuss them.

4.8   Repeat Example 4.4 and look for more than two best projection planes.
        Describe your results. Do you find planes using the moment index
        where the planes exhibit structure other than outliers?

Chapter 5

Finding Clusters

We now turn our attention to the problem of finding groups or clusters in our data, which is an important method in EDA and data mining. We present two of the basic methods in this chapter: agglomerative clustering and *k*-means clustering. Another method for fuzzy clustering based on estimating a finite mixture probability density function is described in the following chapter. Most cluster methods allow users to specify a desired number of groups. We address the problem of assessing the quality of resulting clusters at the end of the chapter, where we describe several statistics and plots that will aid in the analysis, as well as in Chapter 8, where we provide some ways to graphically assess cluster output.

5.1 Introduction

Clustering is the process of organizing a set of data into groups in such a way that observations within a group are more similar to each other than they are to observations belonging to a different cluster. It is assumed that the data represent features that would allow one to distinguish one group from another. An important starting point in the process is choosing a way to represent the objects to be clustered. Many methods for grouping or clustering data can be found in various communities, such as statistics, machine learning, data mining, and computer science. We note, though, that no clustering technique is universally appropriate for finding all varieties of groupings that can be represented by multidimensional data [Jain, Murty and Flynn, 1999]. So in the spirit of EDA, the user should try different clustering methods on a given data set to see what patterns emerge.

Clustering is also known as *unsupervised learning* in the literature. To understand clustering a little better, we will compare it to *discriminant analysis* or *supervised learning*. In supervised learning, the collection of observations has a class label associated with it. Thus, we know the true number of groups in the data, as well as the actual group membership of each data point. We use the data, along with the class labels, to create a classifier.

Then when we encounter a new, unlabeled observation, we can use the classifier to attach a label to it [Duda, Hart and Stork, 2001; Webb, 1999].

However, with clustering, we usually do not have class labels for the observations. Thus, we do not know how many groups are represented by the data, what the group membership or structure is, or even if there are any groups in the first place. As we said earlier, most clustering methods will find some desired number of groups, but what we really want is some meaningful clusters that represent the true phenomena. So, the analyst must look at the resulting groups and determine whether or not they aid in understanding the problem. Of course, nothing prevents us from using clustering methods on data that have class labels associated with the observations. As we will see in some of the examples, knowing the true clustering helps us assess the performance of the methods.

One can group the usual steps of clustering into the following [Jain and Dubes, 1988]:

1. *__Pattern representation__*: This includes much of the preparation and initial work, such as choosing the number of clusters to look for, picking what measurements to use (*feature selection*), determining how many observations to process, and choosing the scaling or other transformations of the data (*feature extraction*). Some of this might be beyond the control of analysts.

2. *__Pattern proximity measure__*: Many clustering methods require a measure of distance or proximity between observations and maybe between clusters. As one might suspect, different distances give rise to different partitions of the data. We discuss various distances and measures of proximity in Appendix A.

3. *__Grouping__*: This is the process of partitioning the data into clusters. The grouping can be *hard*, which means that an observation either belongs to a group or not. In contrast, it can be *fuzzy*, where each data point has a degree of membership in each of the clusters. It can also be *hierarchical*, where we have a nested sequence of partitions.

4. *__Data abstraction__*: This is the optional process of obtaining a simple and compact representation of the partitions. It could be a description of each cluster in words (e.g., one cluster represents lung cancer, while another one corresponds to breast cancer). It might be something quantitative such as a representative pattern, e.g., the centroid of the cluster.

5. *__Cluster Assessment__*: This could involve an assessment of the data to see if it contains any clusters. However, more often, it means an examination of the output of the algorithm to determine whether or not the clusters are meaningful.

In our discussion so far, we've assumed that we know what a cluster actually is. However, several authors have pointed out the difficulty of formally defining such a term [Everitt, Landau and Leese, 2001; Estivill-Castro, 2002]. Most clustering methods assume some sort of structure or model for the clusters (e.g., spherical, elliptical). Thus, they find clusters of that type, regardless of whether they are really present in the data or not.

Humans are quite adept at locating clusters in 2-D scatterplots, as we show in Figure 5.1. Bonner [1964] argued that the meaning of terms like *cluster* and *group* is in the 'eye of the beholder.' We caution the reader that it is usually easy to assign some structure or meaning to the clusters that are found. However, we should keep in mind that the groups might be a result of the clustering method and that we could be *imposing* a pattern rather than *discovering* something that is actually there.

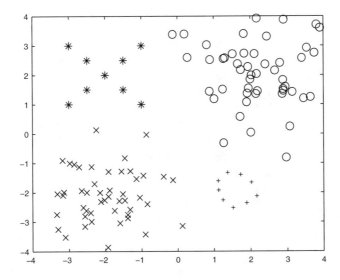

FIGURE 5.1
Here we show an example of some clusters. Keep in mind that what constitutes a cluster is based on one's definition and application.

5.2 Hierarchical Methods

One of the most common approaches to clustering is to use a hierarchical method. This seems to be popular in the areas of data mining and gene expression analysis [Hand, Mannila and Smyth, 2001; Hastie, Tibshirani and Friedman, 2001]. In **hierarchical clustering**, one does not have to know the number of groups ahead of time; i.e., the data are not divided into a predetermined number of partitions. Rather, the result consists of a hierarchy or set of nested partitions. As we point out below, there are several types of hierarchical clustering, and each one can give very different results on a given data set. Thus, there is not one recommended method, and the analyst should use several in exploring the data.

The process consists of a sequence of steps, where two groups are either merged (*agglomerative*) or divided (*divisive*) according to some optimality criterion. In their simplest and most commonly used form, each of these hierarchical methods have all *n* observations in their own group (i.e., *n* total groups) at one end of the process and one group with all *n* data points at the other end. The difference between them is where we start. With agglomerative clustering, we have *n* singleton clusters and end up with all points belonging to one group. Divisive methods are just the opposite; we start with everything in one group and keep splitting them until we have *n* clusters.

One of the issues with hierarchical clustering is that once points are grouped together or split apart, the step cannot be undone. Another issue, of course, is how many clusters are appropriate.

We will not cover the divisive methods in this book, because they are less common and they can be computationally intensive (except in the case of binary variables, see Everitt, Landau and Leese [2001]). However, Kaufman and Rousseeuw [1990] point out that an advantage with divisive methods is that most data sets have a small number of clusters and that structure would be revealed in the beginning of a divisive method, whereas, with agglomerative methods, this does not happen until the end of the process.

Agglomerative clustering requires the analyst to make several choices, such as how to measure the proximity (distance) between data points and how to define the distance between two clusters. Determining what distance to use is largely driven by the type of data one has: continuous, categorical or a mixture of the two, as well as what aspect of the features one wants to emphasize. Please see Appendix A for a description of various distances, including those that are implemented in the MATLAB Statistics Toolbox.

The input required for most agglomerative clustering methods is the $n \times n$ interpoint distance matrix (as was used before in multidimensional scaling); some also require the full data set. The next step is to specify how we will determine what clusters to link at each stage of the method. The usual way is to link the two closest clusters at each stage of the process, where *closest* is defined by one of the linkage methods described below. It should be noted that using different definitions of distance and linkage can give rise to very different cluster structures. We now describe each of the linkage methods that are available in the MATLAB Statistics Toolbox.

We set up some notation before we continue with a description of the various approaches. Given a cluster *r* and a cluster *s*, the number of objects in each cluster is given by n_r and n_s. The distance between cluster *r* and *s* is denoted by $d_c(r,s)$.

Single Linkage

Single linkage is perhaps the method used most often in agglomerative clustering, and it is the default method in the MATLAB **linkage** function, which produces the hierarchical clustering. Single linkage is also called *nearest neighbor*, because the distance between two clusters is given by the smallest distance between objects, where each one is taken from one of the two groups. Thus, we have the following distance between clusters

$$d_c(r, s) = min\{d(x_{ri}, x_{sj})\} \qquad i = 1, \ldots, n_r \, ; j = 1, \ldots, n_s,$$

where $d(x_{ri}, x_{sj})$ is the distance between observation i from group r and observation j from group s. Recall that this is the interpoint distance (e.g., Euclidean, etc.), which is the input to the clustering procedure.

Single linkage clustering suffers from a problem called *chaining*. This comes about when clusters are not well separated, and snake-like chains can form. Observations at opposite ends of the chain can be very dissimilar, but yet they end up in the same cluster. Another issue with single linkage is that it does not take the cluster structure into account [Everitt, Landau and Leese, 2001].

Complete Linkage

Complete linkage is also called the *furthest neighbor* method, since it uses the largest distance between observations, one in each group, as the distance between the clusters. The distance between clusters is given by

$$d_c(r, s) = max\{d(x_{ri}, x_{sj})\} \qquad i = 1, \ldots, n_r \, ; j = 1, \ldots, n_s \, .$$

Complete linkage is not susceptible to chaining, but it does tend to impose a spherical structure on the clusters. In other words, the resulting clusters tend to be spherical, and it has difficulty recovering nonspherical groups. Like single linkage, complete linkage does not account for cluster structure.

Average (Unweighted and Weighted) Linkage

The *average linkage* method defines the distance between clusters as the average distance from all observations in one cluster to all points in another cluster. In other words, it is the average distance between pairs of observations, where one is from one cluster and one is from the other. Thus, we have the following distance

$$d_c(r, s) = \frac{1}{n_r n_s} \sum_{i=1}^{n_r} \sum_{j=1}^{n_s} d(x_{ri}, x_{sj}).$$

This method tends to combine clusters that have small variances, and it also tends to produce clusters with approximately equal variance. It is relatively robust and does take the cluster structure into account. Like single and complete linkage, this method takes the interpoint distances as input.

Version 5 of the Statistics Toolbox has another related type of linkage called *weighted average distance* or WPGMA. The average linkage mentioned above is *unweighted* and is also known as UPGMA.

Centroid Linkage

Another type, called **centroid linkage**, requires the raw data, as well as the distances. It measures the distance between clusters as the distance between their centroids. Their centroids are usually the mean, and these change with each cluster merge. We can write the distance between clusters, as follows

$$d_c(r, \ s) \ = \ d(\bar{x}_r, \ \bar{x}_s) ,$$

where $\bar{x}_r$ is the average of the observations in the r-th cluster, and $\bar{x}_s$ is defined similarly.

The distance between the centroids is usually taken to be Euclidean. The MATLAB **linkage** function for centroid linkage works only when the interpoint distances are Euclidean, and it does not require the raw data as input. A somewhat related method called *median linkage* computes the distance between clusters using weighted centroids. This is now available in the Statistics Toolbox, version 5. A problem with both centroid and median linkage is the possibility of reversals [Morgan and Ray, 1995]. This can happen when the distance between one pair of cluster centroids is less than the distance between the centroid of another pair that was merged earlier. In other words, the fusion values (distances between clusters) are not monotonically increasing. This makes the results confusing and difficult to interpret.

Ward's Method

Ward [1963] devised a method for agglomerative hierarchical clustering where the fusion of two clusters is determined by the size of the incremental sum of squares. It looks at the increase in the total within-group sum of squares when clusters r and s are joined. The distance between two clusters using Ward's method is given by

$$d(r, \ s) \ = \ n_r n_s d_{rs}^2 / (n_r + n_s) \ ,$$

where d_{rs}^2 is the distance between the r-th and s-th cluster as defined in the centroid linkage definition. In other words, to get each merge in the procedure, the within-cluster sum of squares is minimized over all possible partitions that can be obtained by combining two clusters from the current partition.

Ward's method tends to combine clusters that have a small number of observations. It also has a tendency to locate clusters that are of the same size and spherical. Due to the sum of squares criterion, it is sensitive to the presence of outliers in the data set.

Visualizing Hierarchical Clustering Using the Dendrogram

We discuss the dendrogram in more detail in Chapter 8, where we present several ways to visualize the output from cluster analysis. We briefly introduce it here, so we can use the dendrograms to present the results of this chapter to the reader.

A *dendrogram* is a tree diagram that shows the nested structure of the partitions and how the various groups are linked at each stage. The dendrogram can be shown horizontally or vertically, however we will concentrate on the vertical version for right now, since it seems more 'tree-like.' There is a numerical value associated with each stage of the method where the branches (i.e., clusters) join, which usually represents the distance between the two clusters. The scale for this numerical value is shown on the vertical axis.

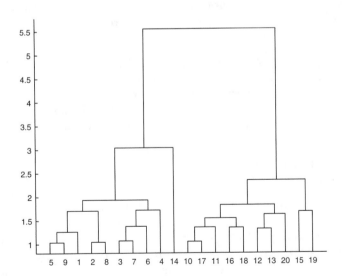

FIGURE 5.2
This is an example of a dendrogram for the two spherical clusters in Figure 5.1, where average linkage has been used to generate the hierarchy. Note that we are showing only 20 leaf nodes. See Chapter 8 for more information on what this means.

We show an example of a dendrogram in Figure 5.2 for a very small data set. Notice that the tree is made up of inverted U-shaped links, where the top of the U represents a fusion between two clusters. In most cases, the fusion levels will be monotonically increasing, yielding an easy to understand dendrogram.

We already discussed the problem of reversals with the centroid and median linkage methods. With reversals, these merge points can decrease, which can make the results confusing. Another problem with some of these methods is the possibility of nonuniqueness of the hierarchical clustering or dendrogram. This can happen when there are ties in the distances between clusters. How these are handled depends on the software, and this information is often left out of the documentation. Morgan and Ray [1995] provide a detailed explanation of the inversion and nonuniqueness problems in hierarchical clustering. This will be explored further in the exercises.

Example 5.1

In this example, we use the **yeast** data to illustrate the procedure in MATLAB for obtaining agglomerative hierarchical clustering. The first step is to load the data and then to get all of the interpoint distances.

```
load yeast
% Get the distances. The output from this function
% is a vector of the n(n-1)/2 interpoint distances.
% The default is Euclidean distance.
Y = pdist(data);
```

The output from the **pdist** function is just the upper triangular portion of the complete $n \times n$ interpoint distance matrix. It can be converted to a full matrix using the function **squareform**, but this is not necessary for the next step, which is to get the hierarchy of partitions. See the **help** on **pdist** for more information on the other distances that are available.

```
% Single linkage (the default) shows chaining.
Z = linkage(Y);
dendrogram(Z);
```

The default for the **linkage** function is single linkage. The output is a matrix **Z**, where the first two columns indicate what groups were linked and the third column contains the corresponding distance or fusion level. The dendrogram for this is shown in Figure 5.3 (top), where we can see the chaining that can happen with single linkage. Now we show how to do the same thing using complete linkage.

```
% Complete linkage does not have the chaining.
Z = linkage(Y,'complete');
dendrogram(Z);
```

This dendrogram is given in Figure 5.3 (bottom), and we see no chaining here. Since the dendrogram shows the entire set of nested partitions, one could say the dendrogram *is* the actual clustering. However, it is useful to know how to get the grouping for any desired number of groups. MATLAB provides the **cluster** function for this purpose. One can specify the number of clusters, as we do below, however there are other options. See the help on **cluster** for other uses.

```
% To get the actual clusters - say based
% on two partitions, use the following
% syntax.
cind = cluster(Z,'maxclust',2);
```

The output argument **cind** is an n-dimensional vector of group labels.
❏

5.3 Optimization Methods - k-Means

The methods discussed in the previous section were all hierarchical, where the output consists of a complete set of nested partitions. Another category of clustering methods consists of techniques that optimize some criterion in order to partition the observations into a *specified* or *predetermined* number of groups. These *partition* or *optimization methods* differ in the nature of the objective function, as well as the optimization algorithm used to come up with the final clustering. One of the issues that must be addressed when employing these methods (as is also the case with the hierarchical methods) is determining the number of clusters in the data set. We will discuss ways to tackle this problem in the next section. However, one of the major advantages of the optimization-based methods is that they require only the data as input (along with some other parameters), not the interpoint distances, as in hierarchical methods. Thus, these are usually more suitable when working with large data sets.

One of the most commonly used optimization-based methods is k-means clustering, which is the only one we will discuss in this book. The reader is referred to Everitt, Landau and Leese [2001] or other books mentioned at the end of the chapter for more information on the other types of partition methods. The MATLAB Statistics Toolbox has a function that implements the k-means algorithm.

The goal of k-means clustering is to partition the data into k groups such that the within-group sum-of-squares is minimized. We start by defining the *within-class scatter matrix* given by

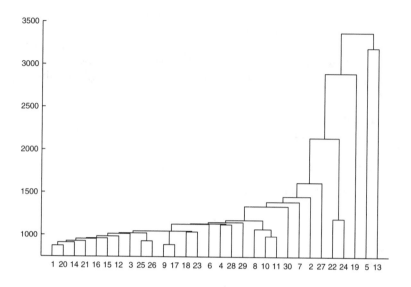

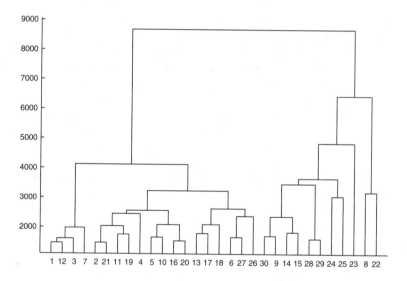

FIGURE 5.3

The first dendrogram shows the results of using Euclidean distance and single linkage on the **yeast** data, and we can see what chaining looks like in the partitions. The second dendrogram is what we obtain using complete linkage.

$$\mathbf{S}_W = \frac{1}{n} \sum_{j=1}^{g} \sum_{i=1}^{n} I_{ij}(\mathbf{x}_i - \bar{\mathbf{x}}_j)(\mathbf{x}_i - \bar{\mathbf{x}}_j)^T,$$

where I_{ij} is one if $\mathbf{x}_i$ belongs to group j and zero otherwise, and g is the number of groups. The criterion that is minimized in k-means is given by the sum of the diagonal elements of $\mathbf{S}_W$, i.e., the trace of the matrix, as follows

$$\mathrm{Tr}(\mathbf{S}_W) = \sum \mathbf{S}_{W_{ii}}.$$

If we minimize the trace, then we are also minimizing the total within-group sum of squares about the group means. Everitt, Landau and Leese [2001] show that minimizing the trace of $\mathbf{S}_W$ is equivalent to minimizing the sum of the squared Euclidean distances between individuals and their group mean. Clustering methods that minimize this criterion tend to produce clusters that have a hyperellipsoidal shape. This criterion can be affected by the scale of the variables, so standardization should be done first.

We briefly describe two procedures for obtaining clusters via k-means. The basic algorithm for k-means clustering is a two step procedure. First, we assign each observation to its closest group, usually using the Euclidean distance between the observation and the cluster centroid. The second step of the procedure is to calculate the new centroids using the assigned observations. These steps are alternated until there are no changes in cluster membership or until the centroids do not change. This algorithm is sometimes referred to as HMEANS [Späth, 1980] or the basic ISODATA method.

Procedure - k-Means

1. Specify the number of clusters k.
2. Determine initial cluster centroids. These can be randomly chosen or the user can specify them.
3. Calculate the distance between each observation and each cluster centroid.
4. Assign every observation to the closest cluster.
5. Calculate the centroid (i.e., the d-dimensional mean) of every cluster using the observations that were just grouped there.
6. Repeat steps 3 through 5 until no more changes are made.

The k-means algorithm could lead to empty clusters, so users should be aware of this possibility. Another issue concerns the optimality of the partitions. With k-means, we are searching for partitions where the within-group sum-of-squares is a minimum. It can be shown [Webb, 1999] that in some cases the final k-means cluster assignment is not optimal, in the sense that moving a single point from one cluster to another may reduce the sum of squared errors. The following procedure that we call the *enhanced k-means* helps address the second problem.

Procedure - Enhanced k-means

1. Obtain a partition of k groups via k-means as described previously.
2. Take each data point $\mathbf{x}_i$ and calculate the Euclidean distance between it and every cluster centroid.
3. Here $\mathbf{x}_i$ is in the r-th cluster, n_r is the number of points in the r-th cluster, and d_{ir}^2 is the Euclidean distance between $\mathbf{x}_i$ and the centroid of cluster r. If there is a group s such that

$$\frac{n_r}{n_r - 1} d_{ir}^2 > \frac{n_s}{n_s + 1} d_{is}^2 ,$$

then move $\mathbf{x}_i$ to cluster s.

4. If there are several clusters that satisfy the above inequality, then move the $\mathbf{x}_i$ to the group that has the smallest value for

$$\frac{n_s}{n_s + 1} d_{is}^2 .$$

5. Repeat steps 2 through 4 until no more changes are made.

We note that there are many algorithms for k-means clustering described in the literature that improve the efficiency, allow clusters to be created and deleted during the process, and other improvements. See Webb [1999] and the other references at the end of the chapter for more information.

Example 5.2

For this example, we turn to a data set that is familiar to most statisticians: the iris data. These data consist of three classes of iris: *Iris setosa, Iris versicolor,* and *Iris virginica*. They were originally analyzed by Fisher [1936], because he was interested in developing a method for discriminating the species of iris based on their sepal length, sepal width, petal length, and petal width. The **kmeans** function in MATLAB requires the data as input, along with the desired number of groups. MATLAB also allows the user to specify a distance measure used in the minimization process. In other words, **kmeans** computes the centroid clusters differently for the different distance measures. We will use the default of squared Euclidean distance.

```
load iris
```

```
% First load up the data and put into one data
% matrix.
data = [setosa; versicolor; virginica];
vars = ['Sepal Length';
         'Sepal Width ';
         'Petal Length';
         'Petal Width '];
 kmus = kmeans(data,3);
```

We illustrate the results in a scatterplot matrix shown in Figure 5.4.[1] The first plot shows the results from k-means, and the second shows the true groups that are in the data. Different symbols correspond to the different groups, and we see that k-means produced reasonable clusters that are not too far off from the truth. We will explore this more in a later example.
❑

The k-means method is dependent on the chosen initial cluster centers. MATLAB allows the user to specify different starting options, such as randomly selecting k data points as centers (the default), uniformly generated p-dimensional vectors over the range of **X**, or user-defined centers. As in many optimization problems that rely on a designated starting point, k-means can get stuck in a locally optimal solution. Thus, k-means should be performed several times with different starting points. MATLAB provides this option also as an input argument to the **kmeans** function. For another implementation of k-means in MATLAB, see Martinez and Martinez [2002].

5.4 Evaluating the Clusters

In this section, we turn our attention to understanding more about the quality of our cluster results and to estimating the 'correct' number of groups in our data. We present the following measures that can be used to address both of these issues. The first is the Rand index that can be used to compare two different groupings of the same data set. Next, we discuss the cophenetic correlation coefficient that provides a way to compare a set of nested partitions from hierarchical clustering with a distance matrix or with another set of nested partitions. We also cover a method due to Mojena [1977] for determining the number of groups in hierarchical clustering based on the fusion levels. We illustrate the silhouette plot and silhouette statistic that can be used to help decide how many groups are present. Finally, we discuss a recently developed method called the gap statistic [Tibshirani, Walther and Hastie, 2001] that seems to be successful at estimating the number of clusters.

---

[1] See the file **Example54.m** for the MATLAB code used to construct the plots.

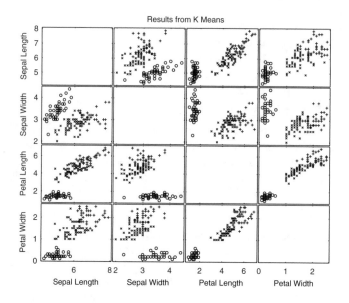

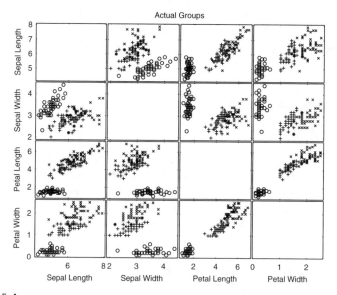

FIGURE 5.4

The first scatterplot matrix shows the results from *k*-means, while the second one shows the true groups in the data. We see that for the most part, *k*-means did a reasonable job finding the groups. See the associated color figure following page 144.

An added benefit of the gap statistic approach is that it addresses the issue of whether or not there are any clusters at all.

5.4.1 Rand Index

Say we have two partitions of the same data set called G_1 and G_2, with g_1 groups and g_2 groups, respectively. This can be represented in a $g_1 \times g_2$ matching matrix $\mathbf{N}$ with elements n_{ij}, where n_{ij} is the number of observations in group i of partition G_1 that are also in group j of partition G_2. Note that the number of groups in each partition do not have to be equal, and that the classifications can be obtained through any method.

The **Rand index** [Rand, 1971] was developed to help analysts answer four questions about their cluster results. These are:

How well does a method retrieve natural clusters?

How sensitive is a method to perturbation of the data?

How sensitive is a method to missing data?

Given two methods, do they produce different results when applied to the same data?

In this book, we are more interested in the last question.

The motivation for the Rand index follows three assumptions. First, clustering is considered to be discrete in that every point is assigned to a specific cluster. Second, it is important to define clusters with respect to the observations they *do not* contain, as well as by the points that they *do* contain. Finally, all data points are equally important in determining the cluster structure.

Since the cluster labels are arbitrary, the Rand index looks at pairs of points and how they are partitioned in groupings G_1 and G_2. There are two ways that data points x_i and x_j can be grouped similarly (i.e., the groupings agree):

1. x_i and x_j are put in the same cluster in G_1 and G_2.
2. x_i and x_j are in different clusters in G_1 and in different clusters in G_2.

There are also two ways that x_i and x_j can be grouped differently:

3. x_i and x_j are in the same cluster in G_1 and different clusters in G_2.
4. x_i and x_j are in different clusters in G_1 and the same cluster in G_2.

The Rand index calculates the proportion of the total of n choose 2 objects that agree between the two groupings. It is given by the following

$$
RI = \frac{\binom{n}{2} + \sum_{i=1}^{g_1}\sum_{j=1}^{g_2} n_{ij}^2 - \frac{1}{2}\sum_{i=1}^{g_1}\left(\sum_{j=1}^{g_2} n_{ij}\right)^2 - \frac{1}{2}\sum_{j=1}^{g_2}\left(\sum_{i=1}^{g_1} n_{ij}\right)^2}{\binom{n}{2}}.
$$

The Rand index is a measure of similarity between the groupings, and it ranges from zero when the two groupings are not similar to a value of one when the groupings are exactly the same.

Fowlkes and Mallows [1983] developed their own index for the case $g_1 = g_2$, which will be presented in the exercises. They point out in that paper that the Rand index increases as the number of clusters increase, and the possible range of values is very narrow. Hubert and Arabie [1985] developed an *adjusted Rand index* that addresses these issues. This is given by $RI_A = N/D$, where

$$
N = \sum_{i=1}^{g_1}\sum_{j=1}^{g_2}\binom{n_{ij}}{2} - \sum_{i=1}^{g_1}\binom{n_{i\cdot}}{2}\sum_{j=1}^{g_2}\binom{n_{\cdot j}}{2} \div \binom{n}{2},
$$

$$
D = \left[\sum_{i=1}^{g_1}\binom{n_{i\cdot}}{2} + \sum_{j=1}^{g_2}\binom{n_{\cdot j}}{2}\right] \div 2 - \sum_{i=1}^{g_1}\binom{n_{i\cdot}}{2}\sum_{j=1}^{g_2}\binom{n_{\cdot j}}{2} \div \binom{n}{2},
$$

and

$$
n_{\cdot j} = \sum_{i=1}^{g_1} n_{ij} \qquad n_{i\cdot} = \sum_{j=1}^{g_2} n_{ij}.
$$

The binomial coefficient $\binom{m}{2}$ is defined as 0 when $m = 0$ or $m = 1$. The adjusted Rand index provides a standardized measure such that its expected value is zero when the partitions are selected at random and one when the partitions match completely.

Example 5.3

We return to the **iris** data of the previous example to illustrate the Rand index, by comparing the k-means results with the true class labels. First we get some of the needed information to construct the matching matrix.

```
% Get some of the preliminary information.
% You can load the data using: load example52
```

```
ukmus = unique(kmus); ulabs = unique(labs);
n1 = length(ukmus); n2 = length(ulabs);
n = length(kmus);
```

Now we find the matrix **N**, noting that it is not necessarily square (the number of partitions do not have to be equal), and it is usually not symmetric.

```
% Now find the matching matrix N
N = zeros(n1,n2);
I = 0;
for i = ukmus(:)'
    I = I + 1;
    J = 0;
    for j = ulabs(:)'
        J = J + 1;
        indI = find(kmus == i);
        indJ = find(labs == j);
        N(I,J) = length(intersect(indI,indJ));
    end
end
nc2 = nchoosek(n,2);
nidot = sum(N);
njdot = sum(N');
ntot = sum(sum(N.^2));
num = nc2+ntot-0.5*sum(nidot.^2)-0.5*sum(njdot.^2);
ri = num/nc2;
```

The resulting Rand index has a value of 0.8797, which indicates good agreement between the classifications. We provide a function called **randind** that implements this code for any two partitions **P1** and **P2**. We also implement the adjusted Rand index in the function **adjrand**. It is used in the following manner.

```
% Now use the adjusted Rand index function.
ari = adjrand(kmus,labs);
```

This yields 0.7302, indicating an agreement above what is expected by chance alone.
□

5.4.2 Cophenetic Correlation

In some applications, we might be interested in comparing the output from two hierarchical partitions. We can use the *cophenetic correlation coefficient* for this purpose. Perhaps the most common use of this measure is in comparing hierarchical clustering results with the proximity data (e.g., interpoint distances) that were used to obtain the partitions. For example, as discussed before, the various hierarchical methods impose a certain structure on the data, and we might want to know if this is distorting the original relationships between the points as specified by their proximities.

We start with the cophenetic matrix, **H**. The *ij*-th element of **H** contains the fusion value where object *i* and *j* were first clustered together. We only need the upper triangular entries of this matrix (i.e., those elements above the diagonal). Say we want to compare this partition to the interpoint distances. Then, the cophenetic correlation is given by the product moment correlation between the values (upper triangular only) in **H** and their corresponding entries in the interpoint distance matrix. This has the same properties as the product moment correlation coefficient. Values close to one indicate a higher degree of correlation between the fusion levels and the distances. To compare two hierarchical clusters, one would compare the upper triangular elements of the cophenetic matrix for each one.

Example 5.4

The cophenetic correlation coefficient is primarily used to assess the results of a hierarchical clustering method by comparing the fusion level of observations with their distance. MATLAB provides a function called **cophenet** that calculates the desired measure. We return to the **yeast** data in Example 5.1 to determine the cophenetic coefficient for single linkage and complete linkage.

```
load yeast
% Get the Euclidean distances.
Y = pdist(data);
% Single linkage output.
Zs = linkage(Y);
% Now get the cophenetic coefficient.
scoph = cophenet(Zs,Y);
```

The cophenetic coefficient is 0.9243, indicating good agreement between the distances and the hierarchical clustering. Now we apply the same procedure to the complete linkage.

```
% Now do the same thing for the complete linkage.
Zc = linkage(Y,'complete');
ccoph = cophenet(Zc,Y);
```

The coefficient in this case is 0.8592, showing less correlation. The **cophenet** function only does the comparison between the clustering and the distances and not the comparison between two hierarchical structures.
❏

5.5.3 Upper Tail Rule

The upper tail rule was developed by Mojena [1977] as a way of determining the number of groups in hierarchical clustering. It uses the relative sizes of

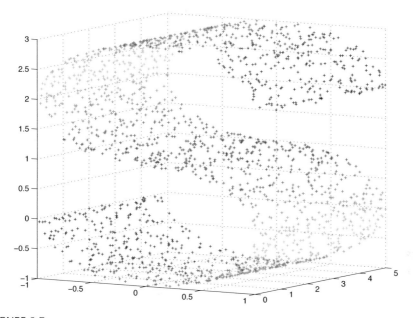

FIGURE 3.7
These points represent a random sample from the S-curve manifold. The colors are mapped to the height of the surface and are an indication of the neighborhood of a point.

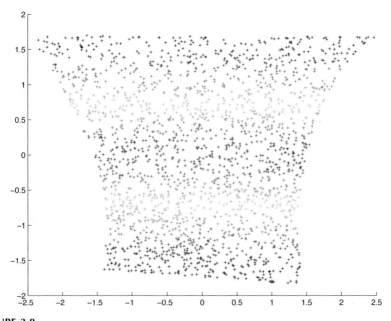

FIGURE 3.8
This is the 2-D embedding recovered using LLE. Note by the color that the neighborhood structure is preserved.

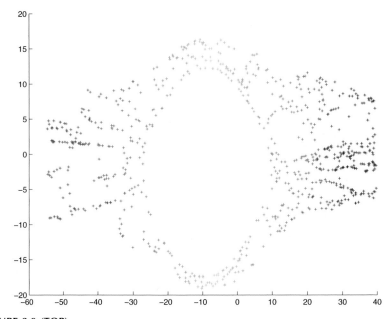

FIGURE 3.9 (TOP)
This is the 2-D embedding from ISOMAP. Note that the correct structure is not obtained.

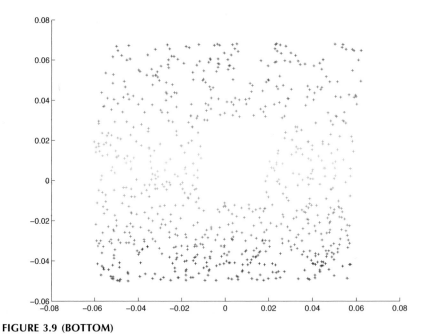

FIGURE 3.9 (BOTTOM)
This scatterplot shows the 2-D embedding from HLLE. Note that it was able to recover the correct embedding.

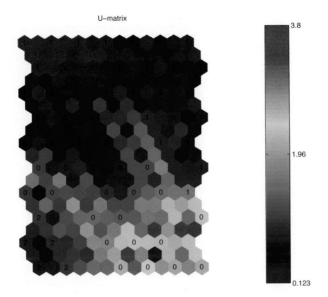

FIGURE 3.10

This is a U-matrix visualization of the SOM for the oronsay data. The distance of a map unit to each of its neighbors is calculated and then visualized using a color scale. We see evidence of a cluster in the lower left corner and the upper part of the map.

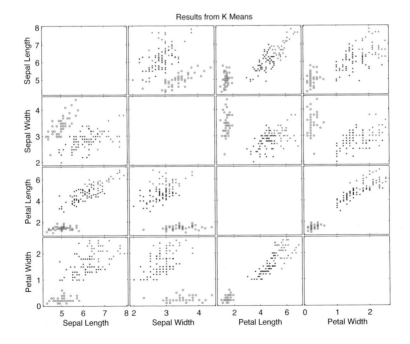

FIGURE 5.4

This is a scatterplot matrix showing the results of applying *k*-means clustering (with $k = 3$) to the iris data.

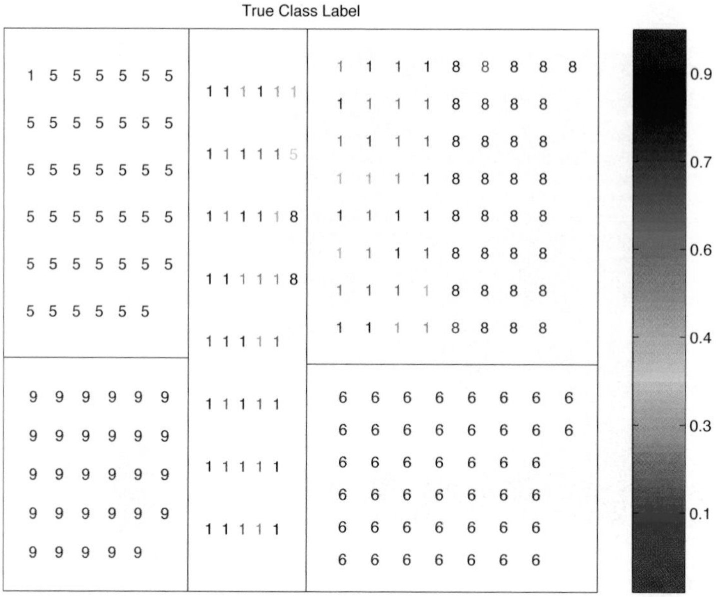

True Class Label

```
1 5 5 5 5 5 5        1 1 1 1 1 1      1 1 1 1 8 8 8 8 8        0.9
5 5 5 5 5 5 5                         1 1 1 1 8 8 8 8
                     1 1 1 1 1 5      1 1 1 1 8 8 8 8          0.7
5 5 5 5 5 5 5                         1 1 1 1 8 8 8 8
5 5 5 5 5 5 5        1 1 1 1 1 8      1 1 1 1 8 8 8 8          0.6
5 5 5 5 5 5 5                         1 1 1 1 8 8 8 8
                     1 1 1 1 1 8      1 1 1 1 8 8 8 8
5 5 5 5 5 5                           1 1 1 1 8 8 8 8          0.4
                     1 1 1 1 1

9 9 9 9 9 9          1 1 1 1 1      6 6 6 6 6 6 6 6
9 9 9 9 9 9                         6 6 6 6 6 6 6 6            0.3
9 9 9 9 9 9          1 1 1 1 1      6 6 6 6 6 6 6
                                   6 6 6 6 6 6 6
9 9 9 9 9 9          1 1 1 1 1      6 6 6 6 6 6 6              0.1
9 9 9 9 9                          6 6 6 6 6 6 6
```

FIGURE 8.7 (TOP)
This ReClus plot shows the cluster configuration for the text data based on the best model chosen from model-based clustering. Here we plot the true class label with the color indicating the probability that the observation belongs to the cluster.

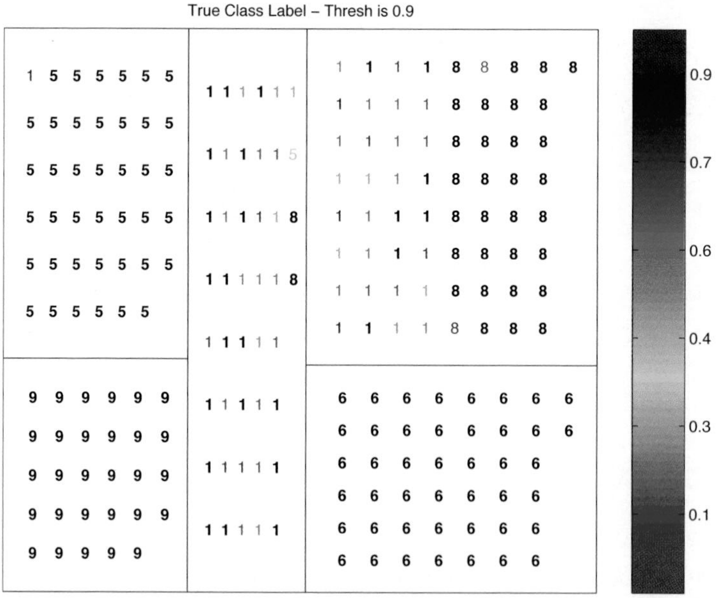

True Class Label – Thresh is 0.9

```
1 5 5 5 5 5 5        1 1 1 1 1 1      1 1 1 1 8 8 8 8 8        0.9
5 5 5 5 5 5 5                         1 1 1 1 8 8 8 8
                     1 1 1 1 1 5      1 1 1 1 8 8 8 8          0.7
5 5 5 5 5 5 5                         1 1 1 1 8 8 8 8
5 5 5 5 5 5 5        1 1 1 1 1 8      1 1 1 1 8 8 8 8          0.6
5 5 5 5 5 5 5                         1 1 1 1 8 8 8 8
                     1 1 1 1 1 8      1 1 1 1 8 8 8 8
5 5 5 5 5 5                           1 1 1 1 8 8 8 8          0.4
                     1 1 1 1 1

9 9 9 9 9 9          1 1 1 1 1      6 6 6 6 6 6 6 6
9 9 9 9 9 9                         6 6 6 6 6 6 6 6            0.3
9 9 9 9 9 9          1 1 1 1 1      6 6 6 6 6 6 6
                                   6 6 6 6 6 6 6
9 9 9 9 9 9          1 1 1 1 1      6 6 6 6 6 6 6              0.1
9 9 9 9 9                          6 6 6 6 6 6 6
```

FIGURE 8.7 (BOTTOM)
This ReClus plot shows the same cluster configuration as in Figure 8.7 (top), except that we now show observations with posterior probabilities greater than 0.9 in black bold font.

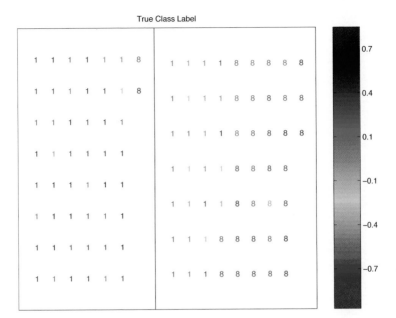

FIGURE 8.8
Here is the ReClus plot when we split the text data into two groups using hierarchical clustering. In this case, the scale corresponds to the silhouette values.

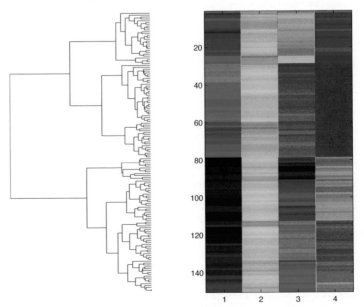

FIGURE 8.9
On the right, we have the data image for the iris data. The colors are mapped to the attribute values. Each row corresponds to an observation, and the columns are the variables. The rows have been re-ordered based on the leaves of the dendrogram at the left. Three groups are now visible in the data image.

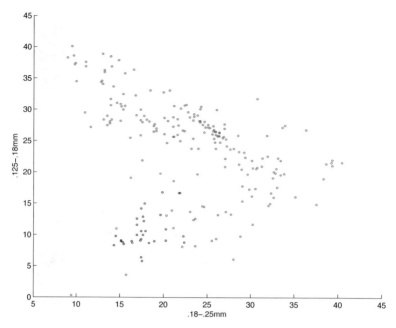

FIGURE 10.3 (TOP)
This figure shows the 2-D scatterplot for two variables of the oronsay data. The color indicates the midden class membership.

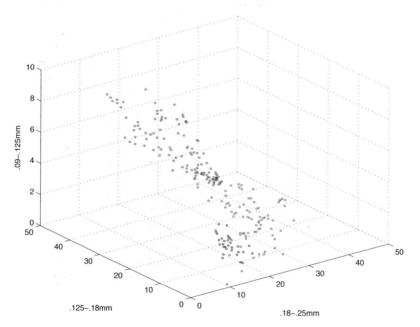

FIGURE 10.3 (BOTTOM)
This is the 3-D scatterplot for three variables of the oronsay data. The color indicates the midden class membership.

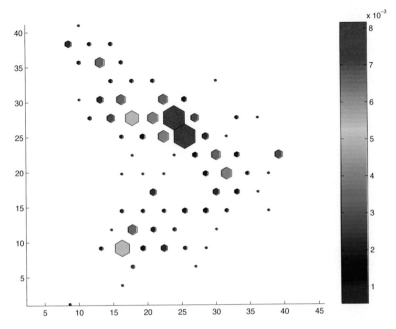

FIGURE 10.5
This shows a scatterplot of the oronsay data based on hexagonal binning. The color of the symbols represents the value of the probability density at that bin.

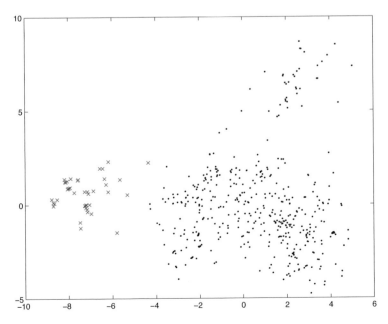

FIGURE 10.7
The red points in this scatterplot were highlighted using the scattergui function.

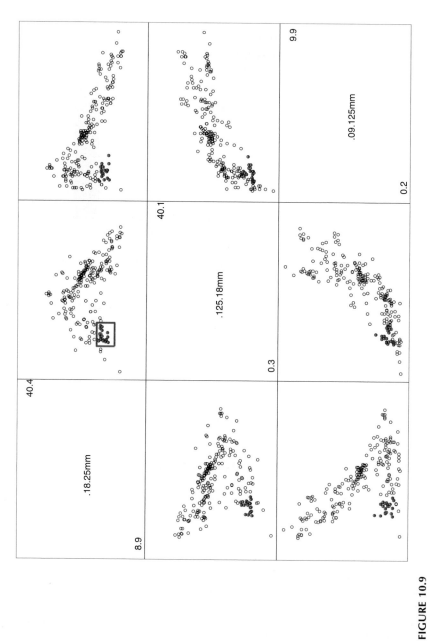

FIGURE 10.9

This is the scatterplot matrix with brushing and linking. This mode is transient, where only points inside the brush are highlighted. Corresponding points are highlighted in all scatterplots.

the different fusion levels in the hierarchy. We let the fusion levels $\alpha_0, \alpha_1, \alpha_2,$ ..., α_{n-1} correspond to the stages in the hierarchy with $n,\ n-1,\ldots,\ 1$ clusters. We also denote the average and standard deviation of the j previous fusion levels by $\bar{\alpha}$ and s_α. To apply this rule, we estimate the number of groups as the first level at which we have

$$\alpha_{j+1} > \bar{\alpha} + cs_\alpha, \tag{5.1}$$

where c is a constant. Mojena suggests a value of c between 2.75 and 3.50, but Milligan and Cooper [1985] offer a value of 1.25 based on their study of simulated data sets. One could also look at a plot of the values

$$\frac{(\alpha_{j+1} - \bar{\alpha})}{s_\alpha} \tag{5.2}$$

against the number of clusters j. A break in the plot is an indication of the number of clusters. Given the dependence on the value of c in Equation 5.1, we recommend the graphical approach of Equation 5.2.

Example 5.5

We now show how to implement the graphical Mojena procedure in a way that makes it similar to the 'elbow' plots of previous applications. We turn to the **lungB** data set for this example, and we use the standardized Euclidean distance, where each coordinate in the sum of squares is inversely weighted by its sample variance.

```
load lungB
% Get the distances and the linkage.
% Use the standardized Euclidean distance.
Y = pdist(lungB','seuclidean');
Z = linkage(Y,'complete');
% Plot dendrogram with fewer leaf nodes.
dendrogram(Z,15);
```

The dendrogram is shown in Figure 5.5 (top). We are going to flip the **Z** matrix to make it easier to work with, and we will find the values in Equation 5.2 for a maximum of 10 clusters.

```
nc = 10;
% Flip the Z matrix - makes it easier.
Zf = flipud(Z);
% Now get the vectors of means
% and standard deviations
for i = 1:nc
    abar(i) = mean(Zf(i:end,3));
    astd(i) = std(Zf(i:end,3));
```

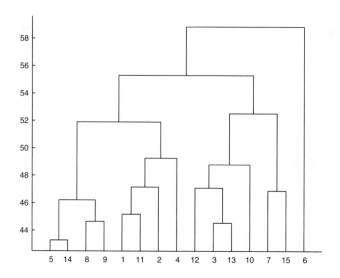

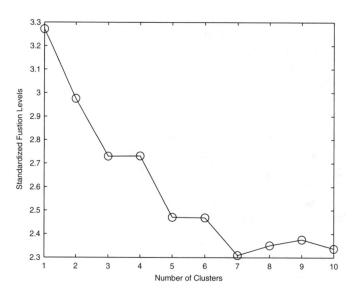

FIGURE 5.5

In the top panel, we show the dendrogram (with 15 leaf nodes) for the **lungB** data set, where standardized Euclidean distance is used with complete linkage. The second panel is the plot of the standardized fusion levels. The 'elbow' in the curve indicates that three clusters is reasonable. However, some other 'elbows' at 5 and 7 might provide interesting clusters, too.

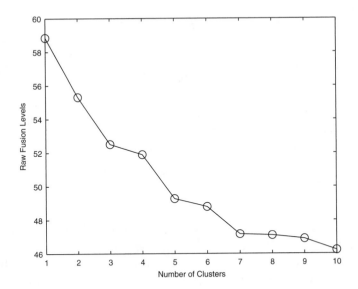

FIGURE 5.6
This is a plot of the raw fusion levels for Example 5.5. Again, we see an elbow at three clusters.

```
    end
    % Get the y values for plotting.
    yv = (Zf(1:nc,3) - abar(:))./astd(:);
    xv = 1:nc;
    plot(xv,yv,'-o')
```

This plot is shown in Figure 5.5 (bottom), where the elbow in the curve seems to indicate that three clusters could be chosen for this application. We could also plot the raw fusion values in a similar plot, as follows.

```
    % We can also plot just the fusion levels
    % and look for the elbow.
    plot(1:nc,Zf(1:nc,3),'o-')
```

This plot is given in Figure 5.6, and we see again that three seems to be a reasonable estimate for the number of clusters. We provide a function called **mojenaplot** that will construct the plot, given the output from **linkage**.
❑

5.5.4 Silhouette Plot

Kaufman and Rousseeuw [1990] present the **silhouette statistic** as a way of estimating the number of groups in a data set. Given observation i, we denote the average dissimilarity to all other points in its own cluster as a_i. For any other cluster c, we let $\bar{d}(i, c)$ represent the average dissimilarity of i to all objects in cluster c. Finally, we let b_i denote the minimum of these average dissimilarities $\bar{d}(i, c)$. The **silhouette width** for the i-th observation is

$$sw_i = \frac{(b_i - a_i)}{\max(a_i, b_i)}.$$ (5.3)

We can find the average silhouette width by averaging sw_i over all observations:

$$\overline{sw} = \frac{1}{n}\sum_{i=1}^{n} sw_i.$$

Observations with a large silhouette width are well clustered, but those with small values tend to be ones that are scattered between clusters. The silhouette width sw_i in Equation 5.3 ranges from –1 to 1. If an observation has a value close to 1, then the data point is closer to its own cluster than a neighboring one. If it has a silhouette width close to –1, then it is not very well-clustered. A silhouette width close to zero indicates that the observation could just as well belong to its current cluster or one that is near to it.

Kaufman and Rousseeuw use the average silhouette width to estimate the number of clusters in the data set by using the partition with two or more clusters that yields the largest average silhouette width. They state that an average silhouette width greater than 0.5 indicates a reasonable partition of the data, and a value of less than 0.2 would indicate that the data do not exhibit cluster structure.

There is also a nice graphical display called a **silhouette plot**, which is illustrated in the next example. This type of plot displays the silhouette values for each cluster, ranking them in decreasing order. This allows the analyst to rapidly visualize and assess the cluster structure.

Example 5.6

MATLAB provides a function called **silhouette** that will construct the silhouette plot and also returns the silhouette values, if desired. We illustrate its functionality using the **iris** data and k-means, where we choose both $k = 3$ and $k = 4$. In this example, we will use **replicates** (i.e., repeating the k-means procedure 5 times) and the **display** option that summarizes information about the replicates.

```
load iris
data = [setosa; versicolor; virginica];
% Get a k-means clustering using 3 clusters,
% and 5 replicates. We also ask MATLAB to
% display the final results for each replicate.
```

```
kmus3 = kmeans(data,3,...
    'replicates',5,'display','final');
```

When we run this code, the following is echoed to the command window. Since this procedure uses a random selection of starting points, you might see different results when implementing this code.

```
5 iterations, total sum of distances = 78.8514
4 iterations, total sum of distances = 78.8514
7 iterations, total sum of distances = 78.8514
6 iterations, total sum of distances = 78.8514
8 iterations, total sum of distances = 142.754
```

From this, we see that local solutions do exist, but the final result from MATLAB will be the one corresponding to the lowest value of our objective function, which is 78.8514. Now we repeat this for four clusters and get the corresponding silhouette plots.

```
% Get a k-means clustering using 4 clusters.
kmus4 = kmeans(data,4,...
    'replicates',5,'display','final');
% Get the silhouette plots and the values.
[sil3, h3] = silhouette(data, kmus3);
[sil4, h4] = silhouette(data, kmus4);
```

The silhouette plots for both $k = 3$ and $k = 4$ are shown in Figure 5.7. We see that we have mostly large silhouette values in the case of three clusters, and all are positive. On the other hand, four clusters yield some with negative values and some with small (but positive) silhouette indexes. To get a one-number summary describing each clustering, we can find the average of the silhouette values.

```
mean(sil3)
mean(sil4)
```

The three cluster solution has an average silhouette value of 0.7357, and the four cluster solution has an average of 0.6714. This indicates that the grouping into three clusters using k-means is better than the one with four groups.
❑

5.5.5 Gap Statistic

Tibshirani, et al. [2001] define a technique called the *gap statistic method* for estimating the number of clusters in a data set. This methodology applies to any technique used for grouping the data. The clustering method could be k-means, hierarchical, etc. This technique compares the within-cluster dispersion with what one might expect given a reference null distribution (i.e., no clusters). Fridlyand and Dudoit [2001] note a type of gap test has been

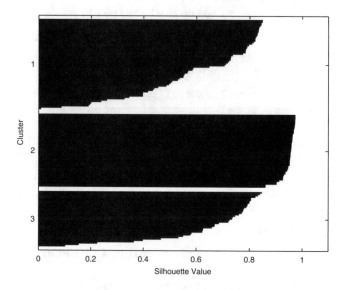

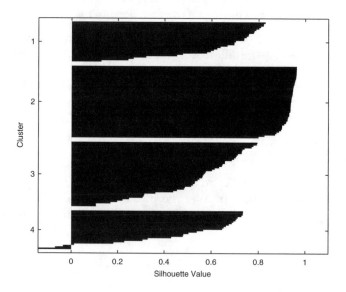

FIGURE 5.7
Here we have the silhouette plots for $k = 3$ and $k = 4$ clusters using the **iris** data. The top one indicates large values for cluster 2 and a few large values for clusters 1 and 3. In the second plot with 4 clusters, we see that there are some negative values in cluster 4, and clusters 1, 3, and 4 have many low silhouette values. These plots indicate that 3 clusters are a better fit to the data.

used by Bock [1985] in cluster analysis to test the null hypothesis of a homogeneous population versus a heterogeneous one, but they are defined somewhat differently.

We start with a set of k clusters $C_1, \dots, C_k$ from some clustering method, where the r-th group has n_r observations. We denote the sum of the pairwise distances for all points in cluster r as

$$D_r = \sum_{i,\, j \in C_r} d_{ij}.$$

We now define W_k as

$$W_k = \sum_{r=1}^{k} \frac{1}{2n_r} D_r. \tag{5.4}$$

Tibshirani, et al. [2001] note that the factor 2 in Equation 5.4 makes this the same as the pooled within-cluster sum of squares around the cluster means, if the distance used is squared Euclidean distance.

The gap statistic method compares the standardized graph of the within-dispersion index $\log(W_k)$, $k = 1,\dots,K$ with its expectation under a reference null distribution. The null hypothesis of a single cluster is rejected in favor of a model with k groups if there is strong evidence for this based on the gap statistic (see Equation 5.5). Their estimate for the number of groups is the value of k where $\log(W_k)$ is the furthest below this reference curve. The reference curve shows the expected values of $\log(W_k)$ and is found using random sampling.

Tibshirani, et al. [2001] show that there are two possible choices for the reference distribution. These are the uniform distribution over the range of observed values for a given variable or a uniform distribution over a box aligned with the principal components of the data set. The second method is preferred when one wants to ensure that the shape of the distribution is accounted for. We first discuss how to generate data from these distributions, then we go on to the describe the entire gap statistic procedure.

Generating Data from Reference Distributions

1. **Gap-Uniform**: For each of the i dimensions (or variables), we generate n one-dimensional variates that are uniformly distributed over the range of x_i^{min} to x_i^{max}, where x_i represents the i-th variable or the i-th column of **X**.

2. **Gap-PC**: We assume that the data matrix **X** has been column-centered. We then compute the singular value decomposition

$$\mathbf{X} = \mathbf{UDV}^{T}.$$

Next we transform $\mathbf{X}$ using

$$\mathbf{X}' = \mathbf{XV}.$$

We generate a matrix of random variates $\mathbf{Z}'$ as in the gap-uniform case, using the range of the columns of $\mathbf{X}'$ instead. We transpose back using

$$\mathbf{Z} = \mathbf{Z}'\mathbf{V}^{T}.$$

The basic gap statistic procedure is to simulate B data sets according to either of the null distributions and then apply the clustering method to each of them. We can calculate the same index $\log(W_k{}^*)$ for each simulated data set. The estimated expected value would be their average, and the estimated gap statistic is given by

$$gap(k) = \frac{1}{B} \sum_{b} \log(W^*_{k,\,b}) - \log(W_k).$$

Tibshirani, et al. [2001] did not offer insights as to what value of B to use, but Fridlyand and Dudoit [2001] use $B = 10$ in their work. We use the gap statistic to decide on the number of clusters as outlined in the following procedure.

Procedure - Gap Statistic Method

1. Cluster the given data set to obtain partitions $k = 1, 2, \dots , K$, using any desired clustering method.
2. For each partition with k clusters, calculate the observed $\log(W_k)$.
3. Generate a random sample of size n using either the gap-uniform or the gap-PC procedure. Call this sample $\mathbf{X^*}$.
4. Cluster the random sample $\mathbf{X^*}$ using the same clustering method as in step 1.
5. Find the within-dispersion measures for this sample, call them $\log(W^*_{k,\,b})$.
6. Repeat steps 3 through 5 for a total of B times. This yields a set of measures $\log(W^*_{k,\,b})$, $k = 1, \dots, K$ and $b = 1, \dots , B$.
7. Find the average of these values, using

$$\overline{W}_k = \frac{1}{B} \sum_{b} \log(W^*_{k,\,b}),$$

and their standard deviation

$$sd_k = \sqrt{\frac{1}{B} \sum_b [\log(W_{k,\,b}^*) - \overline{W}]^2} .$$

8. Calculate the estimated gap statistic

$$gap(k) = \overline{W}_k - \log(W_k) . \qquad (5.5)$$

9. Define

$$s_k = sd_k \sqrt{1 + 1/B} ,$$

and choose the number of clusters as the smallest k such that

$$gap(k) \geq gap(k+1) - s_{k+1} .$$

Example 5.7

We show how to implement the gap statistic method for a uniform null reference distribution using the **lungB** data set. We use agglomerative clustering (with complete linkage) in this application rather than k-means, mostly because we can get up to K clusters without performing the clustering again for a different k. Note also that we standardized the columns of the data matrix.

```
load lungB
% Take the transpose, because the
% columns are the observations.
X = lungB';
[n,p] = size(X);
% Standardize the columns.
for i = 1:p
    X(:,i) = X(:,i)/std(X(:,i));
end
```

We now find the observed $\log(W_k)$ for a maximum of $K = 10$ clusters.

```
% Test for a maximum of 10 clusters.
K = 10;
Y = pdist(X,'euclidean');
Z = linkage(Y,'complete');
% First get the observed log(W_k).
% We will use the squared Euclidean distance
% for the gap statistic.
```

```
% Get the one for 1 cluster first.
W(1) = sum(pdist(X).^2)/(2*n);
for k = 2:K
    % Find the index for k.
    inds = cluster(Z,k);
    for r = 1:k
        indr = find(inds==r);
        nr = length(indr);
        % Find squared Euclidean distances.
        ynr = pdist(X(indr,:)).^2;
        D(r) = sum(ynr)/(2*nr);
    end
    W(k) = sum(D);
end
```

We now repeat the same procedure *B* times, except that we use the uniform reference distribution as our data.

```
% Now find the estimated expected
% values.
B = 10;
% Find the range of columns of X for gap-uniform
minX = min(X);
maxX = max(X);
Wb = zeros(B,K);
% Now do this for the bootstrap.
Xb = zeros(n,p);
for b = 1:B
    % Generate according to the gap-uniform method.
    % Find the min values and max values.
    for j = 1:p
        Xb(:,j) = unifrnd(minX(j),maxX(j),n,1);
    end
    Yb = pdist(Xb,'euclidean');
    Zb = linkage(Yb,'complete');
    % First get the observed log(W_k)
    % We will use the squared Euclidean distance.
    % Get the one for 1 cluster first.
    Wb(b,1) = sum(pdist(Xb).^2)/(2*n);
    for k = 2:K
        % Find the index for k.
        inds = cluster(Zb,k);
        for r = 1:k
            indr = find(inds==r);
            nr = length(indr);
            % Find squared Euclidean distances.
            ynr = pdist(Xb(indr,:)).^2;
```

```
              D(r) = sum(ynr)/(2*nr);
          end
          Wb(b,k) = sum(D);
      end
  end
```

The matrix **Wb** contains our $\log(W_k^*)$ values, one set for each row. The following code gets the gap statistics, as well as the observed and expected $\log(W_k)$.

```
% Find the mean and standard deviation
Wobs = log(W);
muWb = mean(log(Wb));
sdk = (B-1)*std(log(Wb))/B;
gap = muWb - Wobs;
% Find the weighted version.
sk = sdk*sqrt(1 + 1/B);
gapsk = gap - sk;
% Find the lowest one that is larger:
ineq = gap(1:9) - gapsk(2:10);
ind = find(ineq > 0);
khat = ind(1);
```

The estimated number of clusters is two, which corresponds to the correct number of cancer types. This can be compared to the results of Example 5.5, where the Mojena graphical rule indicated three clusters. In any event, the results of the gap statistic method seem to indicate that there is evidence to reject the hypothesis that there is only one group. We plot the gap curve in Figure 5.8, where we can see the strong maximum at $k = 2$.
❑

The gap statistic method relies on B random samples from the reference null distribution to estimate the expected value of $\log(W_k)$, each of which is clustered using the same procedure that was used to get the observed ones. The data analyst should keep in mind that this can be computationally intensive if the sample size is large and a clustering method such as agglomerative clustering is used.

5.5 Summary and Further Reading

Clustering is a technique that has been used in many diverse areas, such as biology, psychiatry, archaeology, geology, marketing, and others [Everitt, Landau and Leese, 2001]. Because of this, clustering has come to be called different things, such as unsupervised learning [Duda and Hart, 1973; Jain,

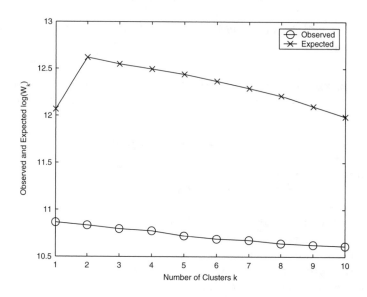

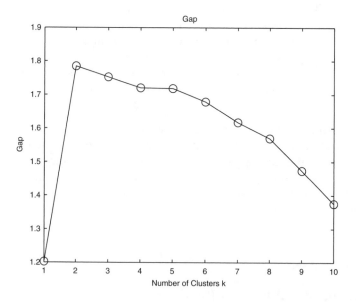

FIGURE 5.8
The upper panel shows the estimated expected and observed values of the log(Wk). The lower plot is the gap statistic curve, where we see a clear maximum at $k = 2$ clusters.

Murty and Flynn, 1999], numerical taxonomy [Sneath and Sokal, 1973], and vector quantization [Hastie, Tibshirani and Friedman, 2001].

In this chapter, we presented examples of the two most commonly-used types of clustering: hierarchical methods and optimization-based methods. Hierarchical methods yield an entire sequence of nested partitions. On the other hand, optimization or partition methods, like k-means, group the data into k nonoverlapping data sets. Hierarchical methods use the $n(n-1)/2$ interpoint distances as inputs (some also require the data), while optimization methods just require the data, making them suitable for large data sets.

In the next chapter, we describe another grouping technique called model-based clustering that is based on estimating finite mixture probability density functions. Note that some of the methods discussed in Chapter 3, such as self-organizing maps, generative topographic maps, and multidimensional scaling, can be considered a type of clustering, where the clusters are sometimes assessed visually. Since clustering methods (in most cases) will yield a grouping of the data, it is important to perform some validation or assessment of the cluster output. To that end, we presented several of these methods in this chapter.

We feel we should mention that, for the most part, we discussed clustering *methods* in this chapter. This can be contrasted with clustering *algorithms*, which are the underlying computational steps to achieve each of the clustering structures. For example, for any given hierarchical or optimization based method, many different algorithms are available that will achieve the desired result.

This is not meant to be an exhaustive treatment of the subject and much is left out, so we offer pointers to some additional resources. An excellent book to consult is Everitt, Landau and Leese [2001], which has been recently updated to include some of the latest developments in clustering based on the classification likelihood and neural networks. It is relatively nonmathematical, and it includes many examples for the student or practitioner. For a good summary and discussion of methods for estimating the number of clusters, as well as for other clustering information, we recommend Gordon [1999]. Most books that focus strictly on clustering are Kaufman and Rousseeuw [1990], Jain and Dubes [1988], Späth [1980], Hartigan [1975], and Anderberg [1973]. Other books on statistical pattern recognition usually include coverage of unsupervised learning or clustering. An excellent book like this is Webb [1999]. It is very readable and the author includes many applications and step-by-step algorithms. Of course, one of the seminal books in pattern recognition is Duda and Hart [1973], which has recently been updated to a new edition [Duda, Hart and Stork, 2001]. For more of the neural network point of view, one could consult Ripley [1996].

Some survey papers on clustering are available. A recent one that provides an overview of clustering from the machine learning point of view is Jain, Murty and Flynn [1999]. The goal in their paper is to provide useful advice on clustering to a broad community of users. Another summary paper written by a *Panel on Discriminant Analysis, Classification, and Clustering* [1989] describes methodological and theoretical aspects of these topics. A presentation of clustering from an EDA point of view is Dubes and Jain [1980]. An early review paper on grouping is by Cormack [1971], where he provides a summary of distances, clustering techniques, and their various limitations. An interesting position paper on clustering algorithms from a data mining point of view is Estivill-Castro [2002].

For a survey and analysis of procedures for estimating the number of clusters, see Milligan and Cooper [1985]. One of the successful ones in their study uses a criterion based on the within-group sum-of-squares objective function, which was developed by Krzanowski and Lai [1988]. Roeder [1994] proposes a graphical technique for this purpose. Tibshirani, et al. [2001] develop a new method called prediction strength for validating clusters and assessing the number of groups [2001], which is based on cross-validation. Another well-known index for measuring the appropriateness of data partitions is by Davies and Bouldin [1979]. Bailey and Dubes [1982] develop something called cluster validity profiles that quantify the interaction between a cluster and its environment in terms of its compactness and isolation, thus providing more information than a single index would. One should apply several cluster methods, along with the appropriate cluster assessment and validation methods with each data set, to search for interesting and informative groups.

Exercises

5.1   Get the Euclidean distances for the **iris** data. Apply centroid linkage and construct the dendrogram. Do you get inversions? Do the same thing with Mahalanobis distance. Do you get similar results? Try some of the other distances and linkages. Do you still get inversions?

5.2   Apply single linkage hierarchical clustering to the following data sets. Do you get chaining? Explore some of the other types of distance/linkage combinations and compare results.

   a. **geyser**

   b. **singer**

   c. **skulls**

   d. **spam**

   e. **sparrow**

   f. **oronsay**

   g. gene expression data sets

5.3   Construct the dendrogram for the partitions found in problem 5.2. Is there evidence of groups or clusters in each of the data sets?

5.4   The inconsistency coefficient can be used to determine the number of clusters in hierarchical clustering. This compares the length of a link in the hierarchy with the average length of neighboring links. If the merge is consistent with those around it, then it will have a low inconsistency coefficient. A higher inconsistency coefficient indicates the merge is inconsistent and thus indicative of clusters. One of the arguments to the **cluster** function can be a threshold for the inconsistency coefficient that corresponds to the **cutoff** argument. The cut point of the dendrogram occurs where links are greater than this value. MATLAB has a separate function called **inconsistent** that returns information in a matrix, where the last column contains the inconsistency coefficients. Generate some bivariate data that contains two well-separated clusters. Apply a suitable hierarchical clustering method and construct the dendrogram. Obtain the output from the **inconsistent** function and use these values to get a threshold for the **cutoff** argument in **cluster**. Knowing that there are only two clusters, does the inconsistency coefficient give the correct result?

5.5   Apply the inconsistent threshold to get partitions for the hierarchical clustering in problem 5.2. Where possible, construct scatterplots or a scatterplot matrix (using **plotmatrix** or **gplotmatrix**) of the resulting groups. Use different colors and/or symbols for the groups. Discuss your results.

5.6   Do a **help** on the **cophenet** function. Write your own MATLAB function that will calculate the cophenetic coefficient comparing two dendrograms.

5.7   Generate 2-D uniform random variables. Apply the gap statistic procedure and plot the expected gap statistic and observed value for each k. Compare the curves. What is the estimated number of clusters?

5.8   Now generate bivariate normal random variables with two well-separated clusters. Apply the gap statistic procedure and plot the expected gap statistic and observed value for each k. What is the estimated number of clusters?

5.9   Write a MATLAB function that implements the gap-PC approach.

5.10   Apply the gap statistic procedure to the cluster results from problem 5.2. Is there evidence for more than one group? How many clusters are present according to the gap statistic? Compare with your results in problem 5.5.

5.11   Apply the upper tail rule and **mojenaplot** to the data and groupings found in problem 5.2. How many clusters are present?

5.12   In what situation would the Rand index be zero? Use **adjrand** and **randind** functions to verify your answer.

5.13   Using the cluster results from one of the data sets in problem 5.2, use the **adjrand** function with the input arguments the same. The value should be one. Do the same thing using **randind**.

5.14   Apply *k*-means and the agglomerative clustering method of your choice to the **oronsay** data set (both classifications), using the correct number of known groups. Use the silhouette plot and average silhouette values. Repeat this process varying the value for *k*. Discuss your results.

5.15   Using the same data and partitions in problem 5.14, use the Rand index and adjusted Rand index to compare the estimated partitions with the true ones.

5.16   Repeat Example 5.4 using different distances and linkages. Discuss your results.

5.17   Repeat Example 5.7 using gap-PC method. Are the results different?

5.18   Calinski and Harabasz [1974] defined an index for determining the number of clusters as follows

$$ch_k = \frac{tr(\mathbf{S}_{B_k})/(k-1)}{tr(\mathbf{S}_{W_k})/(n-k)},$$

where $\mathbf{S}_{W_k}$ is the within-class scatter matrix for *k* clusters. $\mathbf{S}_{B_k}$ is the **between-class scatter matrix** that describes the scatter of the cluster means about the total mean and is defined as

$$\mathbf{S}_{B_k} = \sum_{j=1}^{k} \frac{n_j}{n}(\bar{\mathbf{x}}_j - \bar{\mathbf{x}})(\bar{\mathbf{x}}_j - \bar{\mathbf{x}})^T.$$

The estimated number of clusters is given by the largest value of ch_k for $k \geq 2$. Note that this is not defined for $k = 1$, as is the case with the gap statistic method. Implement this in MATLAB and apply it to the data used in Example 5.7. How does this compare to previous results?

5.19   Hartigan [1985] also developed an index to estimate the number of clusters. This is given by

$$hart_k = \left(\frac{tr(\mathbf{S}_{W_k})}{tr(\mathbf{S}_{W_{k+1}})} - 1\right)(n - k - 1).$$

The estimated number of clusters is the smallest value of *k*, where

$$1 \leq k \leq 10 .$$

Implement this in MATLAB, apply it to the data used in Example 5.7, and compare with previous results.

5.20   The Fowlkes-Mallows [1983] index can be calculated using the matching matrix **N** (which in this case must be $g \times g$, where $g = g_1 = g_2$) as

$$B_g = T_g / \sqrt{P_g Q_g} \ ,$$

where

$$T_g = \sum_{i=1}^{g} \sum_{j=1}^{g} n_{ij}^2 - n$$

$$P_g = \sum_{i=1}^{g} n_{i\cdot}^2 - n$$

$$Q_g = \sum_{j=1}^{g} n_{\cdot j}^2 - n$$

Implement this in a MATLAB function. Use the Fowlkes-Mallows index with the **oronsay** data, as in problem 5.14 and compare with previous results.

Chapter 6

Model-Based Clustering

In this chapter, we present a method for clustering that is based on finite mixture probability models. We first provide an overview of a comprehensive procedure for model-based clustering, so the reader has a framework for the methodology. We then describe the constituent techniques used in model-based clustering, such as finite mixtures, the EM algorithm, and model-based agglomerative clustering. We then put it all together again and include more discussion on its use in EDA and clustering. Finally, we show how to use a GUI tool that generates random samples based on the finite mixture models presented in the chapter.

6.1 Overview of Model-Based Clustering

Recall from the previous chapter that we presented two main types of clustering: hierarchical and partition-based (k-means). The hierarchical method we discussed in detail was agglomerative clustering, where two groups are merged at each step, such that some criterion is optimized. We will also use agglomerative clustering in this chapter, where the objective function is now based on optimizing the classification likelihood function.

We mentioned several issues with the methods in Chapter 5. One problem is that the number of clusters must be specified in k-means or chosen later in agglomerative hierarchical clustering. We already presented several ways to handle this, and the model-based clustering framework is another approach that addresses the issue of choosing the number of groups represented by the data. Another problem we mentioned is that many clustering methods are heuristic and impose a certain structure on the clusters. In other words, using Ward's method tends to produce same size, spherical clusters (as does k-means clustering). Additionally, some of the techniques are sensitive to outliers (e.g., Ward's method), and the statistical properties are generally unknown.

The model-based clustering methodology is based on probability models, such as the finite mixture model for probability densities. The idea of using probability models for clustering has been around for many years. See Bock [1996] for a survey of cluster analysis in a probabilistic and inferential framework. In particular, finite mixture models have been proposed for cluster analysis by Edwards and Cavalli-Sforza [1965], Day [1969], Wolfe [1970], Scott and Symons [1971], and Binder [1978]. Later researchers recognized that this approach can be used to address some of the issues in cluster analysis that we just discussed [Fraley and Raftery, 2002; Everitt, Landau and Leese, 2001; McLachlan and Peel, 2000; McLachlan and Basford, 1988; Banfield and Raftery, 1993].

The finite mixture approach assumes that the probability density function can be modeled as the sum of weighted component densities. As we will see shortly, when we use finite mixtures for cluster analysis, then the clustering problem becomes one of estimating the parameters of the assumed mixture model, i.e., probability density estimation. Each component density corresponds to a cluster and posterior probabilities are used to determine cluster membership.

The most commonly used method for estimating the parameters of a finite mixture probability density is the Expectation-Maximization (EM) algorithm, which is based on maximum likelihood estimation [Dempster, Laird and Rubin, 1977]. In order to apply the finite mixture – EM methodology, several issues must be addressed:

1. We must specify the number of component densities (or groups).[1]
2. The EM algorithm is iterative, so we need initial values of the parameters to get started.
3. We have to assume some form (e.g., multivariate normal, t distribution, etc.) for the component densities.

The model-based clustering framework provides a principled way to tackle all of these problems.

Let's start with the second problem: initializing the EM algorithm. We use model-based agglomerative clustering for this purpose. Model-based agglomerative clustering [Murtagh and Raftery, 1984; Banfield and Raftery, 1993] uses the same general ideas of hierarchical agglomerative clustering, where all observations start out in a single group, and two clusters are merged at each step. However, in the model-based case, two clusters are merged such that the classification likelihood is maximized.[2] Recall that we get a complete set of nested partitions with hierarchical clustering. From this, we can obtain initial estimates of our component parameters based on a given partition from the hierarchical clustering.

This brings us to issue three: the form of the component densities. Banfield and Raftery [1993] devise a general framework for multivariate normal

---

[1] This would be similar to specifying the k in k-means clustering.
[2] The classification likelihood is similar to the mixture likelihood, except that each observation is allowed to belong to only one component density. See Section 6.4 for a discussion.

mixtures based on constraining the component covariance matrices. Imposing these constraints governs certain geometric properties of the clusters, such as orientation, volume, and shape. In model-based clustering, we assume that the finite mixture is composed of multivariate normal densities, where constraints on the component covariance matrices yield different models.

So, why is this called *model*-based clustering? This really pertains to issues (1) and (2), together. We can consider the number of groups, combined with the form of the component densities, as producing different statistical models for the data. Determining the final model is accomplished by using the Bayesian Information Criterion (BIC), which is an approximation to Bayes factors [Schwarz, 1978; Kass and Raftery, 1995]. The model with the optimal BIC value is chosen as the 'best' model. The reader should be aware that we also use the word *model* to represent the type of constraints and geometric properties of the covariance matrices. Hopefully, the meaning of the word *model* will be clear from the context.

The steps of the model-based clustering framework are illustrated in Figure 6.1. First, we choose our constraints on the component covariance matrices (i.e., the model), and then we apply agglomerative model-based clustering. This provides an initial partition of our data for any given number of groups and any desired model. We use this partition to get initial estimates of our component density parameters for use in the EM algorithm. Once we converge to an estimate using the EM, we calculate the BIC. We continue to do this for various models (i.e., number of groups and forms of the covariance matrices). The final model that we choose is the one that produces the highest BIC. We now present more information and detail on the constituent parts of model-based clustering.

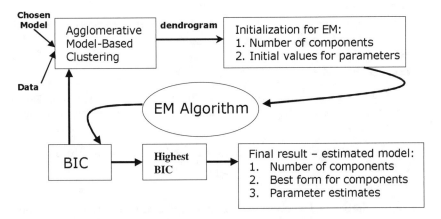

FIGURE 6.1
This shows the flow of the steps in the model-based clustering procedure.

6.2 Finite Mixtures

In this section, we provide more in-depth information regarding finite mixtures, as well as the different forms of the constrained covariance matrices used in model-based clustering. However, first we investigate an example of a *univariate* finite mixture probability density function to facilitate understanding of the multivariate approach.

Example 6.1

In this example, we show how to construct a probability density function that is the weighted sum of two univariate normal densities. First, we set up the various component parameters. We have equal mixing proportions or weights. The means are given by –2 and 2, and the corresponding standard deviations are 0.5 and 1.2. Mathematically, we have

$$f(x;\theta_k) = 0.5 \times \phi(x;-2, 0.5) + 0.5 \times \phi(x; 2, 1.2),$$

where ϕ denotes a normal probability density function. The following MATLAB code assigns the component density parameters.

```
% First the mixing coefficients will be equal.
pie1 = 0.5;
pie2 = 0.5;
% Let the means be -2 and 2.
mu1 = -2;
mu2 = 2;
% The standard deviation of the first term will be 0.5
% and the second one be 1.2.
sigma1 = 0.5;
sigma2 = 1.2;
```

Now we need to get a domain over which to evaluate the density, making sure that we have enough points to encompass the interesting parts of this density.

```
% Now generate a domain over which to evaluate the
% function.
x = linspace(-6, 6);
```

We use the Statistics Toolbox function **normpdf** to evaluate the component probability density functions. To get the finite mixture density, we need to weight these according to their mixing proportions and then add them together.

```
% The following is a Statistics Toolbox function.
```

```
y1 = normpdf(x,mu1,sigma1);
y2 = normpdf(x,mu2,sigma2);
% Now weight and add to get the final curve.
y = pie1*y1 + pie2*y2;
% Plot the final function.
plot(x,y)
xlabel('x'), ylabel('Probability Density Function')
title('Univariate Finite Mixture - Two Terms')
```

The plot is shown in Figure 6.2. The two terms are clearly visible, but this is not always the case. In the cluster analysis context, we would speculate that a group is located at –2 and 2.

❏

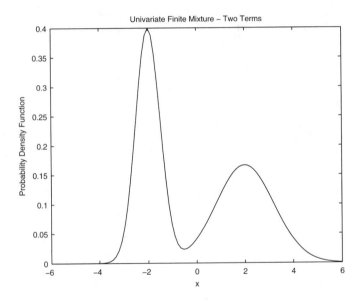

FIGURE 6.2
This shows the univariate probability density function as described in Example 6.1.

6.2.1 Multivariate Finite Mixtures

The *finite mixture* approach to probability density estimation (and cluster analysis) can be used for both univariate and multivariate data. In what follows, we will concern ourselves with the multivariate case only.

Finite mixtures encompass a family of probability density functions that are a weighted sum of component densities. The form of the density is given by

$$f(\mathbf{x}; \pi, \, \theta) \, = \, \sum_{k=1}^{c} \pi_k \, g_k(\mathbf{x}; \, \theta_k). \tag{6.1}$$

The *component density* is denoted by $g_k(\mathbf{x}; \, \theta_k)$ with associated parameters represented by θ_k. Note that θ_k is used to denote any type and number of parameters. The *weights* are given by π_k, with the constraint that they are nonnegative and sum to one. These weights are also called the *mixing proportions* or *mixing coefficients*.

Say we want to use the finite mixture in Equation 6.1 as a model for the distribution that generated our data. Then, to estimate the density, we must know the number of components c, and we need to have a form for the function g_k.

The component densities can be any *bona fide* probability density, but one of the most commonly used ones is the multivariate normal. This yields the following equation for a multivariate Gaussian finite mixture

$$f(\mathbf{x}; \pi, \, \mu_k, \, \Sigma_k) \, = \, \sum_{k=1}^{c} \pi_k \phi \, (\mathbf{x}; \mu_k, \, \Sigma_k), \tag{6.2}$$

where ϕ represents a multivariate normal probability density function given by

$$\phi(\mathbf{x}_i; \mu_k, \, \Sigma_k) \, = \, \frac{\exp\left\{-\frac{1}{2}(\mathbf{x}_i - \mu_k)^T \Sigma_k^{-1} (\mathbf{x}_i - \mu_k)\right\}}{(2\pi)^{p/2} \sqrt{|\Sigma_k|}}.$$

Thus, we have the parameters μ_k, where each one is a p-dimensional vector of means, and Σ_k, where each is a $p \times p$ covariance matrix.

Now that we have a form for the component densities, we know what we have to estimate using our data. We need to estimate the weights π_k, the p-dimensional means for each term, and the covariance matrices. Before we describe how to do this using the EM algorithm, we first look at the form of the multivariate normal densities in a little more detail.

6.2.2 Component Models - Constraining the Covariances

Banfield and Raftery [1993] and Celeux and Govaert [1995] provide the following eigenvalue decomposition of the k-th covariance matrix

$$\Sigma_k = \lambda_k \mathbf{D}_k \mathbf{A}_k \mathbf{D}_k^T. \tag{6.3}$$

The factors in Equation 6.3 are given by the following:

- The volume of the k-th cluster or component density is governed by the λ_k, which is proportional to the volume of the standard deviation ellipsoid. We note that the volume is different from the size of a cluster. The *size* is the number of observations falling into the cluster, while the *volume* is the amount of space encompassed by the cluster.

- $\mathbf{D}_k$ is a matrix with columns corresponding to the eigenvectors of Σ_k. It determines the *orientation* of the cluster.

- $\mathbf{A}_k$ is a diagonal matrix. $\mathbf{A}_k$ contains the normalized eigenvalues of Σ_k along the diagonal. By convention, they are arranged in decreasing order. This matrix is associated with the *shape* of the distribution.

We now describe the various models in more detail, using the notation and classification of Celeux and Govaert. The eigenvalue decomposition given in Equation 6.3 produces a total of 14 models, by keeping factors λ_k, $\mathbf{A}_k$, and $\mathbf{D}_k$ constant across terms and/or restricting the form of the covariance matrices (e.g., restrict it to a diagonal matrix). There are three main families of models: the spherical family, the diagonal family, and the general family.

Celeux and Govaert provide covariance matrix update equations based on these models for use in the EM algorithm. Some of these have a closed form, and others must be solved in an iterative manner. We concern ourselves only with the nine models that have a closed form update for the covariance matrix. Thus, what we present in this text is a subset of the available models. These are summarized in Table 6.1.

Spherical Family

The models in this family are characterized by diagonal matrices, where each diagonal element of the covariance matrix Σ_k has the same value. Thus, the distribution is spherical; i.e., each variable of the component density has the same variance. We have two closed-form cases in this family, each corresponding to a fixed spherical shape.

The first has equal volume across all components, so the covariance is of the form

$$\Sigma_k = \lambda \mathbf{DAD}^T = \lambda \mathbf{I},$$

where $\mathbf{I}$ is the $p \times p$ identity matrix. The other model allows the volumes to vary. In this case, Equation 6.3 becomes

TABLE 6.1

Description of Multivariate Normal Mixture Models - Closed Form Solution to Covariance Matrix Update Equation in EM Algorithm[a]

Model #	Covariance	Distribution	Description
1	$\Sigma_k = \lambda I$	Family: Spherical Volume: Fixed Shape: Fixed Orientation: NA	• Diagonal covariance matrices • Diagonal elements are equal • Covariance matrices are equal • I is a $p \times p$ identity matrix
2	$\Sigma_k = \lambda_k I$	Family: Spherical Volume: Variable Shape: Fixed Orientation: NA	• Diagonal covariance matrices • Diagonal elements are equal • Covariance matrices may vary • I is a $p \times p$ identity matrix
3	$\Sigma_k = \lambda B$	Family: Diagonal Volume: Fixed Shape: Fixed Orientation: Axes	• Diagonal covariance matrices • Diagonal elements may be unequal • Covariance matrices are equal • B is a diagonal matrix
4	$\Sigma_k = \lambda B_k$	Family: Diagonal Volume: Fixed Shape: Variable Orientation: Axes	• Diagonal covariance matrices • Diagonal elements may be unequal • Covariance matrices may vary among components • B is a diagonal matrix
5	$\Sigma_k = \lambda_k B_k$	Family: Diagonal Volume: Variable Shape: Variable Orientation: Axes	• Diagonal covariance matrices • Diagonal elements may be unequal • Covariance matrices may vary among components • B is a diagonal matrix
6	$\Sigma_k = \lambda DAD^T$	Family: General Volume: Fixed Shape: Fixed Orientation: Fixed	• Covariance matrices can have nonzero off-diagonal elements • Covariance matrices are equal
7	$\Sigma_k = \lambda D_k AD_k^T$	Family: General Volume: Fixed Shape: Fixed Orientation: Variable	• Covariance matrices can have nonzero off-diagonal elements • Covariance matrices may vary among components
8	$\Sigma_k = \lambda D_k A_k D_k^T$	Family: General Volume: Fixed Shape: Variable Orientation: Variable	• Covariance matrices can have nonzero off-diagonal elements • Covariance matrices may vary among components
9	$\Sigma_k = \lambda_k D_k A_k D_k^T$	Family: General Volume: Variable Shape: Variable Orientation: Variable	• Covariance matrices can have nonzero off-diagonal elements • Covariance matrices may vary among components

a.   This is a subset of the available models.

$$\Sigma_k = \lambda_k \mathbf{DAD}^T = \lambda_k \mathbf{I}.$$

Celeux and Govaert note that these models are invariant under any isometric transformation.

Example 6.2

We will generate a data set according to model 2 in Table 6.1. Here we have the spherical model, where the covariance matrices are allowed to vary among the component densities. The data set has $n = 250$ data points in 3-D, and we choose to have $c = 2$ component densities. The parameters for the multivariate normal components are given by

$$\mu_1 = [2,\ 2,\ 2]^T \qquad \mu_2 = [-2,\ -2,\ -2]^T$$
$$\Sigma_1 = \mathbf{I} \qquad \Sigma_2 = 2\mathbf{I}$$
$$\pi_1 = 0.7 \qquad \pi_2 = 0.3$$
$$\lambda_1 = 1 \qquad \lambda_2 = 2$$

We use the **genmix** GUI to generate the sample. See the last section of this chapter for information on how to use this tool. Once we generate the data, we can display it in a scatterplot matrix, as in Figure 6.3. Note that the second component centered at $[-2, -2, -2]^T$ has fewer points, but seems to have a larger volume.
❏

Diagonal Family

The models in this family are also diagonal, but now the elements along the diagonal of the covariance matrix are allowed to be different. The covariances are of the form

$$\Sigma = \lambda \mathbf{B} = \lambda \mathbf{DAD}^T,$$

where

$$\mathbf{B} = \text{diag}(b_1, \ldots, b_p),$$

and $|\mathbf{B}| = 1$. The matrix $\mathbf{B}$ determines the shape of the cluster.

The cluster shapes arising in this family of models are elliptical, because the variance in each of the dimensions is allowed to be different. Celeux and Govaert mention that the models in the diagonal family are invariant under any scaling of the variables. However, invariance does not hold for these models under all linear transformations.

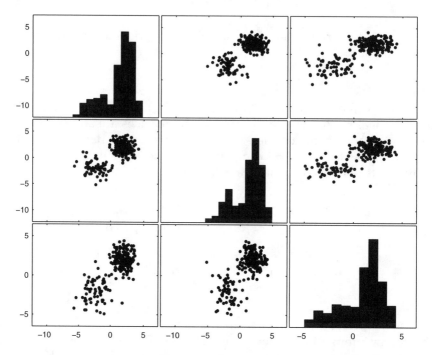

FIGURE 6.3

This shows a scatterplot matrix of the data set randomly generated according to model 2, which is a member of the spherical family.

We include three of the models in this family. The first is where each component covariance matrix is equal, with the same volume and shape:

$$\Sigma_k = \lambda \mathbf{B}.$$

Next we have one where the covariance matrices are allowed to vary in terms of the shape, but not the volume:

$$\Sigma_k = \lambda \mathbf{B}_k.$$

Finally, we allow both volume and shape to vary between the components or clusters:

$$\Sigma_k = \lambda_k \mathbf{B}_k.$$

Example 6.3

To illustrate an example from the diagonal family, we again use **genmix** to generate random variables according to the following model, with $n = 250$, $p = 2$, and $c = 2$. The means, weights, and covariance are given by

$$\mu_1 = [2,\ 2]^T \qquad \mu_2 = [-2,\ -2]^T$$

$$\mathbf{B}_1 = \begin{bmatrix} 3 & 0 \\ 0 & \frac{1}{3} \end{bmatrix} \qquad \mathbf{B}_2 = \begin{bmatrix} \frac{1}{2} & 0 \\ 0 & 2 \end{bmatrix}$$

$$\pi_1 = 0.7 \qquad \pi_2 = 0.3$$

$$\lambda_1 = 1 \qquad \lambda_2 = 1$$

This is model 4: equal volume, but different shapes. A scatterplot matrix of a random sample generated according to this distribution is shown in Figure 6.4.
❑

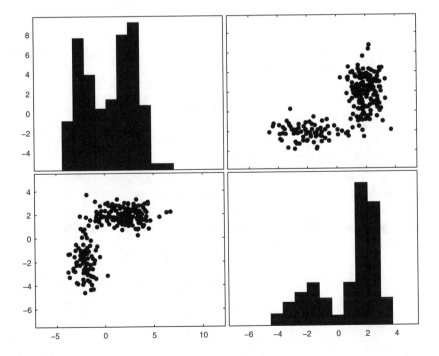

FIGURE 6.4
These data were generated according to a distribution that corresponds to model 4 of the diagonal family.

General Family

This family includes the more general cases for each cluster or component density. The covariances are no longer constrained to be diagonal. In other words, the off-diagonal terms of the matrices can be nonzero. The models in the general family are invariant under any linear transformation of the data.

We include four models from this family. As usual, the first one has all covariances constrained to be equal. This means that the clusters have fixed shape, volume, and orientation:

$$\Sigma_k = \lambda \mathbf{D} \mathbf{A} \mathbf{D}^T.$$

Next we allow the orientation to change, but keep the volume and shape fixed:

$$\Sigma_k = \lambda \mathbf{D}_k \mathbf{A} \mathbf{D}_k^T.$$

Then we allow both shape and orientation to vary, while keeping the volume fixed:

$$\Sigma_k = \lambda \mathbf{D}_k \mathbf{A}_k \mathbf{D}_k^T.$$

Finally, we have the unconstrained version, where nothing is fixed, so shape, volume and orientation are allowed to vary:

$$\Sigma_k = \lambda_k \mathbf{D}_k \mathbf{A}_k \mathbf{D}_k^T.$$

The unconstrained version is the one typically used in finite mixture models with multivariate normal components.

Example 6.4

The general family of models requires a full covariance matrix for each component density. We illustrate this using the following model ($n = 250$, $p = 2$, and $c = 2$):

$$\mu_1 = [2,\ 2]^T \qquad \mu_2 = [-2,\ -2]^T$$

$$\mathbf{A}_1 = \begin{bmatrix} 3 & 0 \\ 0 & \dfrac{1}{3} \end{bmatrix} \qquad \mathbf{A}_2 = \begin{bmatrix} \dfrac{1}{2} & 0 \\ 0 & 2 \end{bmatrix}$$

$$\mathbf{D}_1 = \begin{bmatrix} \cos(\pi \div 8) & -\sin(\pi \div 8) \\ \sin(\pi \div 8) & \cos(\pi \div 8) \end{bmatrix} \qquad \mathbf{D}_2 = \begin{bmatrix} \cos(6\pi \div 8) & -\sin(6\pi \div 8) \\ -\sin(6\pi \div 8) & \cos(6\pi \div 8) \end{bmatrix}$$

$$\pi_1 = 0.7 \qquad\qquad \pi_2 = 0.3$$

$$\lambda_1 = 1 \qquad\qquad \lambda_2 = 1$$

Note that this is model 8 – fixed volume, variable shape, and orientation. We have to multiply these together as in Equation 6.3 to get the covariance matrices for use in the **genmix** tool. These are given below, rounded to the fourth decimal place.

$$\Sigma_1 = \begin{bmatrix} 2.6095 & 0.9428 \\ 0.9428 & 0.7239 \end{bmatrix} \qquad \Sigma_2 = \begin{bmatrix} 1.6667 & -1.3333 \\ -1.3333 & 1.6667 \end{bmatrix}.$$

A scatterplot matrix showing a data set generated from this model is shown in Figure 6.5.
❑

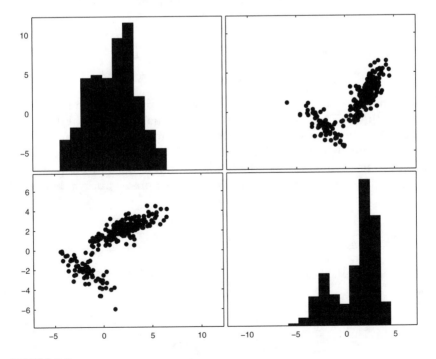

FIGURE 6.5
This shows a data set generated according to model 8 of the general family. These two components have equal volumes, but different shapes and orientations.

Now that we know what models we have for multivariate normal finite mixtures, we need to look at how we can use our data to get estimates of the parameters. The EM algorithm is used for this purpose.

6.3 Expectation-Maximization Algorithm

The problem of estimating the parameters in a finite mixture has been studied for many years. The technique we present here is called the *Expectation-Maximization* (EM) method. This is a general method for optimizing likelihood functions and is useful in situations where data might be missing or simpler optimization methods fail. The seminal paper on this method is by Dempster, Laird and Rubin [1977], where they formalize the EM algorithm and establish its properties. Redner and Walker [1984] apply it to finite mixture probability density estimation. The EM methodology is now a standard tool for statisticians and is used in many applications besides finite mixture estimation.

We wish to estimate the parameters $\theta = \pi_1, \dots, \pi_{c-1}, \mu_1, \dots, \mu_c, \Sigma_1, \dots, \Sigma_c$. Using the maximum likelihood approach, we maximize the log-likelihood given by

$$L(\theta \mid \mathbf{x}_1, \dots, \mathbf{x}_n) = \sum_{i=1}^{n} \ln \left[\sum_{k=1}^{c} \pi_k \phi(\mathbf{x}_i; \mu_k, \Sigma_k) \right]. \tag{6.4}$$

We assume that the components exist in a fixed proportion in the mixture, given by the π_k. Thus, it makes sense to calculate the probability that a particular point $\mathbf{x}_i$ belongs to one of the component densities. It is this component membership (or cluster membership) that is unknown, and why we need to use something like the EM algorithm to maximize Equation 6.4. We can write the *posterior probability* that an observation $\mathbf{x}_i$ belongs to component k as

$$\hat{\tau}_{ik}(\mathbf{x}_i) = \frac{\hat{\pi}_k \phi(\mathbf{x}_i; \hat{\mu}_k, \hat{\Sigma}_k)}{\hat{f}(\mathbf{x}_i; \hat{\pi}_k, \hat{\mu}_k, \hat{\Sigma}_k)}; \qquad k = 1, \dots, c; \ i = 1, \dots, n, \tag{6.5}$$

where

$$\hat{f}(\mathbf{x}_i; \hat{\pi}_k, \hat{\mu}_k, \hat{\Sigma}_k) = \sum_{k=1}^{c} \hat{\pi}_k \phi(\mathbf{x}_i; \hat{\mu}_k, \hat{\Sigma}_k).$$

For those readers not familiar with the notation, the 'hat' or caret above the parameters denotes an estimate. So, the posterior probability in Equation 6.5 is really an estimate based on estimates of the parameters.

Recall from calculus, that to maximize the function given in Equation 6.4, we must find the first partial derivatives with respect to the parameters in θ and then set them equal to zero. These are called the likelihood equations, and they are not provided here. Instead, we provide the solution [Everitt and Hand, 1981] to the likelihood equations as follows

$$\hat{\pi}_k = \frac{1}{n} \sum_{i=1}^{n} \hat{\tau}_{ik} \tag{6.6}$$

$$\hat{\mu}_k = \frac{1}{n} \sum_{i=1}^{n} \frac{\hat{\tau}_{ik} \mathbf{x}_i}{\hat{\pi}_k} \tag{6.7}$$

$$\hat{\Sigma}_k = \frac{1}{n} \sum_{i=1}^{n} \frac{\hat{\tau}_{ik} (\mathbf{x}_i - \hat{\mu}_k)(\mathbf{x}_i - \hat{\mu}_k)^T}{\hat{\pi}_k} . \tag{6.8}$$

The update for the covariance matrix given in Equation 6.8 is for the unconstrained case described in the previous section. This is model 9 in Table 6.1. We do not provide the update equations for the other models in this text, but they are implemented in the model-based clustering function that comes with the EDA Toolbox. Please see Celeux and Govaert [1995] for a complete description of the update equations for all models.

Because the posterior probability (Equation 6.5) is unknown, we need to solve these equations in an iterative manner. It is a two step process, consisting of an E-Step and an M-Step, as outlined below. These two steps are repeated until the estimated values converge.

E-Step

We calculate the posterior probability that the i-th observation belongs to the k-th component, given the current values of the parameters. This is given by Equation 6.5.

M-Step

Update the parameter estimates using the estimated posterior probability and Equations 6.6 through 6.8 (or use the covariance update equation for the specified model).

Note that the E-Step allows us to weight the component parameter updates in the M-Step according to the probability that the observation belongs to that component density.

Hopefully, it is obvious to the reader now why we need to have a way to initialize the EM algorithm, in addition to knowing how many terms or components are in our finite mixture. It is known that the likelihood surface typically has many modes, and the EM algorithm may even diverge, depending on the starting point. However, the EM can provide improved estimates of our parameters if the starting point is a good one [Fraley and Raftery, 2002]. The discovery that partitions based on model-based agglomerative clustering provide a good starting point for the EM algorithm was first discussed in Dasgupta and Raftery [1998].

Procedure - Estimating Finite Mixtures

1. Determine the number of terms or component densities c in the mixture.
2. Determine an initial guess at the component parameters. These are the mixing coefficients, means, and covariance matrices for each multivariate normal density.
3. For each data point x_i, calculate the posterior probability using Equation 6.5 and the current values of the parameters. This is the E-Step.
4. Update the mixing coefficients, the means, and the covariance matrices for the individual components using Equations 6.6 through 6.8. This is the M-Step. Note that in the model-based clustering approach, we use the appropriate covariance update for the desired model (replacing Equation 6.8).
5. Repeat steps 3 through 4 until the estimates converge.

Typically, step 5 is implemented by continuing the iteration until the changes in the estimates at each iteration are less than some pre-set tolerance. Alternatively, one could keep iterating until the likelihood function converges. Note that with the EM algorithm, we use the entire data set to simultaneously update the parameter estimates at each step.

Example 6.5
Since the MATLAB code can be rather complicated for the EM algorithm, we do not provide the code in this example. Instead, we show how to use a function called **mbcfinmix** that returns the weights, means, and covariances given initial starting values. The general syntax for this is

```
[pies,mus,vars]=...
    mbcfinmix(X,muin,varin,wtsin,model);
```

The input argument **model** is the number from Table 6.1. We will use the data generated in Example 6.4. We saved the data to the workspace as variable **X**. The following commands set up some starting values for the EM algorithm.

```
% Need to get initial values of the parameters.
piesin = [0.5, 0.5];
% The musin argument is a matrix of means,
% where each column is a p-D mean.
musin = [ones(2,1), -1*ones(2,1)];
% The varin argument is a 3-D array, where
% each page corresponds to one of the
% covariance matrices.
varin(:,:,1) = 2*eye(2);
varin(:,:,2) = eye(2);
```

Note that our initial values are sensible, but they do need adjusting. We call the EM algorithm with the following:

```
% Now call the function.
[pie,mu,vars]=mbcfinmix(X,musin,varin,piesin,8);
```

The final estimates (rounded to four decimal places) are shown below.

```
pie = 0.7188      0.2812
mu =
      2.1528    -1.7680
      2.1680    -2.2400
vars(:,:,1) =
      2.9582     1.1405
      1.1405     0.7920
vars(:,:,2) =
      1.9860    -1.3329
     -1.3329     1.4193
```

We see that the estimates are reasonable ones. We now show how to construct a surface based on the estimated finite mixture. The resulting plot is shown in Figure 6.6.

```
% Now create a surface for the density.
% Get a domain over which to evaluate the function.
x1 = linspace(-7,7,50);
x2 = linspace(-7,5,50);
[X1,X2] = meshgrid(x1,x2);
% The X1 and X2 are matrices. We need to
% put them as columns into another one.
dom = [X1(:), X2(:)];
% Now get the multivariate normal pdf at
% the domain values - for each component.
% The following function is from the
% Statistics Toolbox.
Y1 = mvnpdf(dom,mu(:,1)',vars(:,:,1));
Y2 = mvnpdf(dom,mu(:,2)',vars(:,:,2));
% Weight them and add to get the final function.
```

```
y = pie(1)*Y1 + pie(2)*Y2;
% Need to reshape the Y vector to get a matrix
% for plotting as a surface plot.
[m,n] = size(X1);
Y = reshape(y,m,n);
surf(X1,X2,Y)
axis([-7 7 -7 5 0 0.12])
xlabel('X_1'),ylabel('X_2')
zlabel('PDF')
```

We compare this to the scatterplot of the random sample shown in Figure 6.5. We see that the surface matches the density of the data.
❑

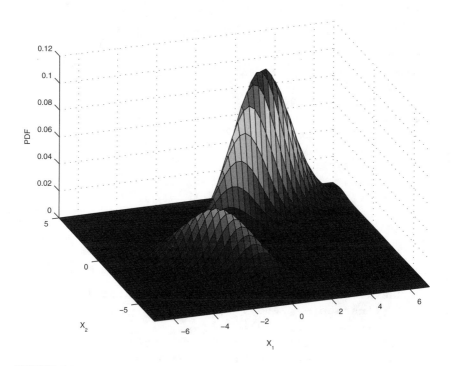

FIGURE 6.6
In Example 6.5, we estimated the density using the data from Figure 6.5 and the EM algorithm. This surface plot represents the estimated function. Compare to the scatterplot of the data in Figure 6.5.

6.4 Hierarchical Agglomerative Model-Based Clustering

We now describe another major component of the model-based clustering process, one that enables us to find initial values of our parameters for any given number of groups. *Agglomerative model-based clustering* works in a similar manner to the hierarchical methods discussed in the previous chapter, except that we do not have any notion of distance. Instead, we work with the *classification likelihood* as our objective function. This is given by

$$\mathcal{L}_{CL}(\theta_k, \gamma_i; \mathbf{x}_i) = \prod_{i=1}^{n} f_{\gamma_i}(\mathbf{x}_i; \theta_{\gamma_i}), \tag{6.9}$$

where γ_i is a label indicating a classification for the i-th observation. We have $\gamma_i = k$, if $\mathbf{x}_i$ belongs to the k-th component. In the mixture approach, the number of observations in each component has a multinomial distribution with sample size of n, and probability parameters given by $\pi_1, \dots, \pi_c$.

Model-based agglomerative clustering is a way to approximately maximize the classification likelihood. We start with singleton clusters consisting of one point. The two clusters producing the largest increase in the classification likelihood are merged at each step. This process continues until all observations are in one group. Note that the form of the objective function is adjusted as in Fraley [1998] to handle singleton clusters.

In theory, we could use any of the nine models in Table 6.1, but previous research indicates that the unconstrained model (number 9) with agglomerative model-based clustering yields reasonable initial partitions for the EM algorithm and any of the models. Thus, our MATLAB implementation of agglomerative model-based clustering includes just the unconstrained model. Fraley [1998] provides efficient algorithms for the four basic models (1, 2, 6, 9) and shows how the techniques developed for these can be extended to the other models.

Example 6.6

The function for agglomerative model-based clustering is called **agmbclust**. It takes the data matrix **X** and returns the linkage matrix **Z**, which is in the same form as the output from MATLAB's **linkage** function. Thus, the output from **agmbclust** can be used with other MATLAB functions that expect this matrix. We will use the familiar **iris** data set for this example.

```
% Load up the data and put into a data matrix.
load iris
X = [setosa; versicolor; virginica];
% Then call the function for agglomerative MBC.
```

```
Z = agmbclust(X);
```

We can show the results in a dendrogram, as in Figure 6.7 (top). Two groups are obvious, but three groups might also be reasonable. We use the silhouette function from the previous chapter to assess the partition for three groups.

```
% We can apply the silhouette procedure for this
% after we find a partition. Use 3 groups.
cind = cluster(Z,3);
[S,H] = silhouette(X,cind);
```

The silhouette plot is shown in Figure 6.7 (bottom). We see one really good cluster (number three). The others have small values and even negative ones. The average silhouette value is 0.7349.
❏

6.5 Model-Based Clustering

The *model-based clustering* framework consists of three major pieces:

1. Initialization of the EM algorithm using partitions from model-based agglomerative clustering.

2. Maximum likelihood estimation of the parameters via the EM algorithm.

3. Choosing the model and number of clusters according to the BIC approximation of Bayes factors.

We have already discussed the first two of these in detail, so we now turn our attention to Bayes factors and the BIC.

The Bayesian approach to model selection started with Jeffreys [1935, 1961]. Jeffreys developed a framework for calculating the evidence in favor of a null hypothesis (or model) using the Bayes factor, which is the posterior odds of one hypothesis when the prior probabilities of the two hypotheses are equal [Kass and Raftery, 1995].

Let's start with a simple two model case. We have our data $\mathbf{X}$, which we assume to have arisen according to either model M_1 or M_2. Thus, we have probability densities $p(\mathbf{X}|M_1)$ or $p(\mathbf{X}|M_2)$ and prior probabilities $p(M_1)$ and $p(M_2)$. From Bayes' Theorem, we obtain the posterior probability of hypothesis M_g given data $\mathbf{X}$,

$$p(M_g|\mathbf{X}) = \frac{p(M_g)p(\mathbf{X}|M_g)}{p(M_1)p(\mathbf{X}|M_1) + p(M_2)p(\mathbf{X}|M_2)}, \qquad (6.10)$$

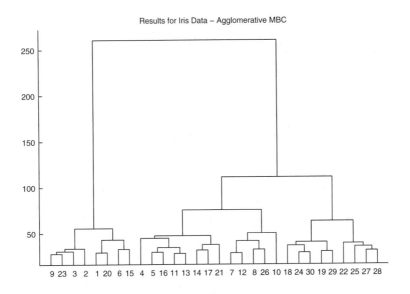

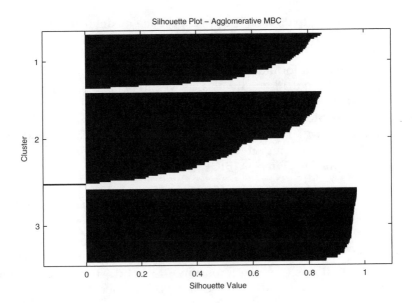

FIGURE 6.7
The top figure is the dendrogram derived from agglomerative model-based clustering of the **iris** data. We partition into three groups and then get their silhouette values. The corresponding silhouette plot is shown in the bottom panel.

for $g = 1, 2$. If we take the ratio of the two probabilities in Equation 6.10 for each of the models, then we get

$$\frac{p(M_1|\mathbf{X})}{p(M_2|\mathbf{X})} = \frac{p(\mathbf{X}|M_1)}{p(\mathbf{X}|M_2)} \times \frac{p(M_1)}{p(M_2)}. \tag{6.11}$$

The **Bayes factor** is the first factor in Equation 6.11:

$$B_{12} = \frac{p(\mathbf{X}|M_1)}{p(\mathbf{X}|M_2)}.$$

If any of the models contain unknown parameters, then the densities $p(\mathbf{X}|M_g)$ are obtained by integrating over the parameters. In this case, the resulting quantity $p(\mathbf{X}|M_g)$ is called the **integrated likelihood** of model M_g. This is given by

$$p(\mathbf{X}|M_g) = \int p(\mathbf{X}|\theta_g, M_g)p(\theta_g|M_g)d\theta_g, \tag{6.12}$$

for $g = 1, \ldots, G$.

The natural thing to do would be to choose the model that is most likely, given the data. If the prior probabilities are equal, then this simply means that we choose the model with the highest integrated likelihood $p(\mathbf{X}|M_g)$. This procedure is relevant for model-based clustering, because it can be applied when there are more than two models, as well as being a Bayesian solution [Fraley and Raftery, 1998, 2002; Dasgupta and Raftery, 1998].

One of the problems with using Equation 6.12 is the need for the prior probabilities $p(\theta_g|M_g)$. For models that satisfy certain regularity conditions, the logarithm of the integrated likelihood can be approximated by the **Bayesian Information Criterion** (BIC), given by

$$p(\mathbf{X}|M_g) \approx BIC_g = 2\log p(\mathbf{X}|\hat{\theta}_g, M_g) - m_g \log(n), \tag{6.13}$$

where m_g is the number of independent parameters that must be estimated for model M_g. It is well known that finite mixture models fail to satisfy the regularity conditions to make Equation 6.13 valid. However, previous applications and studies show the use of the BIC in model-based clustering produces reasonable results [Fraley and Raftery, 1998; Dasgupta and Raftery, 1998].

Now that we have all of the pieces of model-based clustering, we offer the following procedure that puts it all together.

Procedure - Model-Based Clustering

1. Using the unconstrained model, apply the agglomerative model-based clustering procedure to the data. This provides a partition of the data for any given number of clusters.
2. Choose a model, M (see Table 6.1).
3. Choose a number of clusters or component densities, c.
4. Find a partition with c groups using the results of the agglomerative model-based clustering (step 1).
5. Using this partition, find the mixing coefficients, means, and covariances for each cluster. The covariances are constrained according to the model chosen in step 2.
6. Using the chosen c (step 3) and the initial values (step 5), apply the EM algorithm to obtain the final estimates.
7. Calculate the value of the BIC for this value of c and M:

$$BIC_g = 2L_M(\mathbf{X}, \ \hat{\theta}) - m_g \log(n),$$

where L_M is the log likelihood, given the data, the model M, and the estimated parameters $\hat{\theta}$.
8. Go to step 3 to choose another value of c.
9. Go to step 2 to choose another model M.
10. Choose the 'best' configuration (number of clusters c and form for the covariance matrices) that corresponds to the highest BIC.

We see from this procedure that model-based clustering assumes various models for the covariance matrices and numbers of clusters, performs the cluster analysis, and then chooses the most likely clustering. So, it is an exhaustive search over the space of models that are both interesting and available. If we were interested in looking for 1 to C groups using all nine of the models, then we would have to perform the procedure $C \times 9$ times. As one might expect, this can be computationally intensive.

Example 6.7

We are going to use the **iris** data for this example, as well. The function called **mbclust** invokes the entire model-based clustering procedure: agglomerative model-based clustering, finite mixture estimation via the EM algorithm, and evaluating the BIC. The outputs from this function are: (1) a matrix of BIC values, where each row corresponds to a model, (2) a structure that has fields representing the parameters (**pies**, **mus**, and **vars**) for the best model, (3) a structure that contains information about *all* of the models (more on this in the next example), (4) the matrix **z** representing the

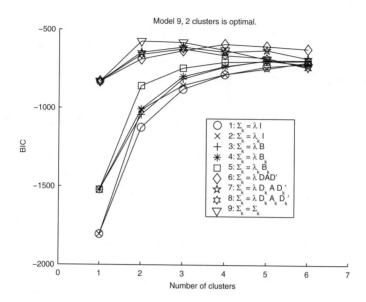

FIGURE 6.8

This shows the BIC curves for the model-based clustering of the **iris** data. We see that the highest BIC corresponds to model 9 and 2 groups.

agglomerative model-based clustering, and (5) a vector of group labels assigned using the best model.

```
load iris
data = [setosa;versicolor;virginica];
% Call the model-based clustering procedure with a
% maximum of 6 clusters.
[bics,bestmodel,allmodel,Z,clabs] = ...
    mbclust(data,6);
```

We can plot the BIC curves using **plotbic**, as follows

```
% Display the BIC curves.
plotbic(bics)
```

This is given in Figure 6.8. We see that the best model yielding the highest BIC value is for two clusters. We know that there are three groups in this data set, but two of the clusters tend to overlap and are hard to distinguish. Thus, it is not surprising that two groups were found using model-based clustering. As stated earlier, the EM algorithm can diverge so the covariance matrix can become singular. In these cases, the EM algorithm is stopped for that model, so you might see some incomplete BIC curves.

❏

The model-based clustering procedure we just outlined takes us as far as the estimation of a finite mixture probability density. Recall, that in this framework for cluster analysis, each component of the density provides a template for a cluster. Once we have the best model (number of clusters and form of the covariance matrices), we use it to group data based on their posterior probability. It makes sense to label an observation according to its highest posterior probability:

$$\{j|\hat{\tau}^*_{ij} = \max_k \hat{\tau}_{ik}\}.$$

In other words, we find the posterior probability that the *i*-th observation belongs to each of the component densities. We then say that it belongs to the cluster with the highest posterior probability. One benefit of using this approach is that we can use the quantity $1 - \max_k \hat{\tau}_{ik}$ as a measure of the uncertainty in the classification [Bensmail, et al. 1997].

Example 6.8

Returning to the results of the previous example, we show how to use some of the other outputs from **mbclust**. We know that the **iris** data has three groups. First, let's see how we can extract the model for three groups, model 9. We can do this using the following syntax to reference it:

```
allmodel(9).clus(3)
```

The variable **allmodel** is a structure with one field called **clus**. It (**allmodel**) has nine records, one for each model. The field **clus** is another structure. It has **maxclus** records and three fields: **pies**, **mus**, and **vars**. To obtain the group labels according to this model, we use the **mixclass** function. This requires the data and the finite mixture parameters for the model.

```
% Extract the parameter information:
pies = allmodel(9).clus(3).pies;
mus = allmodel(9).clus(3).mus;
vars = allmodel(9).clus(3).vars;
```

Now we can group the observations.

```
% The mixclass function returns group labels
% as well as a measure of the uncertainty in the
% classification.
[clabs3,unc] = mixclass(data,pies,mus,vars);
```

We can assess the results, as before, using the **silhouette** function.

```
% As before, we can use the silhouette procedure
% to assess the cluster results.
[S, H] = silhouette(data,clabs3);
```

```
title('Iris Data - Model-Based Clustering')
```

The average silhouette value is 0.6503, and the silhouette plot is shown in Figure 6.9. The reader is asked in the exercises to explore the results of the two cluster case with the best model in a similar manner.
❑

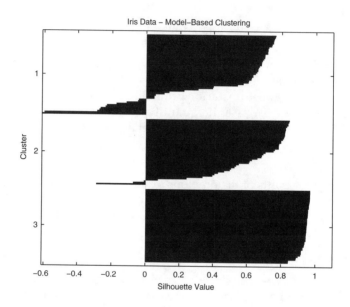

FIGURE 6.9

This is the silhouette plot showing the results of the model-based clustering, model 9, three clusters.

6.6 Generating Random Variables from a Mixture Model

In many situations, one might be interested in generating data from one of the finite mixture models discussed in this chapter. Of course, we could always do this manually using a multivariate normal random number generator (e.g., **mvnrnd** in the MATLAB Statistics Toolbox), but this can be tedious to use in the case of mixtures. So, we include a useful GUI tool called **genmix** that generates data according to a finite mixture.

The GUI is invoked by typing **genmix** at the command line. A screen shot is shown in Figure 6.10. The steps for entering the required information are listed on the left-side of the GUI window. We outline them below and briefly describe how they work.

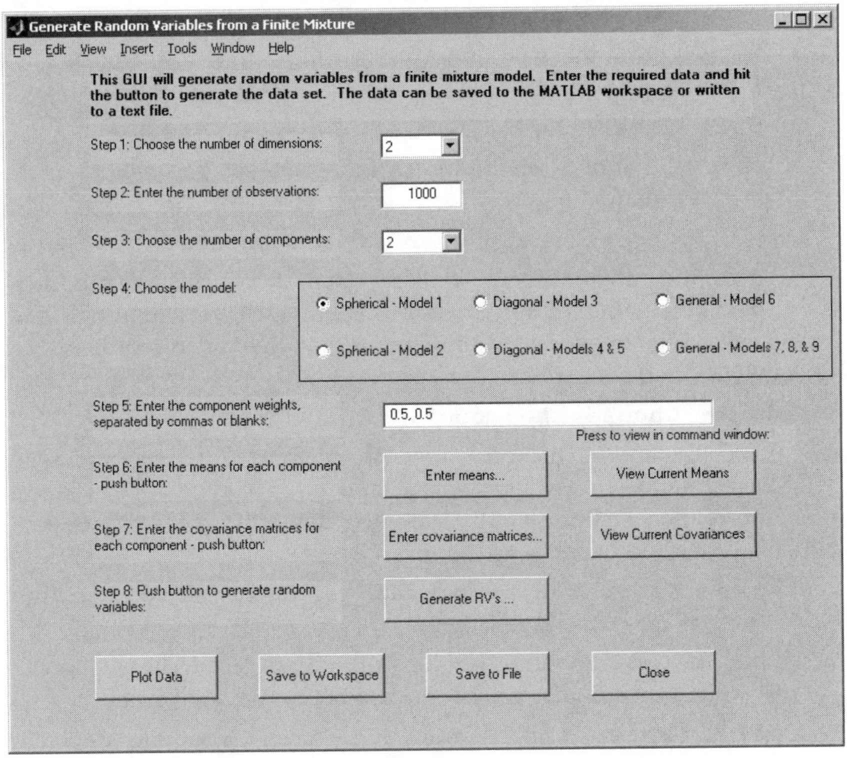

FIGURE 6.10

This shows the **genmix** GUI. The steps that must be taken to enter the parameter information and generate the data are shown on the left side of the window.

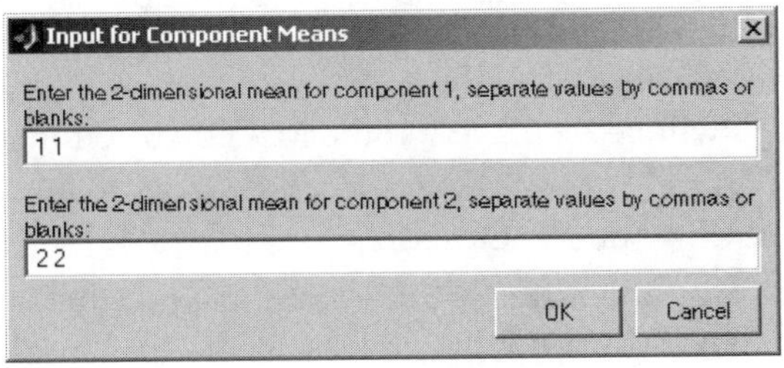

FIGURE 6.11

This shows the pop-up window for entering 2-D means and for the two components or terms in the finite mixture.

Step 1: Choose the number of dimensions.

This is a pop-up menu. Simply select the number of dimensions for the data.

Step 2: Enter the number of observations.

Type the total number of points in the data set. This is the value for n in Equation 6.4.

Step 3: Choose the number of components.

This is the number of terms or component densities in the mixture. This is the value for c in Equations 6.2 and 6.4. Note that one can use this GUI to generate a finite mixture with only one component, which can be useful in other applications.

Step 4: Choose the model.

Select the model for generating the data. The model numbers correspond to those described in Table 6.1. The type of covariance information you are required to enter depends on the model you have selected here.

Step 5: Enter the component weights, separated by commas or blanks.

Enter the corresponding weights (π_k) for each term. These must be separated by commas or spaces, and they must sum to 1.

Step 6: Enter the means for each component-push button.

Click on the button **Enter means...** to bring up a window for entering the p-dimensional means, as shown in Figure 6.11. There will be a different number of text boxes in the window, depending on the number of components selected in **Step 3**. Note that you must have the right number of values in each text box. In other words, if you have dimensionality $p = 3$ (**Step 1**), then each mean requires 3 values. If you need to check on the means that were used, then you can click on the **View Current Means** button. The means will be displayed in the MATLAB command window.

Step 7: Enter the covariance matrices for each component - push button.

Click on the button **Enter covariance matrices...** to activate a pop-up window. You will get a different window, depending on the chosen model (**Step 4**). See Figure 6.12 for examples of the three types of covariance matrix input windows. As with the means, you can push the **View Current Covariances** button to view the covariance matrices in the MATLAB command window.

Step 8: Push the button to generate random variables.

After all of the variables have been entered, push the button labeled **Generate RVs...** to generate the data set.

(a) This shows the pop-up window(s) for the spherical family of models. The only value that must be entered for each covariance matrix is the volume λ.

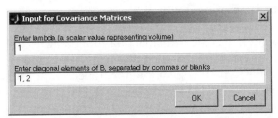

(b) This shows the pop-up window(s) for the diagonal family of models. One needs to enter the volume λ (first text box) and the diagonal elements of the matrix **B** (second text box).

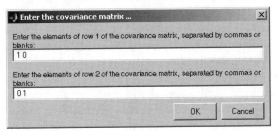

(c) When the general family is selected, one of these pop-up windows will appear for each unique covariance matrix. Each text box corresponds to a *row* of the covariance matrix.

FIGURE 6.12
Here we have examples of the covariance matrix inputs for the three families of models.

Once the variables have been generated, you have several options. The data can be saved to the workspace using the button **Save to Workspace**. When this is activated, a pop-up window appears, and you can enter a variable name in the text box. The data can also be saved to a text file for use in other software applications by clicking on the button **Save to File**. This brings up the usual window for saving files. An added feature to verify the output is the **Plot Data** button. The generated data are displayed in a scatterplot matrix (using the **plotmatrix** function) when this is pushed.

6.7 Summary and Further Reading

In this chapter, we have described another approach to clustering that is based on estimating a finite mixture probability density function. Each component of the density function represents one of the groups, and we can use the posterior probability that an observation belongs to the component density to assign group labels. The model-based clustering methodology consists of three main parts: (1) model-based agglomerative clustering to obtain initial partitions of the data, (2) the EM algorithm for maximum likelihood estimation of the finite mixture parameters, and (3) the BIC for choosing the best model.

Several excellent books are available on finite mixtures. Everitt and Hand [1981] is a short monograph that includes many applications for researchers. McLachlan and Basford [1988] is more theoretical, as is Titterington, Smith and Makov [1985]. McLachlan and Peel [2000] contains a more up-to-date treatment of the subject. They also discuss the application of mixture models to large databases, which is of interest to practitioners in EDA and data mining. Most books on finite mixtures also include information on the EM algorithm, but they also offer other ways to estimate the density parameters. McLachlan and Krishnan [1997] is one book that provides a complete account of the EM, including theory, methodology, and applications. Martinez and Martinez [2002] includes a description of the univariate and multivariate EM algorithm for finite mixtures, as well as a way to visualize the estimation process.

The EM algorithm described in this chapter is sometimes called the *iterative EM*. This is because it operates in batch mode; all data must be available for use in the update equations at each step. This imposes a high computational burden and storage load when dealing with massive data sets. A *recursive* form of the EM algorithm exists and can be used in an on-line manner. See Martinez and Martinez [2002] for more information on this algorithm and its use in MATLAB.

Celeux and Govaert [1995] present the complete set of models and associated procedures for obtaining the parameters in the EM algorithm when the covariances are constrained (multivariate normal mixtures). They also address the restricted case, where mixing coefficients are equal. We looked only at the unrestricted case, where these are allowed to vary.

There are many papers on the model-based clustering methodology, including interesting applications. Some examples include minefield detection [Dasgupta and Raftery, 1998], detecting faults in textiles [Campbell, et al., 1997, 1999], classification of gamma rays in astronomy [Mukherjee, et. al, 1998], and analyzing gene expression data [Yeung, et al., 2001]. One of the issues with agglomerative model-based clustering is the computational load with massive data sets. Posse [2001] proposes a starting partition based on minimal spanning trees, and Solka and Martinez [2004] describe a method for initializing the agglomerative model-based clustering using adaptive mixtures. Further extensions of model-based clustering for large data sets can be found in Wehrens, et al. [2003] and Fraley, Raftery and Wehrens [2003]. Two review papers on model-based clustering are Fraley and Raftery [1998, 2002].

Other methods for choosing the number of terms in finite mixtures have been offered. We provide just a few references here. Banfield and Raftery [1993] use an approximation to the integrated likelihood that is based on the classification likelihood. This is called the AWE, but the BIC seems to perform better. An entropy criterion called the NEC (normalized entropy criterion) has been described by Biernacki and Govaert [1997] and Biernacki, et al. [1999]. Bensmail, et al. [1997] provide an approach that uses exact Bayesian inference via Gibbs sampling. Chapter 6 of McLachlan and Peel [2000] provides a comprehensive treatment of this issue.

Additional software is available for model-based clustering. One such package is MCLUST, which can be downloaded at

http://www.stat.washington.edu/mclust/

MCLUST is compatible with S-Plus and R.[3] Fraley and Raftery [1999, 2002, 2003] provide an overview of the MCLUST software. McLachlan, et al. [1999] wrote a software package called EMMIX that fits mixtures of normals and t-distributions. The website for EMMIX is

http://www.maths.uq.edu.au/~gjm/emmix/emmix.html

Exercises

6.1 Apply model-based agglomerative clustering to the **oronsay** data set. Use the gap statistic to choose the number of clusters. Evaluate the partition using the silhouette plot and the Rand index.

6.2 Use the model-based clustering on the **oronsay** data. Apply the gap method to find the number of clusters for one of the models. Compare the gap estimate of the number of clusters to what one gets from the BIC.

6.3 Repeat the procedure in Example 6.1 for component densities that have closer means. How does this affect the resulting density function?

6.4 Construct another univariate finite mixture density, as in Example 6.1, using a mixture of univariate exponentials.

---

[3] See Appendix B for information on the R language.

6.5   Cluster *size* and *volume* are not directly related. Explain how they are different. In the finite mixture model, what parameter is related to cluster size?

6.6   Write a MATLAB function that will return the normal probability density function for the univariate case:

$$f(x;\mu,\ \sigma)\ =\ \frac{1}{\sigma\sqrt{2\pi}}\exp\left\{-\frac{(x-\mu)^2}{2\sigma^2}\right\}.$$

Use this in place of **normpdf** and repeat Example 6.1.

6.7   Apply model-based agglomerative clustering to the following data sets. Display in a dendrogram. If necessary, first reduce the dimensionality using a procedure from Chapters 2 and 3.

a. **skulls**

b. **sparrow**

c. **oronsay** (both classifications)

d. BPM data sets

e. gene expression data sets

6.8   Generate data according to the remaining six models of Table 6.1 that are not demonstrated in this text. Display the results using **plotmatrix**.

6.9   Return to the **iris** data in Example 6.6. Try two and four clusters from the agglomerative model-based clustering. Assess the results using the silhouette procedure.

6.10   The NEC is given below. Implement this in MATLAB and apply it to the models in Example 6.7. This means that instead of a matrix of BIC values, you will have a matrix of NEC values (i.e., one value for each model and number of clusters). Are the results different with respect to the number of clusters chosen for each model? For $1 \leq k \leq K$ we have,

$$L(K)\ =\ C(K)+E(K)$$

$$C(K)\ =\ \sum_{k=1}^{K}\sum_{i=1}^{n}\hat{\tau}_{ik}\ln[\hat{\pi}_k f(x_i;\hat{\theta}_k)]$$

$$E(K)\ =\ -\sum_{k=1}^{K}\sum_{i=1}^{n}\hat{\tau}_{ik}\ln\hat{\tau}_{ik}\geq 0$$

This is a decomposition of the log-likelihood $L(K)$ into a classification log-likelihood term $C(K)$ and the entropy $E(K)$ of the classification matrix with terms given by $\hat{\tau}_{ik}$. The NEC is given by

$$NEC(K) = \frac{E(K)}{L(K) - L(1)} \qquad K > 1 .$$

$$NEC(1) = 1$$

One chooses the K that corresponds to the minimum value of the NEC.

6.11 Apply the NEC to the clustering found in problem 6.2. How does the NEC estimate of the number of clusters compare with the gap estimate?

6.12 Apply other types of agglomerative clustering (e.g., different distances and linkage) to the **iris** data. Assess the results using three clusters and the **silhouette** function. Compare with the model-based agglomerative clustering results in Example 6.6. Also visually compare your results via the dendrogram.

6.13 Generate random variables according to the different models in Table 6.1. Use $n = 300$, $p = 2$, $c = 2$. Apply the model-based clustering procedure. Is the correct model (number of terms and form of covariance matrix) recovered?

6.14 Repeat Example 6.5 for a larger sample size. Are the parameter estimates closer to the true ones?

6.15 Repeat Example 6.8 using the best model, but the two cluster case. Compare and discuss your results.

6.16 Generate bivariate random samples according to the nine models. Apply model-based clustering and analyze your results.

6.17 Apply the full model-based clustering procedure to the following data sets. If necessary, first reduce the dimensionality using one of the procedures from Chapters 2 and 3. Assess your results using methods from Chapter 5.

a. **skulls**

b. **sparrow**

c. **oronsay** (both classifications)

d. BPM data sets

e. gene expression data sets

Chapter 7

Smoothing Scatterplots

In many applications, we might make distributional and model assumptions about the underlying process that generated the data, so we can use a *parametric* approach in our analysis. The parametric approach offers many benefits (known sampling distributions, lower computational burden, etc.), but can be dangerous if the assumed model is incorrect. At the other end of the data-analytic spectrum, we have the *nonparametric* approach, where one does not make any formal assumptions about the underlying structure or process. When our goal is to summarize the relationship between variables, then *smoothing methods* are a bridge between these two approaches. Smoothing methods make the weak assumption that the relationship between variables can be represented by a smooth curve or surface. In this chapter, we cover the *loess method* for scatterplot smoothing and its various extensions. Spline smoothing is briefly discussed at the end in conjunction with the Curve Fitting Toolbox from The MathWorks.

7.1 Introduction

In most, if not all, scientific experiments and analyses, the researcher must summarize, interpret and/or visualize the data to gain insights, to search for patterns, and to make inferences. For example, with some gene expression data, we might be interested in the distribution of the gene expression values for a particular patient or experimental condition, as this might indicate what genes are most active. We could use a probability density estimation procedure for this purpose. Another situation that often arises is one in which we have a response y and a predictor x, and we want to understand and model the relationship between the y and x variables.

The main goal of smoothing from an EDA point of view is to obtain some insights into how data are related to one another and to search for patterns. This idea has been around for many years, especially in the area of smoothing time series data, where data are measured at equally spaced points in time. Readers might be familiar with some of these methods, such as moving averages, exponential smoothing, and other specialized filtering techniques using polynomials.

The loess procedures described in this chapter obtain the smooths by performing local regression in a moving neighborhood that is analogous to the moving average in time series analysis. The main loess procedures for 2-D and multivariate data are presented first, followed by a robust version to counter the effect of outliers. We then proceed to a discussion of residuals and diagnostic plots that can be used to assess the results of loess smoothing. Cleveland and McGill [1984] developed some extensions of the loess scatterplot smoothing to look for relationships in the bivariate distribution of x and y called pairs of middle smoothings and polar smoothings, which are covered next. Smoothing via loess and other techniques is available in the MATLAB Curve Fitting toolbox, so we finish the chapter with a brief section that describes the functions provided in this optional toolbox.

7.2 Loess

Loess (also called *lowess* in earlier works) is a locally weighted regression procedure for fitting a regression curve (or surface) by smoothing the dependent variable as a function of the independent variable. The framework for loess is similar to what is commonly used in regression. We have n measurements of the dependent variable y_i, and associated with each y_i is a corresponding value of the independent variable x_i. For now, we assume that our dimensionality is $p = 1$; the case of multivariate predictor variables x is covered next.

We assume that the data are generated by

$$y_i = g(x_i) + \varepsilon_i,$$

where the ε_i are independent normal random variables with mean zero and variance σ^2. In the classical regression (or parametric) framework, we would assume that the function g belongs to a class of parametric functions, such as polynomials. With loess or local fitting, we assume that g is a smooth function of the independent variables, and our goal is to estimate this function g. The estimate is denoted as $\hat{y}$, a plot of which can be added to our scattterplot for EDA purposes or it can be used to make predictions of our response variable. The point denoted by the ordered pair $(x_0, \hat{y}_0)$ is called the smoothed point at a point in our domain x_0, and $\hat{y}_0$ is the corresponding fitted value.

The curve obtained from a loess model is governed by two parameters: α and λ. The parameter α is a smoothing parameter that governs the size of the neighborhood; it represents the proportion of points that are included in the local fit. We restrict our attention to values of α between zero and one, where high values for α yield smoother curves. Cleveland [1993] addresses the case where α is greater than one. The second parameter λ determines the degree of the local regression. Usually, a first or second degree polynomial is used, so $\lambda = 1$ or $\lambda = 2$, but higher degree polynomials can be used. Some of the earlier work in local regression used $\lambda = 0$, so a constant function was fit in each neighborhood [Cleveland and Loader, 1996].

The general idea behind loess is the following. To get a value of the curve $\hat{y}_0$ at a given point x_0, we first determine a local neighborhood of x_0 based on α. All points in this neighborhood are weighted according to their distance from x_0, with points closer to x_0 receiving larger weight. The estimate $\hat{y}_0$ at x_0 is obtained by fitting a linear or quadratic polynomial using the weighted points in the neighborhood. This is repeated for a uniform grid of points x in the domain to get the desired curve.

The reason for using a weighting function is that only points close to x_0 will contribute to the regression, since they should be a 'truer' indication of the relationship between the variables in the neighborhood of x_0. The weight function W should have the following properties:

1. $W(x) > 0$ for $|x| < 1$
2. $W(-x) = W(x)$
3. $W(x)$ is a nonincreasing function for $x \geq 0$
4. $W(x) = 0$ for $|x| \geq 1$

The basic idea is, for each point x_0 where we want to get a smoothed value $\hat{y}_0$, we define weights using the weight function W. We center the function W at x_0 and scale it so that W first becomes zero at the k-th nearest neighbor of x_0. As we will see shortly, the value for k is governed by the parameter α.

We use the tri-cube weight function in our implementation of loess. Thus, the weight $w_i(x_0)$ at x_0 for the i-th data point x_i is given by the following

$$w_i(x_0) = W\left(\frac{|x_0 - x_i|}{\Delta_k(x_0)}\right), \tag{7.1}$$

with

$$W(u) = \begin{cases} (1 - u^3)^3; & 0 \leq u < 1 \\ 0; & \text{otherwise.} \end{cases} \tag{7.2}$$

The denominator $\Delta_k(x_0)$ is defined as the distance from x_0 to the k-th nearest neighbor of x_0, where k is the greatest integer less than or equal to $\alpha \times n$. We denote the neighborhood of x_0 as $N(x_0)$. The tri-cube weight function is illustrated in Figure 7.1.

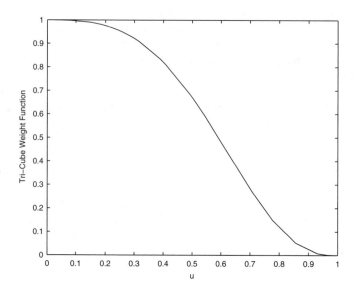

FIGURE 7.1
This is the tri-cube weight function.

In step 6 of the loess procedure outlined below, one can fit either a straight line to the weighted points (x_i , y_i), for x_i in the neighborhood $N(x_0)$, or a quadratic polynomial can be used (in our implementation). If a line is used as the local model, then $\lambda = 1$. The values of β_0 and β_1 are found such that the following is minimized

$$\sum_{i=1}^{k} w_i(x_0)(y_i - \beta_0 - \beta_1 x_i)^2, \tag{7.3}$$

for (x_i , y_i), with x_i in $N(x_0)$. Letting $\hat{\beta}_0$ and $\hat{\beta}_1$ be the values that minimize Equation 7.3, the loess fit at x_0 is given by

$$\hat{y}(x_0) = \hat{\beta}_0 + \hat{\beta}_1 x_0. \tag{7.4}$$

When $\lambda = 2$, then we fit a quadratic polynomial using weighted least-squares, again using only those points in $N(x_0)$. In this case, we find the values for the β_i that minimize

$$\sum_{i=1}^{k} w_i(x_0)(y_i - \beta_0 - \beta_1 x_i - \beta_2 x_i^2)^2. \tag{7.5}$$

Similar to the linear case, if $\hat{\beta}_0$, $\hat{\beta}_1$, and $\hat{\beta}_2$ minimize Equation 7.5, then the loess fit at x_0 is

$$\hat{y}(x_0) = \hat{\beta}_0 + \hat{\beta}_1 x_0 + \hat{\beta}_2 x_0^2. \qquad (7.6)$$

For more information on weighted least squares see Draper and Smith [1981].

We describe below the steps for obtaining a loess curve, which is illustrated in Figure 7.2. Using a set of generated data, we show the loess fit for a given point x_0. The top panel shows the linear fit in the neighborhood of x_0, and the bottom panel shows the quadratic fit. The open circle on the respective curves is the smoothed value at that point.

Procedure - Loess Curve Construction

1. Let x_i denote a set of n values for a predictor variable and let y_i represent the corresponding response.
2. Choose a value for α such that $0 < \alpha < 1$. Let $k = \lfloor \alpha \times n \rfloor$, where k is the greatest integer less than or equal to $\alpha \times n$.
3. For each x_0 where we want to obtain an estimate of the smooth $\hat{y}_0$, find the k points x_i in the data set that are closest to x_0. These x_i comprise a neighborhood of x_0, and this set is denoted by $N(x_0)$.
4. Compute the distance of the x_i in $N(x_0)$ that is furthest away from x_0 using

$$\Delta_k(x_0) = \max_{x_i \in N_0} |x_0 - x_i|.$$

5. Assign a weight to each point (x_i, y_i), x_i in $N(x_0)$, using the tri-cube weight function (Equations 7.1 and 7.2).
6. Obtain the value $\hat{y}_0$ of the curve at the point x_0 for the given λ using a weighted least squares fit of the points x_i in the neighborhood $N(x_0)$. (See Equations 7.3 through 7.6.)
7. Repeat steps 3 through 6 for all x_0 of interest.

We illustrate the loess procedure for univariate data in Example 7.1 using a well-known data set discussed in the loess literature [Cleveland and Devlin, 1988; Cleveland and McGill, 1984; Cleveland, 1993]. This data set contains 111 measurements of four variables, representing ozone and other meteorological data. They were collected during May 1 and September 30, 1973 at various sites in the New York City region. The goal is to describe the relationship between ozone (PPB) and the meteorological variables (solar radiation measured in Langleys, temperature in degrees Fahrenheit, and wind speed in MPH) so one might predict ozone concentrations.

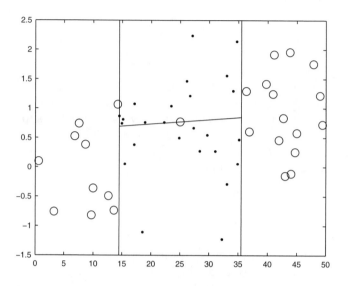

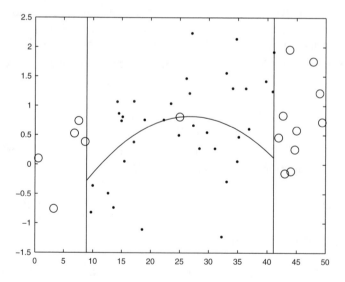

FIGURE 7.2
The top panel shows the local fit at a point $x_0 = 25$, with $\lambda = 1$ and $\alpha = 0.5$. The vertical lines indicate the limits of the neighborhood. The second panel shows the local fit at x_0, where $\lambda = 2$ and $\alpha = 0.7$. The point $(x_0, \hat{y}_0)$ is the open circle on the curves.

Example 7.1

We illustrate the univariate loess procedure using the ozone concentration as our response variable (y) and the temperature as our predictor variable (x). The next few lines of MATLAB code load the data set and display the scatterplot shown in Figure 7.3.

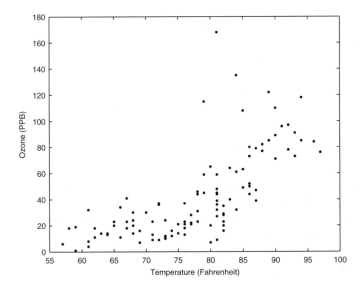

FIGURE 7.3
This is the scatterplot for ozone as it depends on temperature. We see that, in general, the ozone increases as temperature increases.

```
load environmental
% This has all four variables. We will
% use the ozone as the response and
% temperature as the predictor.
% First do a scatterplot
plot(Temperature,Ozone,'.')
xlabel('Temperature (Fahrenheit)')
ylabel('Ozone (PPB)')
```

We see in the scatterplot that the ozone tends to increase when the temperature increases, but the appropriate type of relationship for these data is not clear. We show how to use the loess procedure to find the estimate of the ozone level for a given temperature of 78 degrees Fahrenheit. First, we find some of the parameters.

```
n = length(Temperature); % Number of points
% Find the estimate at this point:
x0 = 78;
```

```
% Set up the other constants:
alpha = 2/3;
lambda = 1;
k = floor(alpha*n);
```

Next, we find the neighborhood at $x_0 = 78$.

```
% First step is to get the neighborhood.
dist = abs(x0 - Temperature);
% Find the closest distances.
[sdist,ind] = sort(dist);
% Get the points in the neighborhood.
Nx = Temperature(ind(1:k));
Ny = Ozone(ind(1:k));
% Maximum distance of neighborhood:
delxo = sdist(k);
```

We now (temporarily) delete all of the points outside the neighborhood and use the remaining points as input to the tri-cube weight function.

```
% Delete the ones outside the neighborhood.
sdist((k+1):n) = [];
% These are the arguments to the weight function.
u = sdist/delxo;
% Get the weights for all points in the neighborhood.
w = (1 - u.^3).^3;
```

The next section of code prepares the matrix for use in the weighted least squares regression (see Draper and Smith [1981]). In other words, we have the values for the x_i, but we need a matrix where the first column contains ones, the second contains the x_i, and the third contains x_i^2 (in the case of $\lambda = 2$).

```
% Now using only those points in the neighborhood,
% do a weighted least squares fit of degree lambda.
% We will follow the procedure in 'polyfit'.
x = Nx(:); y = Ny(:); w = w(:);
% Get the weight matrix
W = diag(w);
% Get the right matrix form: 1, x, x^2.
A = vander(x);
A(:,1:length(x)-lambda-1) = [];
V = A'*W*A;
Y = A'*W*y;
[Q,R] = qr(V,0);
p = R\(Q'*Y);
% The following is to fit MATLAB's convention
% for polynomials
p = p';
```

Now that we have the polynomial for the fit, we can use the MATLAB function **polyval** to get the value of the loess smooth at 78 degrees.

```
% This is the polynomial model for the local fit.
% To get the value at that point, use polyval.
yhat0 = polyval(p,x0);
```

We obtain a value of 33.76 PPB at 78 degrees. The following lines of code produce a loess smooth over the range of temperatures and superimposes the curve on the scatterplot. The **loess** function is included with the EDA Toolbox.

```
% Now call the loess function and plot the result.
% Get a domain over which to evaluate the curve.
X0 = linspace(min(Temperature),max(Temperature),50);
yhat = loess(Temperature,Ozone,X0,alpha,lambda);
% Plot the results.
plot(Temperature,Ozone,'.',X0,yhat)
xlabel('Temp (Fahrenheit)'),ylabel('Ozone (PPB)')
```

The resulting scatterplot with loess smooth is shown in Figure 7.4. It should be noted that we could use the loess function to get the estimated value of ozone at 78 degrees only, using

```
yhat0 = loess(Temperature,Ozone,78,alpha,lambda);
```

❏

Some readers might wonder where the word *loess* comes from. In geology, *loess* is defined as a deposit of fine clay or silt in river valleys. If one takes a vertical cross-section of the earth in such a place, then a *loess* would appear as curved strata running through the cross-section. This is similar to what one sees in the plot of the loess smooth in a scatterplot.

We now turn our attention to the multivariate case, where our predictor variables **x** have $p > 1$ dimensions. The procedure is essentially the same, but the weight function is defined in a slightly different way. We require a distance function in the space of independent variables, which in most cases is taken to be Euclidean distance. As discussed in Chapter 1, it might make sense to divide each of the independent variables by an estimate of its scale before calculating the distance.

Now that we have the distance, we define the weight function for a p-dimensional point $\mathbf{x}_0$ as

$$w_i(\mathbf{x}_0) = W\!\left(\frac{d(\mathbf{x}_0,\ \mathbf{x}_i)}{\Delta_k(\mathbf{x}_0)}\right),$$

where $d(\bullet)$ represents our distance function and $\Delta_k(\mathbf{x}_0)$ is the distance between the k-th nearest neighbor to $\mathbf{x}_0$ using the same definition of distance. W is the tri-cube function, as before. Once we have the weights, we construct

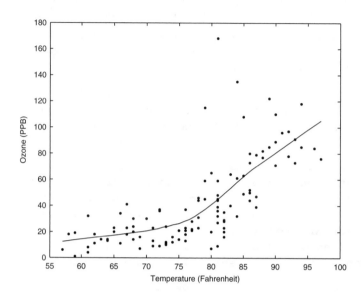

FIGURE 7.4
This shows the scatterplot of **ozone** and **temperature** along with the accompanying loess smooth.

either a multivariate linear or multivariate quadratic fit in the neighborhood of x_0. Linear fits are less of a computational burden, but quadratic fits perform better in applications where the regression surface has a lot of curvature. The Visualizing Data Toolbox (may be downloaded for free) described in Appendix B contains a function for producing loess smooths for bivariate predictors. We illustrate its use in the next example.

Example 7.2

For this example, we use the **galaxy** data set that was analyzed in Cleveland and Devlin [1988]. Buta [1987] measured the velocity of the NGC 7531 spiral galaxy in the Southern Hemisphere at a set of points in the celestial sphere that covered approximately 200 arc sec in the north/south direction and around 135 arc sec in the east/west direction. As usual, we first load the data and set some parameters.

```
% First load the data and set some parameters.
load galaxy
% The following is for the contour plot.
contvals = 1420:30:1780;
% These are the parameters needed for loess.
alpha = 0.25;
lambda = 2;
```

Next we get some (x , y) points in the domain. We will get the estimated surface using loess at these points. Then we call the **loess2** function, which was downloaded as part of the Data Visualization Toolbox.

```
% Now we get the points in the domain.
% The loess surface will be evaluated
% at these points.
XI = -25:2:25;
YI = -45:2:45;
[newx,newy] = meshgrid(XI,YI);
newz = loess2(EastWest,NorthSouth,...
    Velocity,newx,newy,alpha,lambda,1);
```

To plot this in a contour plot, we use the following. The plot is shown in Figure 7.5.

```
% Now do the contour plot and add labels.
[cs,h] = contour(newx,newy,newz,contvals);
clabel(cs,h)
xlabel('East-West Coordinate (arcsec)')
ylabel('North-South Coordinate (arcsec)')
```

❑

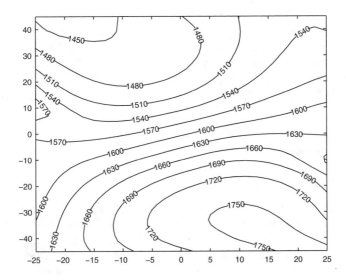

FIGURE 7.5
This is the contour plot showing the loess surface for the **galaxy** data.

We now discuss some of the issues for choosing the parameters for loess. These include the order λ of the polynomial that is estimated at each point of the smooth, the weight function W, and the smoothing parameter α. We have already touched on some of the issues with the degree of the polynomial. Any degree polynomial $\lambda = 0, 1, 2, \ldots$ can be used. $\lambda = 0$ provides a constant fit, but this seems too restrictive and the resulting curves/surfaces could be too rough. $\lambda = 1$ is adequate in most cases and is better computationally, but $\lambda = 2$ should be used in situations where there is a lot of curvature or many local maxima and minima.

As for the function W, we return to the four conditions specified previously. The first condition states that the weights must be positive, because negative weights do not make sense. The second requirement says that the weight function must be symmetric and that points to any side of x_0 should be treated the same way. The third property provides greater weight for those that are closer to x_0. The last property is not required, although it makes things simpler computationally. We save on computations if the weights outside the neighborhood are zero, because those observations do not have to be included in the least squares (see the indices on the summations in Equations 7.3 and 7.5 – they only go up to k, not n). We could use a weight function that violates condition four, such as the normal probability density function, but then we would have to include all n observations in the least squares fit at every point of the smooth.

Perhaps the hardest parameter to come up with is the smoothing parameter α. If α is small, then the loess smooth tends to be wiggly and to overfit the data (i.e., the bias is less). On the other hand, if α is large, then the curve is smoother (i.e., the variance is less). In EDA, where the main purpose of the smooth is to enhance the scatterplot and to look for patterns, the choice of α is not so critical. In these situations, Cleveland [1979] suggests that one choose α in the range 0.2 to 0.8. Different values for α (and λ) could be used to obtain various loess curves. Then the scatterplot with superimposed loess curve and residuals plots (discussed in Section 7.4) can be examined to determine whether or not the model adequately describes the relationship.

7.3 Robust Loess

The loess procedure we described in the previous section is not robust, because it relies on the method of least squares for the local fits. A method is called *robust* if it performs well when the associated underlying assumptions (e.g., normality) are not satisfied [Kotz and Johnson, Vol. 8, 1986]. There are many ways in which assumptions can be violated. A common one is the presence of *outliers* or extreme values in the response data. These are points in the sample that deviate from the pattern of the other observations. Least squares regression is vulnerable to outliers, and it takes only one extreme value to unduly influence the result. The nonrobustness of least squares will be illustrated and explored in the exercises.

Cleveland [1993, 1979] and Cleveland and McGill [1984] present a method for smoothing a scatterplot using a robust version of loess. This technique uses the bisquare method [Hoaglin, Mosteller and Tukey, 1983; Mosteller and Tukey, 1977; Huber, 1973; Andrews, 1974] to add robustness to the weighted least squares step in loess. The idea behind the bisquare is to re-weight points based on their residuals. If the residual for a given point in the neighborhood is large (i.e., it has a large deviation from the model), then the weight for that point should be decreased, since large residuals tend to indicate outlying observations. Alternatively, if the point has a small residual, then it should be weighted more heavily.

Before showing how the bisquare method can be incorporated into loess, we first describe the general bisquare least squares procedure. First a linear regression is used to fit the data, and the residuals $\hat{\varepsilon}_i$ are calculated from

$$\hat{\varepsilon}_i = y_i - \hat{y}_i . \tag{7.7}$$

The residuals are used to determine the weights from the bisquare function given by

$$B(u) = \begin{cases} (1 - u^2)^2; & |u| < 1 \\ 0; & \text{otherwise.} \end{cases} \tag{7.8}$$

The robustness weights are obtained from

$$r_i = B\left(\frac{\hat{\varepsilon}_i}{6\hat{q}_{0.5}}\right), \tag{7.9}$$

where $\hat{q}_{0.5}$ is the median of $|\hat{\varepsilon}_i|$. A weighted least squares regression is performed using weights adjusted by r_i.

To add bisquare to loess, we first fit the loess smooth, using the same procedure as before. We then calculate the residuals using Equation 7.7 and determine the robust weights from Equation 7.9. The loess procedure is repeated using weighted least squares, but the weights are now $r_i w_i(x_0)$. Note that the points used in the fit are the ones in the neighborhood of x_0, as before. This is an iterative process and is repeated until the loess curve converges or stops changing. Cleveland and McGill [1984] suggest that two or three iterations are sufficient to get a reasonable model.

Procedure - Robust Loess

1. Fit the data using the loess procedure with weights w_i.
2. Calculate the residuals, $\hat{\varepsilon}_i = y_i - \hat{y}_i$, for each observation.

3. Determine the median of the absolute value of the residuals, $\hat{q}_{0.5}$.

4. Find the robustness weights from

$$r_i = B\left(\frac{\hat{\varepsilon}_i}{6\hat{q}_{0.5}}\right),$$

using the bisquare function in Equation 7.8.

5. Repeat the loess procedure using weights of $r_i w_i$.

6. Repeat steps 2 through 5 until the loess curve converges.

In essence, the robust loess iteratively adjusts the weights based on the residuals. We illustrate the robust loess procedure in the next example, noting that while our example for robust loess involves only one predictor variable, we can easily apply it to the multivariate case.

Example 7.3

We now illustrate the robust loess using an example from Simonoff [1996]. The data represent the size of the annual spawning stock (x values) and the corresponding production of new fish of catchable size, called recruits (y values). The observations (in thousands of fish) were taken for the Skeena River sockeye salmon from 1940 to 1967. We provide a function called **loessr** that implements the procedure outlined above, and its use is shown below.

```
% Load the data and set the values
% for x and y.
load salmon
x = salmon(:,1);
y = salmon(:,2);
% Obtain a domain over which to get
% the loess curve.
xo = linspace(min(x),max(x));
% Get both the regular loess.
yhat = loess(x,y,xo,0.6,2);
% Get the robust loess curve.
yhatr = loessr(x,y,xo,0.6,2);
% Plot both curves.
plot(xo,yhat,'-',xo,yhatr,':',x,y,'o')
legend({'Loess';'Robust Loess'})
xlabel('Spawners')
ylabel('Recruits')
```

The curves are shown in Figure 7.6, where we see a potential outlier in the upper right area of the plot. The robust loess curve is different in the area influenced by this observation.
❑

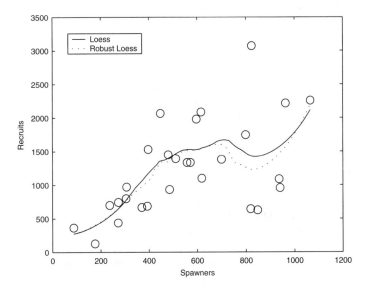

FIGURE 7.6
This shows the regular and robust loess curves for the **salmon** data. We fit a locally quadratic polynomial with α = 0.6. Note the potential outlier in the upper right corner. The robust loess fit downweights the effect of this observation.

7.4 Residuals and Diagnostics

In this section, we address several ways to assess the output from the loess scatterplot smooth, which can also be used to guide the analyst in choosing the values for α and λ. Since we are concerned with EDA methods in this text, we will cover only the graphical methods, such as residual plots, spread smooths, and upper/lower loess smooths. Cleveland and Devlin [1988] describe several statistics that are defined analogously with those used in fitting parametric functions by least squares, so some of the familiar techniques for making inferences in that setting can also be used in loess. Among other things, Cleveland and Devlin describe distributions of residuals, fitted values, and residual sum of squares.

7.4.1 Residual Plots

It is well known in regression analysis that checking the assumptions made about the residuals is important [Draper and Smith, 1981]. This is equally important in smoothing using loess, and we can use similar diagnostic plots. Recall that the true error ε_i is assumed to be normally distributed with mean zero and equal variance, and that the estimated residuals (or errors) are given by

$$\hat{\varepsilon}_i = y_i - \hat{y}_i .$$

We can construct a *normal probability plot* to determine whether the normality assumption is reasonable. We describe normal probability plots in more detail in Chapter 9 and just mention them here in the context of loess. The normal probability plot can be used when we have just one predictor or many predictors. One could also construct a histogram of the residuals to visually assess the distribution.

To check the assumption of constant variance, we can plot the absolute value of the residuals against the fitted values or $\hat{y}_i$. Here we expect to see a horizontal band of points with no patterns or trends.

Finally, to see if there is bias in our estimated curve, we can graph the residuals against the independent variables, where we would also expect to see a horizontal band of points. This is called a *residual dependence plot* [Cleveland, 1993]. If we have multiple predictors, then we can construct one of these plots for each variable. As the next example shows, we can enhance the diagnostic power of these scatterplots by superimposing a loess smooth.

Example 7.4

We turn to the software inspection data that was described in Chapter 1 to illustrate the various residual plots. Here we are interested in determining the relationship between the number of defects found as a function of the time spent inspecting the code or document. We have data that consists of 491 observations, with the x value representing the preparation or inspection time per page, and the response y is the number of defects found per page. After loading the data, we transform it because both variables are skewed. The initial scatterplot of the transformed data is shown in Figure 7.7, where we see that the relationship between the variables seems to be roughly linear.

```
load software
% Transform the data.
X = log(prepage);
Y = log(defpage);
% Get an initial plot.
plot(X,Y,'.')
xlabel('Log [ PrepTime (mins) / Page ]')
```

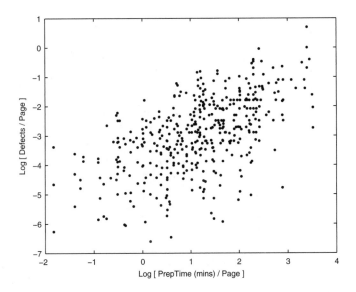

FIGURE 7.7

This is the scatterplot of observations showing the number of defects found per page versus the time spend inspecting each page. We see that the relationship is approximately linear.

```
ylabel('Log [ Defects / Page ]')
```

Next we set up the parameters ($\alpha = 0.5$, $\lambda = 2$) for a loess smooth and show the smoothed scatterplot in Figure 7.8.

```
% Set up the parameters.
alpha = 0.5;
lambda = 2;
% Do the loess on this.
x0 = linspace(min(X),max(X));
y0 = loess(X,Y,x0,alpha,lambda);
% Plot the curve and scatterplot.
plot(X,Y,'.',x0,y0)
xlabel('Log[PrepTime (mins)/Page]')
ylabel('Log[Defects/Page]')
```

We can assess our results by looking at the residual plots. First we find the residuals and plot them in Figure 7.9 (top), where we see that they are roughly symmetric about zero. Then we plot the absolute value of the residuals against the fitted values (Figure 7.9 (bottom)). A loess smooth of these observations show that the variance does not seem to be dependent on the fitted values.

```
% Get the residuals.
% First find the loess values at the observed X values.
```

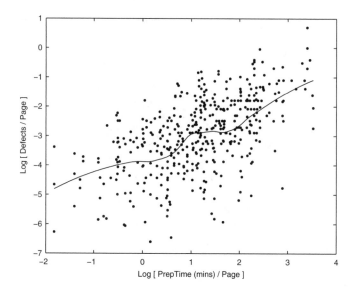

FIGURE 7.8
After we add the loess curve ($\alpha = 0.5$, $\lambda = 2$), we see that the relationship is not completely linear.

```
yhat = loess(X,Y,X,alpha,lambda);
resid = Y - yhat;
% Now plot the residuals.
plot(1:length(resid),resid,'.')
ax = axis;
axis([ax(1), ax(2), -4 4])
xlabel('Index')
ylabel('Residuals')
% Plot the absolute value of the residuals
% against the fitted values.
r0 = linspace(min(yhat),max(yhat),30);
rhat = loess(yhat,abs(resid),r0,0.5,1);
plot(yhat,abs(resid),'.',r0,rhat)
xlabel('Fitted Values')
ylabel('| Residuals |')
```

The following code constructs a residual dependence plot for this loess smooth. We include a loess smooth for this scatterplot to better understand the results. This is shown in Figure 7.10; we do not see any indication of bias.

```
% Now plot the residuals on the vertical
% and the independent values on the
% horizontal. This is the residual
% dependence plot. Include a loess curve.
```

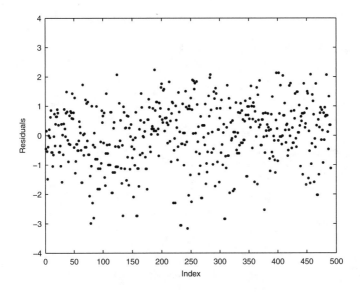

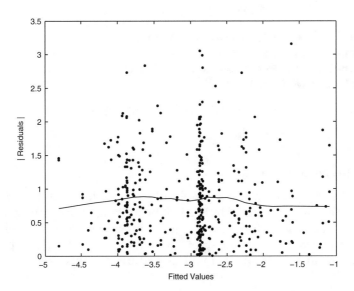

FIGURE 7.9
The upper plot shows the residuals based on the loess curve from Figure 7.8, and we see a nice horizontal band of points. In the lower panel, we have the absolute value of the residuals versus the fitted values. While not perfect, these indicate that the variance is approximately constant.

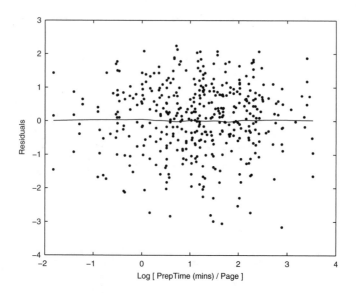

FIGURE 7.10

This shows the residual dependence plot with a superimposed loess curve. We do not see any indication of bias in the estimated curve.

```
rhat = loess(X,resid,x0,.5,1)
plot(X,resid,'.',x0,rhat)
xlabel('Log[PrepTime (mins)/Page]')
ylabel('Residuals')
```

We continue our analysis of these results in the next example.
❏

7.4.2 Spread Smooth

It might be important in certain applications to understand the spread of y given x. We can try to ascertain this just by looking at the scatterplot of the variables, but as we have seen, it is sometimes hard to judge these relationships just from a scatterplot. Cleveland and McGill [1984] describe *spread smoothing* as a way of graphically addressing this issue.

Procedure - Spread Smooths

1. Compute the fitted values $\hat{y}_i$, using loess or some other appropriate estimation procedure.
2. Calculate the residuals $\hat{\varepsilon}_i$ using Equation 7.7.
3. Plot $|\hat{\varepsilon}_i|$ against x_i in a scatterplot.

4. Smooth the scatterplot using loess and add the curve to the plot.

The smoothed values found in step 4 comprise the spread smoothing. We illustrate its use in Example 7.5.

Example 7.5

We show the spread smooth using the same data and residuals as in the previous example. Note that this is similar to the plot we have in Figure 7.9 (bottom), but this time we fit the absolute value of the residuals to the *observed predictor* values. The scatterplot with loess curve given in Figure 7.11 shows that the variance is fairly constant for the observed values of x.

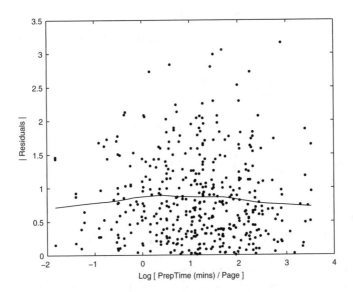

FIGURE 7.11
Here we have the spread smooth plot for the residuals found in Example 7.4. We see that the variance is fairly constant.

```
% The y values in this plot will be
% the absolute value of the residuals.
% Superimpose a loess curve to better
% assess results.
r0 = linspace(min(X),max(X),30);
rhat = loess(X,abs(resid),r0,0.5,1);
plot(X,abs(resid),'.',r0,rhat)
xlabel('Log [ PrepTime (mins) / Page ]')
ylabel('| Residuals |')
```

❏

7.4.3 Loess Envelopes - Upper and Lower Smooths

The loess smoothing method provides a model of the middle of the distribution of y given x. This can be extended to give us upper and lower smooths [Cleveland and McGill, 1984], where the distance between the upper and lower smooths indicates the spread. This shows similar information to the spread smooth, but in a way that is more in keeping with the familiar error bar plots. The procedure for obtaining the upper and lower smooths follows.

Procedure - Upper and Lower Smooths (Loess)

1. Compute the fitted values $\hat{y}_i$ using loess or robust loess.

2. Calculate the residuals $\hat{\varepsilon}_i = y_i - \hat{y}_i$.

3. Find the positive residuals $\hat{\varepsilon}_i^+$ and the corresponding x_i and $\hat{y}_i$ values. Denote these pairs as $(x_i^+, \hat{y}_i^+)$.

4. Find the negative residuals $\hat{\varepsilon}_i^-$ and the corresponding x_i and $\hat{y}_i$ values. Denote these pairs as $(x_i^-, \hat{y}_i^-)$.

5. Smooth the $(x_i^+, \hat{\varepsilon}_i^+)$ and add the fitted values from that smooth to $\hat{y}_i^+$. This is the upper smoothing.

6. Smooth the $(x_i^-, \hat{\varepsilon}_i^-)$ and add the fitted values from this smooth to $\hat{y}_i^-$. This is the lower smoothing.

Example 7.6

We do not show all of the MATLAB code to implement the upper and lower envelopes. Instead, we include a function that will provide them and just show how to use it in this example. We return to the same software inspection data used in the previous examples. The following code invokes the **loessenv** function and plots the curves.

```
% Get the envelopes and plot.
[yhat,ylo,xlo,yup,xup] = loessenv(X,Y,x0,0.5,2,1);
plot(X,Y,'.',x0,y0,xlo,ylo,xup,yup)
xlabel('Log [ PrepTime (mins) / Page ]')
ylabel('Log [ Defects / Page ]')
```

The loess curve with envelopes is given in Figure 7.12. The lower, middle, and upper smooths indicate that the distribution of y given x is symmetric at most values of x and that the variance is fairly constant.

❑

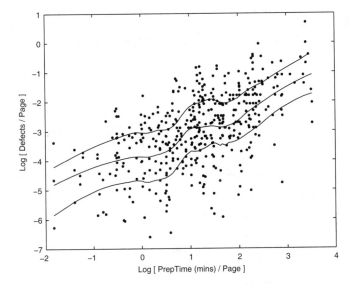

FIGURE 7.12

The lower, middle, and upper smooths indicate that the variance is constant and that the distribution of y given x is symmetric.

7.5 Bivariate Distribution Smooths

We now discuss some smoothings that can be used to graphically explore and summarize the distribution of two variables. Here we are not only looking to see how y depends on x, but also how x depends on y. Plotting pairs of loess smoothings on a scatterplot is one way of understanding this relationship.[1] Polar smoothing can also be used to understand the bivariate distribution between x and y by smoothing the edges of the point cloud.

7.5.1 Pairs of Middle Smoothings

In our previous examples of loess, we were in a situation where y was our response variable and x was the predictor variable, and we wanted to model or explore how y depends on the predictor. However, there are many situations where none of the variables is a factor or a response, and the goal is simply to understand the bivariate distribution of x and y.

---

[1] There are other ways to convey and understand a bivariate distribution, such as probability density estimation. The finite mixture method is one way (Chapter 6), and histograms is another (Chapter 9).

We can use pairs of loess curves to address this situation. The idea is to smooth *y* given *x*, as before, and also smooth *x* given *y* [Cleveland and McGill, 1984; Tukey, 1977]. Both of these smooths are plotted simultaneously on the scatterplot. We illustrate this in Example 7.7, where it is applied to the software data.

Example 7.7

We now look at the bivariate relationships between three variables: number of defects per SLOC, preparation time per SLOC, and meeting time per SLOC (single line of code). First we get some of the parameters needed for obtaining the loess smooths and for constructing the scatterplot matrix. Notice also that we are transforming the data first using the natural logarithm, because the data are skewed.

```
% Get some things needed for plotting.
vars = ['log Prep/SLOC';...
   ' log Mtg/SLOC';' log Def/SLOC'];
% Transform the data using logs.
X = log(prepsloc);
Y = log(mtgsloc);
Z = log(defsloc);
% Set up the parameters.
alpha = 0.5;
lambda = 2;
n = length(X);
```

Next we obtain all pairs of smooths; there are six of them.

```
% Get the pairs of middle smoothings.
% There should be 6 unique cases of these.
% First get domains.
x0 = linspace(min(X),max(X),50);
y0 = linspace(min(Y),max(Y),50);
z0 = linspace(min(Z),max(Z),50);
% Now get the curves.
xhatvy = loess(Y,X,y0,alpha,lambda);
yhatvx = loess(X,Y,x0,alpha,lambda);
xhatvz = loess(Z,X,z0,alpha,lambda);
zhatvx = loess(X,Z,x0,alpha,lambda);
yhatvz = loess(Z,Y,z0,alpha,lambda);
zhatvy = loess(Y,Z,y0,alpha,lambda);
```

Finally, we construct the scatterplot matrix. We use MATLAB's Handle Graphics to add the lines to each plot.

```
% Now do the plotmatrix.
data = [X(:),Y(:),Z(:)];
% gplotmatrix is in the Statistics Toolbox.
```

```
[H,AX,BigAx] = gplotmatrix(data,[],[],'k','.',...
    0.75,[],'none',vars,vars);
% Use Handle Graphics to construct the lines.
axes(AX(1,2));
line(y0,xhatvy);line(yhatvx,x0,'LineStyle','--');
axes(AX(1,3));
line(z0,xhatvz);line(zhatvx,x0,'LineStyle','--');
axes(AX(2,1));
line(x0,yhatvx);line(xhatvy,y0,'LineStyle','--');
axes(AX(2,3));
line(z0,yhatvz);line(zhatvy,y0,'LineStyle','--');
axes(AX(3,1));
line(x0,zhatvx);line(xhatvz,z0,'LineStyle','--');
axes(AX(3,2));
line(y0,zhatvy);line(yhatvz,z0,'LineStyle','--');
```

The results are shown in Figure 7.13. Note that we have the smooths of y given x shown as solid lines, and the smooths of x given y plotted using dashed lines. In the lower left plot, we see an interesting relationship between the preparation time and number of defects found, where we see a local maximum, possibly indicating a higher rate of defects found.
❑

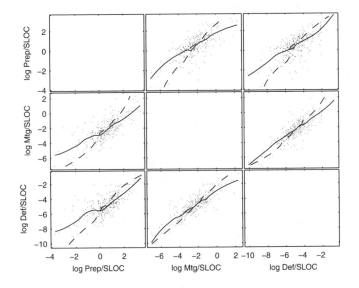

FIGURE 7.13
This shows the scatterplot matrix with superimposed loess curves for the software data, where we look at inspection information per SLOC. The solid lines indicate the smooths for y given x, and the dashed lines are the smooths for x given y.

7.5.2 Polar Smoothing

The goal of *polar smoothing* is to convey where the central portion of the bivariate point cloud lies. This summarizes the position, shape, and orientation of the cloud of points. Another way to achieve this is to find the *convex hull* of the points, which is defined as the smallest convex polygon that completely encompasses the data. However, the convex hull is sensitive to outliers because it must include them, so it can overstate the area covered by the data cloud. Polar smoothing by loess does not suffer from this problem.

We now describe the procedure for polar smoothing. The first three steps implement one of the many ways to center and scale x and y. The smooth is done using polar coordinates based on a form of these pseudovariates. At the end, we transform the results back to the original scale for plotting.

Procedure - Polar Smoothing

1. Normalize x_i and y_i using

$$x_i^* = (x_i - \text{median}(x)) \div \text{MAD}(x)$$
$$y_i^* = (y_i - \text{median}(y)) \div \text{MAD}(y) \ '$$

   where MAD is the median absolute deviation.

2. Calculate the following values:

$$s_i = y_i^* + x_i^*$$
$$d_i = y_i^* - x_i^*$$

3. Normalize the s_i and d_i as follows:

$$s_i^* = s_i \div \text{MAD}(s)$$
$$d_i^* = d_i \div \text{MAD}(d)$$

4. Convert the (s_i^*, d_i^*) to polar coordinates (θ_i, m_i).

5. Transform the m_i as follows

$$z_i = m_i^{2/3} .$$

6. Smooth z_i as a function of θ_i. This produces the fitted values $\hat{z}_i$.

7. Find the fitted values for m_i by transforming the $\hat{z}_i$ using

$$\hat{m}_i = \hat{z}_i^{3/2} \, .$$

8. Convert the coordinates $(\theta_i, \hat{m}_i)$ to Cartesian coordinates $(\hat{s}_i^*, \hat{d}_i^*)$.

9. Transform these coordinates back to the original x and y scales by the following:

$$\hat{s}_i = \hat{s}_i^* \times \text{MAD}(s)$$

$$\hat{d}_i = \hat{d}_i^* \times \text{MAD}(d)$$

$$\hat{x}_i = [(\hat{s}_i - \hat{d}_i) \div 2] \times \text{MAD}(x) + \text{median}(x)$$

$$\hat{y}_i = [(\hat{s}_i + \hat{d}_i) \div 2] \times \text{MAD}(y) + \text{median}(y)$$

10. Plot the $(\hat{x}_i, \hat{y}_i)$ coordinates on the scatterplot by connecting each point by a straight line and then closing the polygon by connecting the first point to the n-th point.

More information on the smooth called for in Step 6 is in order. Cleveland and McGill [1984] suggest the following way to use the regular loess procedure to get a circular smooth. Recall that θ_i is an angle for $i = 1, ..., n$. We order the θ_i in ascending order, and we let $j = n/2$ be rounded *up* to an integer. We then give the following points to loess

$$(-2\pi + \theta_{n-j+1}, \; z_{n-j+1}), \; \ldots, \; (-2\pi + \theta_n, \; z_n),$$
$$(\theta_1, \; z_1), \; \ldots, \; (\theta_n, \; z_n),$$
$$(2\pi + \theta_1, \; z_1), \; \ldots, \; (2\pi + \theta_j, \; z_j).$$

The actual smoothed values that we need are those in the second row.

Example 7.8

We use the BPM data to illustrate polar smoothing. Results from Martinez [2002] show that there is some overlap between topics 8 (death of North Korean leader) and topic 11 (helicopter crash in North Korea). We apply the polar smoothing to these two topics to explore this issue. We use ISOMAP nonlinear dimensionality reduction with the IRad dissimilarity matrix to produce bivariate data. The scatterplot for the data after dimensionality reduction is shown in Figure 7.14.

```
load L1bpm
% Pick the topics we will look at.
t1 = 8;
t2 = 11;
% Reduce the dimensionality using Isomap.
```

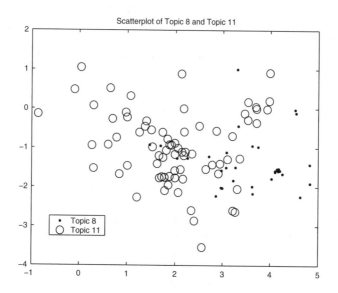

FIGURE 7.14

This is the scatterplot of topics 8 and 11 after we use ISOMAP to reduce the data to 2-D. We see some overlap between the two topics.

```
options.dims = 1:10;        % These are for ISOMAP.
options.display = 0;
[Yiso, Riso, Eiso] = isomap(L1bpm, 'k', 7, options);
% Get the data out.
X = Yiso.coords{2}';
% Get the data for each one and plot
ind1 = find(classlab == t1);
ind2 = find(classlab == t2);
plot(X(ind1,1),X(ind1,2),'.',X(ind2,1),X(ind2,2),'o')
title('Scatterplot of Topic 8 and Topic 11')
```

Now we do the polar smoothing for the first topic.

```
% First let's do the polar smoothing just for
% the first topic. Get the x and y values.
x = X(ind1,1);
y = X(ind1,2);
% Step 1.
% Normalize using the median absolute deviation.
% We will use the Matlab 'inline' functionality.
md = inline('median(abs(x - median(x)))');
xstar = (x - median(x))/md(x);
ystar = (y - median(y))/md(y);
% Step 2.
```

```
s = ystar + xstar;
d = ystar - xstar;
% Step 3. Normalize these values.
sstar = s/md(s);
dstar = d/md(d);
% Step 4. Convert to polar coordinates.
[th,m] = cart2pol(sstar,dstar);
% Step 5. Transform radius m.
z = m.^(2/3);
% Step 6. Smooth z given theta.
n = length(x);
J = ceil(n/2);
% Get the temporary data for loess.
tx = -2*pi + th((n-J+1):n);
% So we can get the values back, find this.
ntx = length(tx);
tx = [tx; th];
tx = [tx; th(1:J)];
ty = z((n-J+1):n);
ty = [ty; z];
ty = [ty; z(1:J)];
tyhat = loess(tx,ty,tx,0.5,1);
% Step 7. Transform the values back.
% Note that we only need the middle values.
tyhat(1:ntx) = [];
mhat = tyhat(1:n).^(3/2);
% Step 8. Convert back to Cartesian.
[shatstar,dhatstar] = pol2cart(th,mhat);
% Step 9. Transform to original scales.
shat = shatstar*md(s);
dhat = dhatstar*md(d);
xhat = ((shat-dhat)/2)*md(x) + median(x);
yhat = ((shat+dhat)/2)*md(y) + median(y);
% Step 10. Plot the smooth.
% We use the convex hull to make it easier
% for plotting.
K = convhull(xhat,yhat);
plot(X(ind1,1),X(ind1,2),'.',X(ind2,1),X(ind2,2),'o')
hold on
plot(xhat(K),yhat(K))
```

We wrote a function called **polarloess** that includes these steps. We use it to find the polar smooth for the second topic and add that smooth to the plot.

```
% Now use the polarloess function to get the
% other one.
[xhat2,yhat2] = ...
```

```
    polarloess(X(ind2,1),X(ind2,2),0.5,1);
  plot(xhat2,yhat2)
  hold off
```

The scatterplot with polar smooths is shown in Figure 7.15, where we get a better idea of the overlap between the groups.
❏

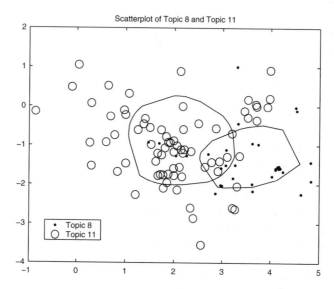

FIGURE 7.15
This is the scatterplot of the two topics with the polar smooths superimposed.

7.6 Curve Fitting Toolbox

We now briefly describe the Curve Fitting Toolbox[2] in this section for completeness. The reader is not required to have this toolbox for MATLAB functionality described outside this section. The Curve Fitting Toolbox is a collection of graphical user interfaces (GUIs) and M-file functions that are written for the MATLAB environment, just like other toolboxes.

The toolbox has the following major features for curve fitting:

- One can perform data preprocessing, such as sectioning, smoothing, and removing outliers.

---

[2] This toolbox is available from The MathWorks, Inc.

- The analyst can fit curves to points using parametric and nonparametric methods. The parametric approaches include polynomials, exponential, rationals, sums of Gaussians, and custom equations. Nonparametric fits include spline smoothing and other interpolation methods.
- It does standard least squares, weighted least squares, and robust fitting procedures.
- Statistics indicating the goodness-of-fit are also available.

The toolbox can only handle fits between y and x; multivariate predictors are not supported.

We now give a brief description of *smoothing splines*. The word spline comes from the engineering community, where draftsmen used long thin strips of wood called splines. They used these splines to draw a smooth curve between points; different global curves are produced when the positions of the points are changed. As we will see, a smoothing spline is a solution to a constrained optimization problem, where we optimally trade-off fidelity of the fit to the data with smoothness of the estimate. In particular, smoothing splines fit piecewise polynomials between the joins, which are called *knots*. Determining the optimal knots can also be part of the optimization problem, but often the knots coincide with the data points for simplicity.

The residual sum of squares is a familiar way to measure fidelity to the data, but we also need to include a penalty for roughness, so the resulting curve will be smooth. The Curve Fitting Toolbox uses the usual roughness penalty approach given by

$$\delta \sum_{i} w_i (y_i - s(x_i))^2 + (1 - \delta) \int \left(\frac{d^2 s}{dx^2} \right)^2 dx \qquad (7.10)$$

to find the smoothing spline s. The first term in Equation 7.10 represents the residual sum of squares, with optional weights w_i, and the second term provides a penalty for roughness. The smoothing parameter is denoted by δ, and it ranges between zero and one. A value of zero performs a least squares straight line fit to the data, and $\delta = 1$ produces a cubic spline interpolant.

The MATLAB toolbox also has a stand-alone function for getting various smooths called **smooth**. The general syntax for this function is

```
x = smooth(x,y,span,method);
```

where span governs the size of the neighborhood. There are six methods available, as outlined below:

```
'moving'    - Moving average (default)
'lowess'    - Lowess (linear fit)
'loess'     - Loess (quadratic fit)
```

```
'sgolay'    - Savitzky-Golay
'rlowess'   - Robust Lowess (linear fit)
'rloess'    - Robust Loess (quadratic fit)
```

One possible inconvenience with the **smooth** function (as well as the GUI) is that it provides values of the smooth only at the observed x values.

7.7 Summary and Further Reading

Several excellent books on smoothing and nonparametric regression (a related approach) are available. Perhaps the most comprehensive one on loess is called *Visualizing Data* by Cleveland [1993]. This book also includes extensive information on visualization tools such as contour plots, wire frames for surfaces, coplots, multiway dot plots, and many others. For smoothing methods in general, we recommend Simonoff [1996]. It surveys the use of smoothing methods in statistics. The book has an applied focus, and it includes univariate and multivariate density estimation, nonparametric regression, and categorical data smoothing. A compendium of contributions in the area of smoothing and regression can be found in Schimek [2000]. A monograph on nonparametric smoothing in statistics was written by Green and Silverman [1994]; it emphasizes methods rather than the theory.

Loader [1999] provides an overview of local regression and likelihood, including theory, methods, and applications. It is easy to read, and it uses S-Plus code to illustrate the concepts. Efromovich [1999] provides a comprehensive account of smoothing and nonparametric regression, and he includes time series analysis. Companion software in S-Plus for the text is available over the internet, but he does not include any code in the book itself. Another smoothing book with S-Plus is Bowman and Azzalini [1997]. For the kernel smoothing approach, see Wand and Jones [1995].

Generalized additive models is a way to handle multivariate predictors in a nonparametric fashion. In classical (or parametric) regression, one assumes a linear or some other parametric form for the predictors. In generalized additive models, these are replaced by smooth functions that are estimated by a scatterplot smoother, such as loess. The smooths are found in an iterative procedure. A monograph by Hastie and Tibshirani [1990] that describes this approach is available, and it includes several chapters on scatterplot smoothing, such as loess, splines, and others. There is also a survey paper on this same topic for those who want a more concise discussion [Hastie and Tibshirani, 1986].

For an early review of smoothing methods, see Stone [1977]. The paper by Cleveland [1979] describes the robust form of loess, and includes some information on the sampling distributions associated with locally weighted regression. Next we have Cleveland and McGill [1984], which is a wonderful paper that includes descriptions of many tools for scatterplot enhancements. Cleveland, Devlin and Grosse [1988] discusses methods and computational algorithms for local regression. Titterington [1985] provides an overview of various smoothing techniques used in statistical practice and provides a common unifying structure. We also have Cleveland and Devlin [1988] that shows how loess can be used for exploratory data analysis, diagnostic checking of parametric models, and multivariate nonparametric regression. For an excellent summary and history of smoothing methods, see Cleveland and Loader [1996]. As for other smoothing methods, the reader can refer to Scott [1992] and Hastie and Loader [1993]. Finally, there is a lot of literature on spline methods. For a nice survey article that synthesizes much of the work on splines in statistics, see Wegman and Wright [1983].

Exercises

7.1 First consult the **help** files on the MATLAB functions **polyfit** and **polyval**, if you are unfamiliar with them. Next, for some given domain of points **x**, find the **y** values using **polyval** and degree 3. Add some normally distributed random noise (use **normrnd** with 0 mean and some small σ) to the **y** values. Fit the data using polynomials of degrees 1, 2, and 3 and plot these curves along with the data. Construct and plot a loess curve. Discuss the results.

7.2 Generate some data using **polyval** and degree 1. Add some small random noise to the points using **randn**. Use **polyfit** to fit a polynomial of degree 1. Add one outlying point at either end of the range of the **x** values. Fit these data to a straight line. Plot both of these lines, along with a scatterplot of the data, and comment on the differences.

7.3 Do a scatterplot of the data generated in problem 7.1. Activate the Tools menu in the Figure window and click on the Basic Fitting option. This brings up a GUI that has some options for fitting data. Explore the capabilities of this GUI.

7.4 Load the **abrasion** data set. Construct loess curves for abrasion loss as a function of tensile strength and abrasion loss as a function of hardness. Comment on the results. Repeat the process of Example 7.7 using this data set and assess the results.

7.5 Repeat the process outlined in Example 7.7 using the **environmental** data. Comment on your results.

7.6 Construct a sequence of loess curves for the **votfraud** data set. Each curve should have a different value of $\alpha = 0.2, 0.5, 0.8$. First do this for $\lambda = 1$ and then repeat for $\lambda = 2$. Discuss the differences in the curves. Just by looking at the curves, what α and λ would you choose?

7.7   Repeat problem 7.6, but this time use the residual plots to help you choose the values for α and λ.

7.8   Repeat problems 7.6 and 7.7 for the **calibrat** data.

7.9   Verify that the tri-cube weight functions satisfies the four properties of a weight function.

7.10   Using the data in Example 7.8, plot the convex hulls (use the function **convhull**) of each of the topics. Compare these with the polar smooth.

7.11   Choose some other topics from the BPM and repeat the polar smooths of Example 7.8.

7.12   Using one of the gene expression data sets, select an experiment (tumor, patient, etc.), and apply some of the smoothing techniques from this chapter to see how the genes are related to that experiment. Choose one of the genes and smooth as a function of the experiments. Construct the loess upper and lower smooths to assess the variation. Comment on the results.

7.13   Construct a normal probability plot of the residuals in the **software** data analysis. See Chapter 9 for information or the MATLAB Statistics Toolbox function **normplot**. Construct a histogram (use the **hist** function). Comment on the distributional assumptions for the errors.

7.14   Repeat problem 7.13 for the analyses in problems 7.6 and 7.8.

7.15   Choose two of the dimensions of the **oronsay** data. Apply polar smoothing to summarize the point cloud for each class. Do this for both classifications of the **oronsay** data. Comment on your results.

7.16   Do a 3-D scatterplot (see **scatter3**) and a scatterplot matrix (see **plotmatrix**) of the **galaxy** data of Example 7.2. Does the contour plot in Figure 7.5 match the scatterplots?

Part III

Graphical Methods for EDA

Chapter 8

Visualizing Clusters

In Chapters 5 and 6, we presented various methods for clustering, including agglomerative clustering, k-means clustering, and model-based clustering. In the process of doing that, we showed some visualization techniques such as dendrograms for visualizing hierarchical structures and scatterplots. We now turn our attention to other methods that can be used for visualizing the results of clustering. These include a space-filling version of dendrograms called treemaps, an extension of treemaps called rectangle plots, a new rectangle-based method for visualizing nonhierarchical clustering called ReClus, and data images that can be used for viewing clusters, as well as outlier detection. We begin by providing more information on dendrograms.

8.1 Dendrogram

The *dendrogram* (also called the *tree diagram*) is a mathematical, as well as a visual representation of a hierarchical procedure, which can be divisive or agglomerative. Thus, we often refer to the results of the hierarchical clustering as the dendrogram itself.

We start off by providing some terminology for a dendrogram; the reader can refer to Figure 8.1 for an illustration. The tree starts at the *root*, which can either be at the top for a vertical tree or on the left side for a horizontal tree. The *nodes* of the dendrogram represent clusters, and they can be *internal* or *terminal*. The internal nodes contain or represent all observations that are grouped together based on the type of linkage and distance used. In most dendrograms, terminal nodes contain a single observation. We will see shortly that this is not always the case in MATLAB's implementation of the dendrogram. Additionally, terminal nodes usually have labels attached to them. These can be names, letters, numbers, etc. The MATLAB **dendrogram** function labels the terminal nodes with numbers.

The *stem* or *edge* shows children of internal nodes and the connection with the clusters below it. The length of the edge represents the distances at which clusters are joined. The dendrograms for hierarchical clustering are *binary trees,* so they have two edges emanating from each internal node. The *topology* of the tree refers to the arrangement of stems and nodes.

The dendrogram illustrates the process of constructing the hierarchy, and the internal nodes describe particular partitions, once the dendrogram has been cut at a given level. Data analysts should be aware that the same data and clustering procedure can yield 2^{n-1} dendrograms, each with a different appearance depending on the order used to display the nodes. Software packages choose the algorithm for drawing this automatically, and they usually do not specify how they do this. Some algorithms have been developed for optimizing the appearance of dendrograms based on various objective functions [Everitt, Landau and Leese, 2001]. We will see in the last section where we discuss the data image that this can be an important consideration.

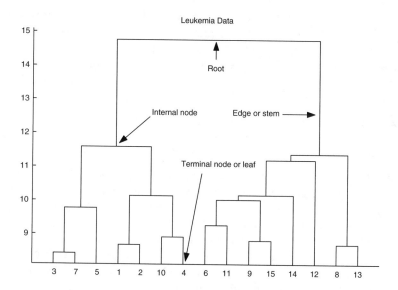

FIGURE 8.1
We applied agglomerative clustering to the **leukemia** data using Euclidean distance and complete linkage. The dendrogram with 15 leaf nodes is shown here. If we cut this dendrogram at level 10.5 (on the vertical axis), then we would obtain 5 clusters or groups.

Example 8.1

We first show how to construct the dendrogram in Figure 8.1 using the **leukemia** data set. The default for the **dendrogram** function is to display a maximum of 30 nodes. This is to prevent the displayed leaf nodes from being too cluttered. The user can specify the number to display, as we do below, or one can display all nodes by using **dendrogram(Z,0)**.

```
% First load the data.
```

```
load leukemia
[n,p] = size(leukemia);
x = zeros(n,p);
% Standardize each row (gene) to be mean
% zero and standard deviation 1.
for i = 1:n
    sig = std(leukemia(i,:));
    mu = mean(leukemia(i,:));
    x(i,:) = (leukemia(i,:) - mu)/sig;
end
% Do hierarchical clustering.
Y = pdist(x);
Z = linkage(Y,'complete');
% Display with only 15 nodes.
% Output arguments are optional.
[H,T] = dendrogram(Z,15);
title('Leukemia Data')
```

Note that the leaf nodes in this example of a MATLAB dendrogram do not necessarily represent one of the original observations; they likely contain several observations. We can (optionally) request some output variables from the **dendrogram** function to help us determine what observations are in the individual nodes. The output vector **T** contains the leaf node number for each object in the data set and can be used to find out what is in node 6 as follows:

```
ind = find(T==6)
ind =

    26
    28
    29
    30
    46
```

Thus, we see that terminal node 6 on the dendrogram in Figure 8.1 contains the original observations 26, 28, 29, 30, and 46.
❑

8.2 Treemaps

Dendrograms are very familiar to data analysts working in hierarchical clustering applications, and they are easy to understand because they match our concept of how trees are laid out in a physical sense with branches and leaves (except that the tree root is in the wrong place!). Johnson and

Shneiderman [1991] point out that the dendrogram does not efficiently use the existing display space, since most of the display consists of white space with very little ink. They proposed a space-filling (i.e., the entire display space is used) display of hierarchical information called *treemaps*, where each node is a rectangle whose area is proportional to some characteristic of interest [Shneiderman, 1992].

The original application and motivation for treemaps was to show the directory and file structure on hard drives. It was also applied to the visualization of organizations such as the departmental structure at a university. Thus, the treemap visualization can be used for an arbitrary number of splits or branches at internal nodes, including the binary tree structure that one gets from hierarchical clustering.

Johnson and Shneiderman [1991] note that hierarchical structures contain two types of information. First, they contain structural or organizational information that is associated with the hierarchy. Second, they have content information associated with each node. Dendrograms can present the structural information, but do not convey much about the leaf nodes other than a text label. Treemaps can depict both the structure and content of the hierarchy.

The treemap displays hierarchical information and relationships by a series of nested rectangles. The parent rectangle (or root of the tree) is given by the entire display area. The treemap is obtained by recursively subdividing this parent rectangle, where the size of each sub-rectangle is proportional to the size of the node. The size could be representative of the size in bytes of the file or the number of employees in an organizational unit. In the case of clustering, the size would correspond to the number of observations in the cluster. We continue to subdivide the rectangles horizontally, vertically, horizontally, etc., until a given leaf configuration (e.g., number of groups in the case of clustering) is obtained.

We show an example of the treemap in Figure 8.2, along with its associated tree diagram. Note that the tree diagram is not the dendrogram that we talked about previously, because it has an arbitrary number of splits at each node. The root node has four children: a leaf node of size 12, a leaf node of size 8, an interior node with four children, and an interior node with three children. These are represented by the first vertical splits of the parent rectangle. We now split horizontally at the second level. The interior node with four children is split into four sub-rectangles proportional to their size. Each one of these is a terminal node, so no further subdivisions are needed. The next interior node has two children that are leaf nodes and one child that is an interior node. Note that this interior node is further subdivided into two leaf nodes of size 6 and 11 using a vertical split.

When we apply the treemap visualization technique to hierarchical clustering output, we must specify the number of clusters. Note also that there is no measure of distance or other objective function associated with the clusters, as there is with the dendrogram. Another issue with the treemap (as well as the dendrogram) is the lack of information about the original data,

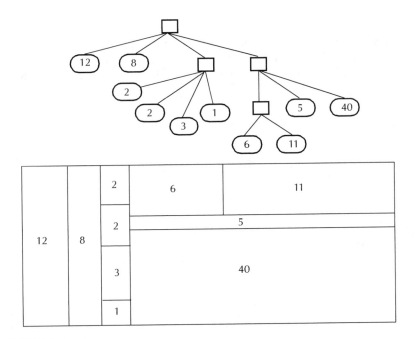

FIGURE 8.2
At the top of this figure, we show a tree diagram with nodes and links. Each leaf node has a number that represents the size of the nodes. An alternative labeling might be just the node number or some text label. The corresponding treemap diagram is shown below. Note that the divisions from the root node are shown as vertical splits of the parent rectangle, where the size of each sub-rectangle is proportional to the total size of all children. The next split is horizontal, and we continue alternating the splits until all leaf nodes are displayed [Shneiderman, 1992].

because the rectangles are just given labels. We implemented the treemap technique in MATLAB; its use is illustrated in the next example.

Example 8.2

In this example, we show how to use the **treemap** provided with the EDA Toolbox. We return to the hierarchical clustering of the **leukemia** data displayed as a dendrogram in Figure 8.1. The function we provide implements the treemap display for binary hierarchical information only and requires the output from the MATLAB **linkage** function. The inputs to the function **treemap** include the **Z** matrix (output from **linkage**) and the desired number of clusters. The default display is to show the leaf nodes with the same labels as in the dendrogram. There is an optional third argument that causes the treemap to display without any labels. The following syntax constructs the treemap that corresponds to the dendrogram in Figure 8.1.

```
% The matrix Z was calculated in
% Example 8.1.
```

```
treemap(Z,15);
```

The treemap is given in Figure 8.3. Notice that we get an idea of the size of the nodes with the treemap, whereas this is not evident in the dendrogram (Figure 8.1). However, as noted before, we lose the concept of distance with treemaps; thus it is perhaps easier to see the number of groups one should have in a dendrogram.

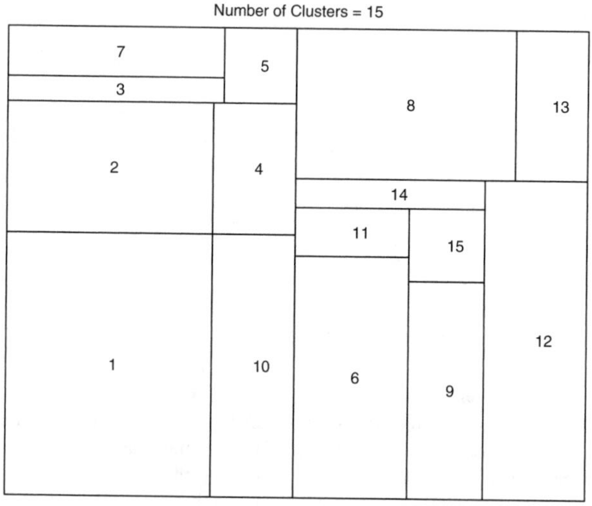

❑

FIGURE 8.3
Here we have the treemap that corresponds to the dendrogram in Figure 8.1.

8.3 Rectangle Plots

Recall that in the dendrogram, the user can specify a value along the axis, and different clusters or partitions are obtained depending on what value is specified. Of course, we do not visualize this change with the dendrogram; i.e., the display of the dendrogram is not dependent on the chosen number of clusters or cutoff distance. However, it is dependent on the number of leaf nodes chosen (in the MATLAB implementation). If one specifies a different number of leaf nodes, then the dendrogram must be completely redrawn, and node labels change. This can significantly change the layout of the dendrogram, as well as the understanding that is gained from it.

To display the hierarchical information as a treemap, the user must specify the number of clusters (or one can think of this as number of leaf nodes) rather than the cutoff point for the distance. If the user wants to explore other cluster configurations by specifying a different number of clusters, then the display is redrawn, as it is with the dendrogram. As stated previously, there is no measure of distance associated with the treemap display, and there is a lack of information about the original data. It would be useful to know what cases are clustered where. The next cluster visualization method attempts to address these issues.

Recall from Chapters 5 and 6 that one of the benefits of hierarchical clustering is that the output can provide a partition of the data for any given number of clusters or, alternatively, for any given level of dissimilarity. This property is an advantage in EDA, because we can run the algorithm once on a large set of data and then explore and visualize the results in a reasonably fast manner. To address some of the issues with treemaps and to take advantage of the strengths of hierarchical clustering, Wills [1998] developed the *rectangle visualization method*. This method is similar to the treemap, but displays the points as glyphs and determines the splits in a different way.

To construct a rectangle plot, we split the rectangles along the longest side, rather than alternating vertical and horizontal splits as in the treemap method. The alternating splits in treemaps are good at showing the depth of the tree, but it has a tendency to create long skinny rectangles, if the trees are unbalanced [Wills, 1998]. The splits in the rectangle plot provide rectangles that are more square.

We keep splitting rectangles until we reach a leaf node or until the cutoff distance is reached. If a rectangle does not have to be split because it reaches this cutoff point, but there is more than one observation in the rectangle, the algorithm continues to split rectangles until it reaches a leaf node. However, it does not draw the rectangles. It uses this leaf-node information to determine the layout of the points or glyphs, where each point is now in its own rectangle. The advantage to this method is that other configurations (i.e., number of clusters or a given distance) can be shown without redisplaying the glyphs. Only the rectangle boundaries are redrawn.

We illustrate the rectangle plot for a simulated data set in Figures 8.4 and 8.5. The data set contains randomly generated bivariate data ($n = 30$) comprising two clusters, one centered at $(-2 , 2)^T$ and the other at $(2 , 2)^T$. In the top part of Figure 8.4, we have the dendrogram with all 30 nodes displayed. We see that the node labels are difficult to distinguish. The corresponding rectangle plot for 30 clusters is shown in the bottom of Figure 8.4, where we see each observation in its own rectangle or cluster.

We show another dendrogram and rectangle plot in Figure 8.5 using the same data set and hierarchical clustering information. We plot a dendrogram requesting 10 leaf nodes and show the results in the top of the figure. The leaf nodes are now easier to read, but they no longer represent the observation numbers. The rectangle plot for 10 clusters is given in the lower half of the figure. When we compare Figures 8.4 and 8.5, we see that the dendrogram changed a lot, whereas only the bounding boxes change in the rectangle plot.

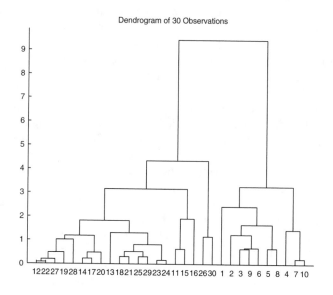

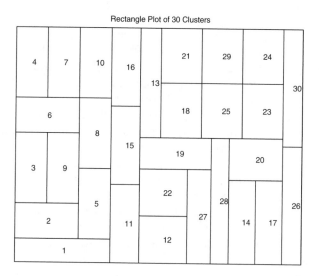

FIGURE 8.4

The top portion of this figure shows the dendrogram (Euclidean distance and complete linkage) for a randomly generated bivariate data set containing two clusters. All $n = 30$ leaf nodes are shown, so we see over plotting of the text labels. The rectangle plot that corresponds to this is shown in the bottom half, where we plot each observation number in its own rectangle or cluster. The original implementation in Wills [1998] plots the observations as dots or circles.

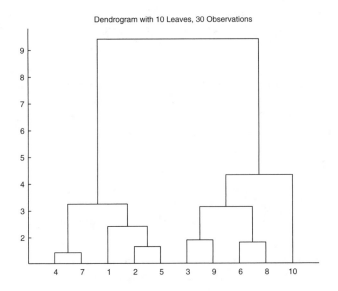

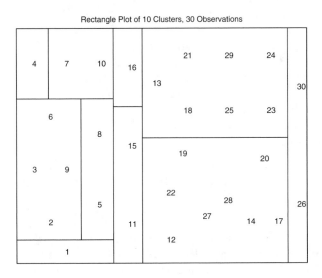

FIGURE 8.5
Using the same information that produced Figure 8.4, we now show the dendrogram when only 10 leaf nodes are requested. We see that the dendrogram has been completely redrawn when we compare it with the one in Figure 8.4. The rectangle plot for 10 clusters is given below the dendrogram. When the rectangle plots in Figures 8.4 and 8.5 are compared, we see that the positions of the glyphs have not changed; only the bounding boxes for the rectangles are different. We also see information about the original data via the observation numbers.

Example 8.3

We continue with the same **leukemia** data set for this example. First we show how to use the **rectplot** function to get a rectangle plot showing observation numbers. For comparison with previous results, we plot 15 clusters.

```
% Use same leukemia data set and matrix Z
% from Example 8.1. The second argument is the
% number of clusters. The third argument is a string
% specifying what information the second
% argument provides.
rectplot(Z,15,'nclus')
```

The plot is shown in the top of Figure 8.6. Next we show how to use the optional input argument to use the true class labels as the glyphs. First we have to convert the class labels to numbers, since the input vector must be numeric.

```
% We now show how to use the optional
% class labels, using the cancer type.
% The argument must be numeric, so we
% convert strings to numbers.
% First set all indices to 0 - this will
% be class ALL.
labs = zeros(length(cancertype),1);
% Now find all of the AML cancers.
% Set them equal to 1.
inds = strmatch('AML',cancertype);
labs(inds) = 1;
% Now do the rectangle plot.
rectplot(Z,15,'nclus',labs)
```

This plot is shown in the lower half of Figure 8.6. The observations labeled 0 are the ALL cancers, and those plotted with a 1 correspond to the AML cancer type.
❑

In our MATLAB implementation of this technique, we plot each point with its observation number or its true class label. This can cause some over plotting with large data sets. A future implementation will include other plot symbols, thus saving on display space. The user can also specify the number of clusters by providing a cutoff dissimilarity based on the dendrogram for the second input to the function. In this case, the third argument to **rectplot** is **'dis'**.

Wills' original motivation for the rectangle plot was to include the notion of distance in a treemap-like display. He did this by providing a supplemental line graph showing the number of clusters on the horizontal axis, and the dissimilarity needed to do the next split on the vertical axis. The

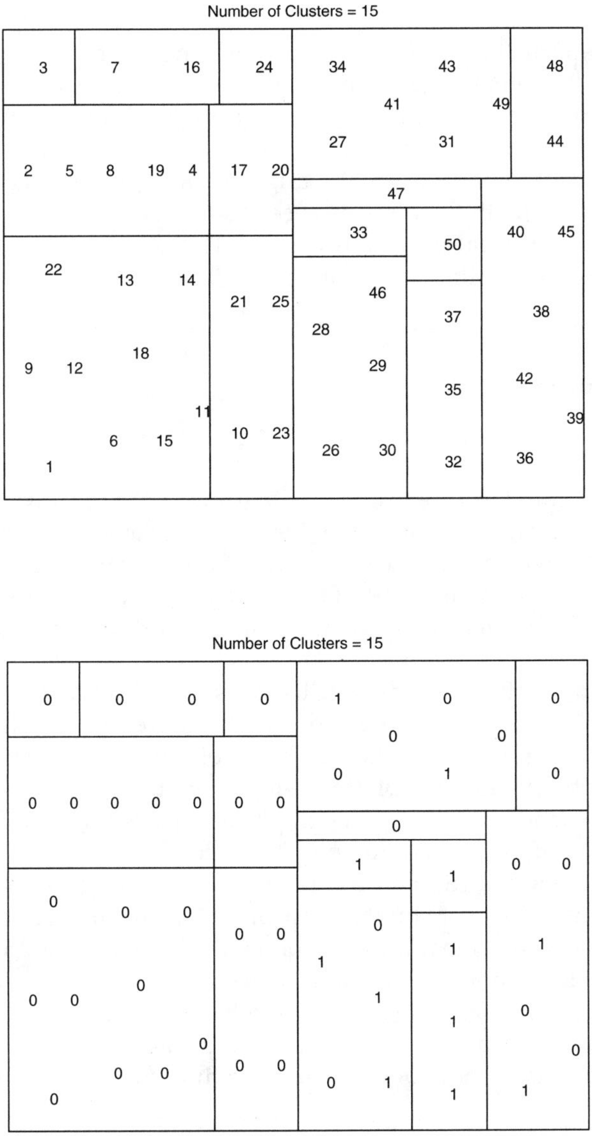

FIGURE 8.6
A rectangle plot for the `leukemia` data is shown in the top of this figure. Here we plot the observation numbers for a specified number of clusters or partitions. The rectangle plot shown in the bottom of the figure plots the observations using the true cancer labels. Class 0 corresponds to ALL and Class 1 is AML.

user can interact with the display by dragging the mouse over the line graph and seeing the corresponding change in the number of clusters shown in the rectangle plot.

The rectangle method is also suitable for linking and brushing applications (see Chapter 10), where one can highlight an observation in one plot (e.g., a scatterplot) and see the same observation highlighted in another (e.g., a rectangle plot). A disadvantage with the rectangle plot is that some of the nesting structure seen in treemaps might be lost in the rectangle display. Another problem with the plots discussed so far is their applicability to the display of hierarchical information only. The plot we show next can be applied to non-hierarchical clustering procedures.

8.4 ReClus Plots

The *ReClus method* was developed by Martinez [2002] as a way to view the output of nonhierarchical clustering methods, such as k-means, model-based clustering, etc., that is reminiscent of the treemap and rectangle displays. We note that ReClus (standing for RectangleClusters) can also be used to convey the results of any hierarchical clustering method once we have a given partition.

As in the previous methods, ReClus uses the entire display area as the parent rectangle. This is then partitioned into sub-rectangles, where the area of each one is proportional to the number of observations that belong to that cluster. The observations are plotted using either the observation number or the true class label, if known. The glyphs are plotted in a systematic way, either by order of the observation number or the class label.

There are some additional options. If the output is from model-based clustering, then we can obtain the probability that an observation belongs to the cluster. This additional information is displayed via the font color. For faster and easier comprehension of the cluster results, we can set a threshold so that the higher probabilities are shown in bold black type. We provide a similar capability for other cluster methods, such as k-means, using the silhouette values. We now outline the procedure for constructing a ReClus plot.

Procedure - ReClus Plot

1. Set up the parent rectangle. We will subdivide the rectangle along the longer side of the parent rectangle according to the proportion of observations that are in each group.

2. Find all of the points in each cluster and the corresponding proportion.

3. Order the proportions in ascending order.

4. Partition the proportions into two groups. If there is an odd number of clusters, then put more of the clusters into the 'left/lower' group.

5. Based on the total proportion in each group, split the longer side of the parent rectangle. We now have two children. Note that we have to re-normalize the proportions based on the current parent rectangle.

6. Repeat steps 4 through 5 until all rectangles represent only one cluster.

7. Find the observations in each cluster and plot, either as the case label or the true class label.

We illustrate the use of ReClus in the next example.

Example 8.4

For this example, we use the **L1bpm** interpoint distance matrix derived from the BPMS of the 503 documents discussed in Chapter 1. We first use ISOMAP to reduce the dimensionality and get our derived observations. We are going to use only five of the sixteen topics, so we also set up the indices to extract them.

```
load L1bpm
% Get just those topics on 5 through 11
ind = find(classlab == 5);
ind = [ind; find(classlab == 6)];
ind = [ind; find(classlab == 8)];
ind = [ind; find(classlab == 9)];
ind = [ind; find(classlab == 11)];
% Change the class labels for class 11
% for better plotting.
clabs = classlab(ind);
clabs(find(clabs == 11)) = 1;
% First do dimensionality reduction - ISOMAP
[Y, R, E] = isomap(L1bpm,'k',10);
% Choose 3 dimensions, based on residual variance.
XX = Y.coords{3}';
% Only need those observations in classes of interest.
X = XX(ind,:);
```

Next we do model-based clustering specifying a maximum of 10 clusters.

```
% Now do model-based clustering.
[bics,bestm,allm,Z,clabsmbc] = mbclust(X,10);
```

We see that model-based clustering found the correct number of groups (five), and model seven is the best fit (according to the BIC). Now that we have the cluster information, we need to find the probability that an observation belongs to the cluster or the silhouette values if using some other clustering procedure. In the case of model-based clustering, we use the function **mixclass** to get the probability that an observation does *not* belong to the cluster, based on the model.

```
% Must get the probability that an observation does
% not belong to the cluster.
[clabsB,uncB] = mixclass(X,bestm.pies,...
    bestm.mus,bestm.vars);
% Plot with true class labels.
% The function requires the posterior probability
% that it belongs to the cluster.
reclus(clabsB,clabs,1 - uncB)
```

Note that **mixclass** returns the *uncertainty* in the classification, so we must subtract this from one in the argument to **reclus**. The resulting plot is shown in the top of Figure 8.7. Note that topics 8 and 1 (formerly topic 11) are both about North Korea, and we see some mistakes in how these topics are grouped. However, topics 5, 6, and 9 are clustered together nicely, except for one document from topic 1 that is grouped with topic 5. The color coding of the posterior probabilities makes it easier to see the level of uncertainty. In some applications, we might gain more insights regarding the clusters by looking at a more 'binary' application of color. In other words, we want to see those observations that have a high posterior probability separated from those with a lower one. An optional argument to **reclus** specifies a threshold, where we plot posterior probabilities above this value in bold black font. This makes it easier to get a quick overall view of the quality of the clusters.

```
% Look at the other option in reclus.
% Plot points with posterior probability above
% 0.9 in bold, black font.
reclus(clabsB,clabs,1 - uncB,.9)
```

This ReClus plot is given in the bottom of Figure 8.7, where we see that documents in topics 5, 6, and 9 have posterior probabilities above the threshold. The one exception is the topic 1 document that was grouped with topic 5. We see that this document has a lower posterior probability that was not readily apparent in Figure 8.7 (top). We also see that the documents in the two groups with topics 1 and 8 mixed together have a large number of members with posterior probabilities below the threshold. It would be interesting to explore these two clusters further to see if this grouping is an indication of sub-topics.
❑

The next example shows how to use ReClus with the information from hierarchical clustering, in addition to some of the other options that are

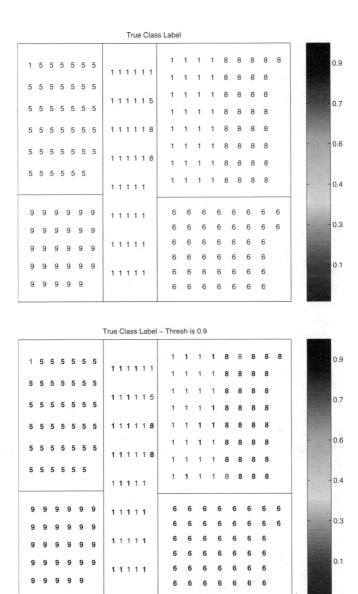

FIGURE 8.7
The ReClus plot at the top shows the cluster configuration based on the best model chosen from model-based clustering. Here we plot the true class label with the color indicating the probability that the observation belongs to that cluster. The next ReClus plot is for the same data and model-based clustering output, but this time we request that probabilities above 0.9 be shown in bold black font. See the associated color figure following page 144.

available with the **reclus** function. One of the advantages of ReClus is the ability to rapidly visualize the 'strength' of the clustering for the observations when the true class membership is known. It would be useful to be able to find out which observations correspond to interesting ones in the ReClus plot. For instance, in Figure 8.7 (top), we might want to locate those stories that were in the two mixed-up groups to see if the way they were grouped makes sense. Or, we might want to find the topic 1 document that was mis-grouped with topic 5.

Example 8.5

We use the same data from the previous example, but now we only look at two topics: 8 and 11. These are the two that concern North Korea, and there was some confusion in the clustering of these topics using model-based clustering. We will use the ReClus plot to help us assess how hierarchical clustering works on these two topics. First we extract the observations that we need and get the labels.

```
% Continuing with same data used in
% Example 8.4.
ind = find(classlab == 8);
ind = [ind; find(classlab == 11)];
clabs = classlab(ind);
% Change the class labels for class 11
% for better plotting.
clabs = classlab(ind);
clabs(find(clabs == 11)) = 1;
% Only need those observations in classes of interest.
X = XX(ind,:);
```

Next we perform hierarchical clustering using Euclidean distance and complete linkage. We use the **cluster** function to request two groups, and then get the silhouette values. Note that this syntax for the **silhouette** function suppresses the plot and only returns the silhouette values.

```
% Get the hierarchical clustering.
Y = pdist(X);
Z = linkage(Y,'complete');
dendrogram(Z);
cids = cluster(Z,'maxclust',2);
% Now get the silhouette values.
S = silhouette(X,cids);
```

Our initial ReClus plot uses the observation number as the plotting symbol. We have to include the true class labels as an argument, so they can be mapped into the same position in the second ReClus plot. The plots are shown in Figure 8.8, and the code to construct these plots is given below.

```
% The following plots the observation
```

```
% numbers, with 'pointers' to the
% true class labels plotted next.
reclus(cids, clabs)
% Now plot same thing, but with true class labels.
reclus(cids,clabs,S)
```

We see from the plots that this hierarchical clustering does not provide a 'good' clustering, as measured by the silhouette values, since we have several negative values in the right-hand cluster.
❑

8.5 Data Image

The *data image* is a technique for visualizing high-dimensional data as if they comprised an image. We have seen an example of this already in Chapter 1, Figure 1.1, where we have the gene expression data displayed as a gray-scale image. The basic idea is to map the data into an image framework, using the gray-scale values or colors (if some other color map is desired) to indicate the magnitude of each variable for each observation. Thus, the data image for a data set of size n with p variables will be $n \times p$.

An early version of the data image can be found in Ling [1973], where the data are plotted as a matrix of characters with different amounts of ink indicating the gray scale value of the observation. However, Ling first plots the interpoint dissimilarity (or similarity) matrix in its 'raw' form. He then reorders rows and columns of the dissimilarity matrix using the cluster labels after some clustering method has been applied. In other words, the original sequence of observations has been arranged such that the members of every cluster lie in consecutive rows and columns of the permuted dissimilarity matrix. Clearly defined dark (or light, depending on the gray scale) squares along the diagonal indicate compact clusters that are well separated from neighboring points. If the data do not contain significant clusters, then this is readily seen in the image.

Wegman [1990] also describes a version of the data image, but he calls it a *color histogram*. He suggests coloring pixels using a binned color gradient and presenting this as an image. Sorting the data based on one variable enables one to observe positive and negative associations. One could explore the data in a tour-like manner by performing this sort for each variable.

Minnotte and West [1998] use binning on a fine scale, so they coined the term 'data image' to be more descriptive of the output. Rather than sorting on only one variable, they suggest finding an ordering such that points that are close in high-dimensional space are close to one another in the image. They suggest that this will help the analyst better visualize high-dimensional structure. We note that this is reminiscent of multidimensional scaling. In

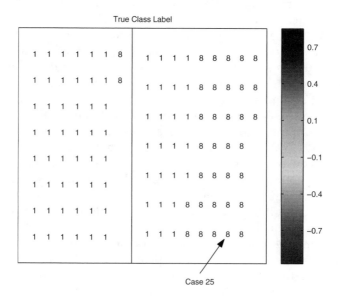

FIGURE 8.8

In Figure 8.8 (top), we show the ReClus plot for topics 8 and 1 (topic 11) based on hierarchical clustering. The positions of these symbols corresponds to the glyphs in Figure 8.8 (bottom). This makes it easy to see what case belongs to observations of interest. The scale in the lower ReClus plot corresponds to the silhouette values. See the associated color figure following page 144.

particular, they apply this method to understand the cluster structure in high-dimensional data, which should appear as vertical bands within the image.

Minnotte and West propose two methods for ordering the data. One approach searches for the shortest path through the cloud of high-dimensional points using traveling salesman type algorithms [Cook, et al., 1998]. Another option is to use an ordering obtained from hierarchical clustering, which is what we do in the next example.

Example 8.6

We use the familiar **iris** data set for this example. We put the three classes of iris into one matrix and randomly reorder the rows. The data image for this arrangement is shown in Figure 8.9 (top). We see the four variables as columns or vertical bands in the image, but we do not see any horizontal bands indicating groups of observations.

```
load iris
% Put into one matrix.
data = [setosa;versicolor;virginica];
% Randomly reorder the data.
data = data(randperm(150),:);
% Construct the data image.
imagesc(-1*data)
colormap(gray(256))
```

We now cluster the data using agglomerative clustering with complete linkage. To get our ordering, we plot the dendrogram with n leaf nodes. The output argument **perm** from the function **dendrogram** provides the order of the observations left to right or bottom to top, depending on the orientation of the tree. We use this to rearrange the data points.

```
% Now get the ordering using hierarchical
% clustering and the dendrogram.
Y = pdist(data);
Z = linkage(Y,'complete');
% Plot both the dendrogram and the data
% image together.
figure
subplot(1,2,1)
[H, T, perm] = dendrogram(Z,0,'orientation','left');
axis off
subplot(1,2,2)
% Need to flip the matrix to show as an image.
imagesc(flipud(-1*data(perm,:)))
colormap(gray(256))
```

The data image for the reordered data and the associated dendrogram are given in Figure 8.9 (bottom). We can now see three horizontal bands that would indicate the presence of three groups. However, we know that each class of iris has 50 observations and that some are likely to be incorrectly clustered together.
❏

The data image displayed along with the dendrogram, as we saw in Example 8.6, is used extensively in the gene expression analysis literature. However, it is not generally known by that name. We now show how to apply the data image concept to locate clusters in a data set using Ling's method.

Example 8.7

We use the data (topics 6 and 9) from Example 8.4 to illustrate how the data image idea can be applied to the interpoint dissimilarity matrix. After extracting the data, we find the distance matrix using **pdist** and **squareform**. The image of this is shown in Figure 8.10 (top). We randomly reordered the data; thus it is difficult to see any cluster structure or clusters. The code to do this follows.

```
% Continuing with same data used in Example 8.4.
% Use just Topics 6 and 9.
ind = find(classlab == 6);
ind = [ind; find(classlab == 9)];
n = length(ind);
clabs = classlab(ind);
data = XX(ind,:);
% Randomly reorder the data and view the
% interpoint distance matrix as an image.
data = data(randperm(n),:);
Y = pdist(data);
Ys = squareform(Y);
imagesc(Ys);
colormap(gray(256))
title('Interpoint Distance Matrix - Random Order')
axis off
```

We now have to obtain some clusters. We will use hierarchical clustering and request cluster labels based on a partition into two groups. Then we order the distances such that observations in the same group are adjacent and replot the distance matrix as an image.

```
% Now apply Ling's method. First need to
% get a partition or clustering.
Z = linkage(Y,'complete');
% Now get the ordering based on the cluster scheme.
T = cluster(Z,'maxclust',2);
```

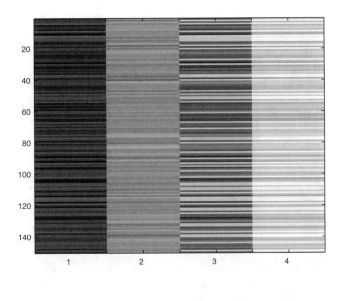

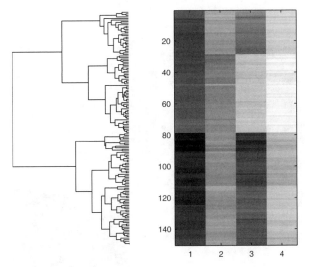

FIGURE 8.9

The data image at the top is for the **iris** data, where the observations have been randomly ordered. Clusters or horizontal bands are not obvious. We apply the data image concept by reordering the data according to the results of hierarchical clustering. The data image of the rearranged data and the corresponding dendrogram are shown at the bottom. Three horizontal bands or clusters are now readily visible. See the associated color figure following page 144.

Interpoint Distance Matrix – Random Order

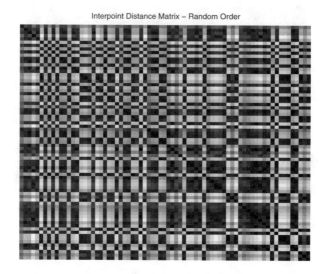

Interpoint Distance Matrix – Cluster Order

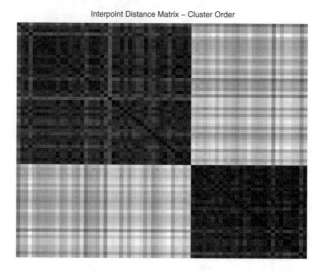

FIGURE 8.10

The top image shows the interpoint distance matrix for topic 6 and topic 9 when the data are in random order. The presence of any groups or clusters is difficult to discern. Next we cluster the data according to our desired method and rearrange the points such that those observations in the same group are adjacent in the distance matrix. We show this re-ordered matrix in the bottom half of the figure, and the two groups can be clearly seen.

```
% Sort these, so points in the same cluster
% are adjacent.
[Ts,inds] = sort(T);
% Sort the distance matrix using this and replot.
figure
imagesc(Ys(inds,inds))
title('Interpoint Distance Matrix - Cluster Order')
colormap(gray(256))
axis off
```

This image is shown in Figure 8.10 (bottom). We can now clearly see the two clusters along the diagonal.

❑

8.6 Summary and Further Reading

One of the main purposes of clustering is to organize the data, so graphical tools to display the results of these methods are important. These displays should enable the analyst to see whether or not the grouping of the data illustrates some latent structure. Equally important, if real structure is not present in the data, then the cluster display should convey this fact. The cluster visualization techniques presented in this chapter should enable the analyst to better explore, assess, and understand the results from hierarchical clustering, model-based clustering, *k*-means, etc.

Some enhancements to the dendrogram have been proposed. One is a generalization of dendrograms called *espaliers* [Hansen, Jaumard and Simeone, 1996]. In the case of a vertical diagram, espaliers use the length of the horizontal lines to encode another characteristic of the cluster, such as the diameter. Other graphical aids for assessing and interpreting the clusters can be found in Cohen, et al. [1977].

Johnson and Shneiderman [1991] provide examples of other types of treemaps. One is a *nested treemap*, where some space is provided around each sub-rectangle. The nested nature is easier to see, but some display space is wasted in visualizing the borders. They also show a *Venn diagram treemap*, where the groups are shown as nested ovals.

The treemap display is popular in the computer science community as a tool for visualizing directory structures on disk drives. Several extensions to the treemap have been developed. These include *cushion treemaps* [Wijk and Wetering, 1999] and *squarified treemaps* [Bruls, Huizing and Wijk, 2000]. Cushion treemaps keep the same basic structure of the space-filling treemaps, but they add surface height and shading to provide additional insight into the hierarchical structure. The squarified treemap methodology attempts to construct squares instead of rectangles (as much as possible) to prevent long, skinny rectangles, but this sacrifices the visual understanding of the nested structure to some extent.

Marchette and Solka [2003] use the data image to find outliers in a data set. They apply this concept to the interpoint distance matrix by treating the columns (or alternatively the rows) of the matrix as observations. These observations are clustered using some hierarchical clustering scheme, and the rows and columns are permuted accordingly. Outliers show up as a dark *v* or cross (depending on the color scale).

Others in the literature have discussed the importance of ordering the data to facilitate exploration and understanding. A recent discussion of this can be found in Friendly and Kwan [2003]. They outline a general framework for ordering information in visual displays, including tables and graphs. They show how these *effect-ordered data displays* can be used to discover patterns, trends, and anomalies in the data.

Exercises

8.1 Do a **help** on **dendrogram** and read about the optional input argument **'colorthreshold'**. Repeat Example 8.1 using this option.

8.2 Compare and contrast the dendrogram, treemap, rectangle plots, and ReClus cluster visualization techniques. What are the advantages and disadvantages of each? Comment on the usefulness of these methods for large data sets.

8.3 Find the ReClus plot (without class labels and with class labels) for the **leukemia** partition with 15 clusters obtained in Example 8.1. Compare to the previous results using the hierarchical visualization techniques.

8.4 Repeat Example 8.3 using the distance input argument to specify the number of clusters displayed in the rectangle plot.

8.5 For the following data sets, use an appropriate hierarchical clustering method and visualize using the methods described in the chapter. Analyze the results.

a. **geyser**

b. **singer**

c. **skulls**

d. **sparrow**

e. **oronsay**

f. gene expression data sets

8.6 Repeat Example 8.7 using all of the data in the matrix **XX** (reduced from ISOMAP) from Examples 8.4 and 8.5. Use hierarchical clustering or *k*-means and ask for 16 groups. Is there any evidence of clusters?

8.7 Apply the methodology of Example 8.7 to the **iris** data.

8.8 Repeat Examples 8.4, 8.5 and 8.7 using other BPM data sets and report on your results.

8.9 Repeat Example 8.4 using the silhouette values for the model-based clustering classification.

8.10 Repeat Example 8.5 using other types of hierarchical clustering. Compare your results.

8.11 For the following data sets, use *k*-means or model-based clustering. Use the ReClus method for visualization. Analyze your results.

    a. **skulls**

    b. **sparrow**

    c. **oronsay** (both classifications)

    d. BPM data sets

    e. gene expression data sets

8.12 Looking at the data image in Figure 8.9, comment on what variables seem most useful for classifying the species of iris.

Chapter 9

Distribution Shapes

In this chapter, we show various methods for visualizing the shapes of distributions. The ability to visualize the distribution shape in exploratory data analysis is important for several reasons. First, we can use it to summarize a data set to better understand general characteristics such as shape, spread, or location. In turn, this information can be used to suggest transformations or probabilistic models for the data. Second, we can use these methods to check model assumptions, such as symmetry, normality, etc. We present several techniques for visualizing univariate and bivariate distributions. These include 1-D and 2-D histograms, boxplots, quantile-based plots, and bagplots.

9.1 Histograms

A *histogram* is a way to graphically summarize or describe a data set by visually conveying its distribution using vertical bars. They are easy to create and are computationally feasible, so they can be applied to massive data sets. In this section, we describe several varieties of histograms. These include the frequency and relative frequency histogram, and what we are calling the density histogram.

9.1.1 Univariate Histograms

A *frequency histogram* is obtained by first creating a set of bins or intervals that cover the range of the data set. It is important that these bins do not overlap and that they have equal width. We then count the number of observations that fall into each bin. To visualize this information, we place a bar at each bin, where the height of the bar corresponds to the frequency. *Relative frequency histograms* are obtained by mapping the height of the bin to the relative frequency of the observations that fall into the bin.

The basic MATLAB package has a function for calculating and plotting a univariate frequency histogram called **hist**. This function is illustrated in the example given below, where we show how to construct both types of histograms.

Example 9.1

In this example, we look at the two univariate histograms showing relative frequency and frequency. We can obtain a simple histogram in MATLAB using these commands:

```
load galaxy
% The 'hist' function can return the
% bin centers and frequencies.
% Use the default number of bins - 10.
[n, x] = hist(EastWest);
% Plot and use the argument of width = 1
% to get bars that touch.
bar(x,n,1,'w');
title('Frequency Histogram - Galaxy Data')
xlabel('Velocity')
ylabel('Frequency')
```

Note that calling the **hist** function with no output arguments will find the pieces necessary to construct the histogram based on a given number of bins (default is 10 bins) and will also produce the plot. We chose to use the option of first extracting the bin locations and bin frequencies so we could get the relative frequency histogram using the following code:

```
% Now create a relative frequency histogram.
% Divide each box by the total number of points.
% We use bar to plot.
bar (x,n/140,1,'w')
title('Relative Frequency Histogram - Galaxy Data')
xlabel('Velocity')
ylabel('Relative Frequency')
```

These plots are shown in Figure 9.1. Notice that the shapes of the histograms are the same in both types of histograms, but the vertical axes are different. From the shape of the histograms, it seems reasonable to assume that the data are normally distributed (for this bin configuration).
❑

One problem with using a frequency or relative frequency histogram is that they do not represent meaningful probability densities, because the total area represented by the bars does not equal one. This can be seen by superimposing a corresponding normal distribution over the relative frequency histogram as shown in Figure 9.2. However, they are very useful for gaining a quick picture of the distribution of the data.

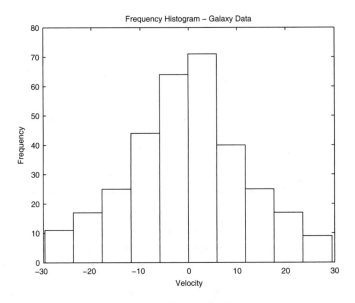

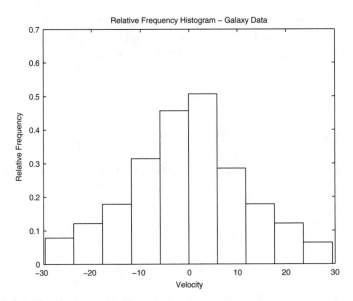

FIGURE 9.1

The top histogram shows the number of observations that fall into each bin, while the bottom histogram shows the relative frequency. Note that the shape of the histograms is the same, but the vertical axes represent different quantities.

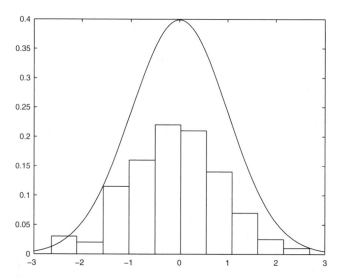

FIGURE 9.2
This shows a relative frequency histogram for some data generated from a standard normal distribution. Note that the curve is higher than the histogram, indicating that the histogram is not a valid probability density function.

A ***density histogram*** is a histogram that has been normalized so the area under the curve (where the curve is represented by the heights of the bars) is one. A density histogram is given by the following equation

$$\hat{f}(x) = \frac{v_k}{nh} \qquad x \text{ in } B_k, \tag{9.1}$$

where B_k denotes the k-th bin, v_k represents the number of data points that fall into the k-th bin, and h represents the width of the bins.

Since our goal is to estimate a *bona fide* probability density, we want to have an estimate that is nonnegative and satisfies the constraint that

$$\int \hat{f}(x)dx = 1.$$

It is left as an exercise to the reader to show that Equation 9.1 satisfies this condition.

The density histogram depends on two parameters: 1) an origin t_0 for the bins and 2) a bin width h. These two parameters define the mesh over which the histogram is constructed. The bin width h is sometimes referred to as the ***smoothing parameter***, and it fulfills a similar purpose as that found in the chapter on smoothing scatterplots. The bin width determines the smoothness of the histogram. Small values of h produce histograms with a lot of variation in the heights of the bins, while larger bin widths yield smoother histograms. This phenomenon is illustrated in Figure 9.3, where we show histograms of the same data, but with different bin widths.

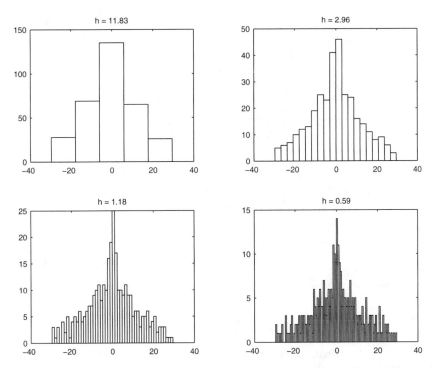

FIGURE 9.3
These are histograms for the **galaxy** data used in Example 9.1. Notice that for the larger bin widths, we have only one peak. As the smoothing parameter gets smaller, the histogram displays more variation and spurious peaks appear in the histogram estimate.

We now look at how we can choose the bin width h, in an attempt to minimize our estimation error. It can be shown that setting h small to reduce the bias increases the variance in our estimate. On the other hand, creating a smoother histogram reduces the variance, at the expense of worsening the bias. This is the familiar trade-off between variance and bias discussed previously. We now present some of the common methods for choosing the bin width, most of which are obtained by trying to minimize the squared error [Scott, 1992] between the true density and the estimate.

Histogram Bin Widths

   1. Sturges' Rule

$$k = 1 + \log_2 n \ , \tag{9.2}$$

   where k is the number of bins. The bin width h is obtained by taking
   the range of the sample data and dividing it into the requisite
   number of bins, k [Sturges, 1926].

   2. Normal Reference Rule - 1-D Histogram

$$h^* = \left(\frac{24\sigma^3 \sqrt{\pi}}{n} \right)^{1/3} \approx 3.5 \times \sigma \times n^{-1/3} \ . \tag{9.3}$$

   Scott [1979, 1992] proposed the sample standard deviation as an
   estimate of σ in Equation 9.3 to get the following bin width rule.

   3. Scott's Rule

$$\hat{h}^* = 3.5 \times s \times n^{-1/3} \ .$$

   4. Freedman-Diaconis Rule

$$\hat{h}^* = 2 \times IQR \times n^{-1/3} \ .$$

   This robust rule developed by Freedman and Diaconis [1981] uses
   the interquartile range (IQR) instead of the sample standard deviation.

   It turns out that when the data are skewed or heavy-tailed, the bin widths
are too large using the Normal Reference Rule. Scott [1979, 1992] derived the
following correction factor for skewed data:

$$\text{skewness factor} = \frac{2^{1/3} \sigma}{e^{5\sigma^2/4}(\sigma^2 + 2)^{1/3}(e^{\sigma^2} - 1)^{1/2}} \ . \tag{9.4}$$

If one suspects the data come from a skewed distribution, then the Normal
Reference Rule bin widths should be multiplied by the factor given in
Equation 9.4.
   So far, we have discussed histograms from a visualization standpoint only.
We might also need an estimate of the density at a given point x, as we will
see in the next section on boxplots. We can find a value for our density
estimate for a given x, using Equation 9.1. We obtain the value $\hat{f}(x)$ by taking

the number of observations in the data set that fall into the same bin as x and multiplying by $1/(nh)$.

Example 9.2

In this example, we provide MATLAB code that calculates the estimated value $\hat{f}(x)$ for a given x. We use the same data from the previous example and Sturges' Rule for estimating the number of bins.

```
load galaxy
n = length(EastWest);
% Use Sturges' Rule to get the number of bins.
k = round(1 + log2(n));
% Bin the data.
[nuk,xk] = hist(EastWest,k);
% Get the width of the bins.
h = xk(2) - xk(1);
% Plot as a density histogram.
bar(xk, nuk/(n*h), 1, 'w')
title('Density Histogram - Galaxy Data')
xlabel('Velocity')
```

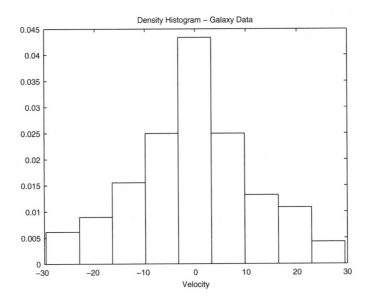

FIGURE 9.4
This shows the density histogram for the **galaxy** data.

The histogram produced by this code is shown in Figure 9.4. Note that we had to adjust the output from **hist** to ensure that our estimate is a *bona fide* density. Let's get the estimate of our function at a point $x_0 = 0$.

```
% Now return an estimate at a point xo.
xo = 0;
% Find all of the bin centers less than xo.
ind = find(xk < xo);
% xo should be between these two bin centers:
b1 = xk(ind(end));
b2 = xk(ind(end)+1);
% Put it in the closer bin.
if (xo-b1) < (b2-xo)    % then put it in the 1st bin
    fhat = nuk(ind(end))/(n*h);
else
    fhat = nuk(ind(end)+1)/(n*h);
end
```

Our result is **fhat = 0.0433**. Looking at Figure 9.4, we see that this is the correct estimate.
❏

9.1.2 Bivariate Histograms

We can easily extend the univariate density histogram to multivariate data, but we restrict our attention in this text to the bivariate case. The bivariate histogram is defined as

$$\hat{f}(\mathbf{x}) = \frac{v_k}{nh_1h_2}; \qquad \mathbf{x} \text{ in } B_k, \tag{9.5}$$

where v_k denotes the number of observations falling into the bivariate bin B_k, and h_i is the width of the bin along the i-th coordinate axis. Thus, the estimate of the probability density would be given by the number of observations falling into that same bin divided by the sample size and the bin widths.

As before, we must determine what bin widths to use. Scott [1992] provides the following multivariate Normal Reference Rule.

Normal Reference Rule - Multivariate Histograms

$$h_i^* \approx 3.5 \times \sigma_i \times n^{\frac{-1}{2+p}}; \qquad i = 1, \ \ldots \ , \ p. \tag{9.6}$$

Notice that this reduces to the same univariate Normal Reference Rule (Equation 9.3) when $p = 1$. As before, we can use a suitable estimate for σ_i, based on our data.

Example 9.3

We return to the data used in Example 7.8 to illustrate the bivariate histogram. Recall that we first reduced the BPM data to 2-D using ISOMAP and the L_1 proximity measure. The code to do this is repeated below.

```
load L1bpm
% Reduce the dimensionality using Isomap.
options.dims = 1:10;     % These are for ISOMAP.
options.display = 0;
[Yiso, Riso, Eiso] = isomap(L1bpm, 'k', 7, options);
% Get the data out.
XX = Yiso.coords{2}';
inds = find(classlab==8 | classlab==11);
x = [XX(inds,:)];
[n,p] = size(x);
```

We use the normal reference rule to find the density histogram of the data.

```
% Need bin origins.
bin0 = floor(min(x));
% The bin width h, for p = 2:
h = 3.5*std(x)*n^(-0.25);
% Find the number of bins
nb1 = ceil((max(x(:,1))-bin0(1))/h(1));
nb2 = ceil((max(x(:,2))-bin0(2))/h(2));
% Find the bin edges.
t1 = bin0(1):h(1):(nb1*h(1)+bin0(1));
t2 = bin0(2):h(2):(nb2*h(2)+bin0(2));
[X,Y] = meshgrid(t1,t2);
% Find bin frequencies.
[nr,nc] = size(X);
vu = zeros(nr-1,nc-1);
for i = 1:(nr-1)
    for j = 1:(nc-1)
        xv = [X(i,j) X(i,j+1) X(i+1,j+1) X(i+1,j)];
        yv = [Y(i,j) Y(i,j+1) Y(i+1,j+1) Y(i+1,j)];
        in = inpolygon(x(:,1),x(:,2),xv,yv);
        vu(i,j) = sum(in(:));
    end
end
% Find the proper height of the bins.
Z = vu/(n*h(1)*h(2));
% Plot using bars.
```

```
bar3(Z,1,'w')
```

The plot for histogram is shown in Figure 9.5. We used some additional MATLAB code to get axes labels that make sense. Please refer to the M-file for this example to see the code to do that. The Statistics Toolbox, Version 5, has a function called **hist3** for constructing histograms of bivariate data. ❑

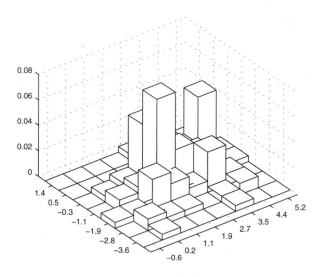

FIGURE 9.5
This is the histogram for the data representing topics 8 and 11. The normal reference rule was used for the bin widths. Please see Figure 7.14 for the corresponding scatterplot.

9.2 Boxplots

Boxplots (sometimes called *box-and-whisker diagrams*) have been in use for many years [Tukey, 1977]. They are an excellent way to visualize summary statistics such as the median, to study the distribution of the data, and to supplement multivariate displays with univariate information. Benjamini [1988] outlines the following characteristics of the boxplot that make them useful:

1. Statistics describing the data are visualized in a way that readily conveys information about the location, spread, skewness, and longtailedness of the sample.

2. The boxplot displays information about the observations in the tails, such as potential outliers.

3. Boxplots can be displayed side-by-side to compare the distribution of several data sets.

4. The boxplot is easy to construct.

5. The boxplot is easily explained to and understood by users of statistics.

In this section, we first describe the basic boxplot. This is followed by several enhancements and variations of the boxplot. These include variable-width boxplots, the histplot, and the box-percentile plot.

9.2.1 The Basic Boxplot

Before we describe the boxplot, we need to define some terms. In essence, three statistics from a data set are needed to construct all of the pieces of the boxplot. These are the *sample quartiles*: $q(0.25)$, $q(0.5)$, $q(0.75)$. The sample quartiles are based on *sample quantiles*, which are defined next [Kotz and Johnson, 1986].

Given a data set $x_1, \ldots, x_n$, we order the data from smallest to largest. These are called the *order statistics*, and we denote them as

$$x_{(1)}, \ldots, x_{(n)}.$$

The u ($0 < u < 1$) quantile $q(u)$ of a random sample is a value belonging to the range of the data such that a fraction u (approximately) of the data are less than or equal to u.

The quantile denoted by $q(0.25)$ is also called the *lower quartile*, where approximately 25% of the data are less than or equal to this number. The quantile $q(0.5)$ is the *median*, and $q(0.75)$ is the *upper quartile*. We need to define a form for the u in quantile $q(u)$. For a random sample of size n, we let

$$u_i = \frac{i - 0.5}{n}. \tag{9.7}$$

The form for u_i given in Equation 9.7 is somewhat arbitrary [Cleveland, 1993] and is defined for u_i, $i = 1, \ldots, n$. This definition can be extended for all values of u, $0 < u < 1$ by interpolation or extrapolation, given the values of u_i and $q(u_i)$. We will study the quantiles in more detail in the next section.

Definitions of quartiles can vary from one software package to another. Frigge, Hoaglin and Iglewicz [1989] describe a study on how quartiles are implemented in some popular statistics programs such as Minitab, S, SAS, SPSS, and others. They provide eight definitions for quartiles and show how this can affect the appearance of the boxplots. We will use the one defined in Tukey [1977] called standard *fourths* or *hinges*.

Procedure - Finding Quartiles

1. Order the data from smallest to largest.
2. Find the median $q(0.5)$, which is in position $(n+1)/2$ in the list of ordered data:
   a. If n is odd, then the median is the middle data point.
   b. If n is even, then the median is the average of the two middle points.
3. Find the lower quartile, which is the median of the data that lie at or below the median:
   a. If n is odd, then $q(0.25)$ is the median of the ordered data in positions 1 through $(n+1)/2$.
   b. If n is even, then $q(0.25)$ is the median of the ordered data in positions 1 through $n/2$.
4. Find the upper quartile, which is the median of the data that lie at or above the median:
   a. If n is odd, then $q(0.75)$ is the median of the ordered data in positions $(n+1)/2$ through n.
   b. If n is even, then $q(0.75)$ is the median of the ordered data in positions $n/2 + 1$ through n.

Thus, we see that the lower quartile is the median of the lower half of the data set, and the upper quartile is the median of the upper half of the sample. We show how to find these using MATLAB in the next example.

Example 9.4

We will use the **geyser** data to illustrate the MATLAB code for finding the quartiles. These data represent the time (in minutes) between eruptions of the Old Faithful geyser at Yellowstone National Park.

```
load geyser
% First sort the data.
geyser = sort(geyser);
% Get the median.
q2 = median(geyser);
% First find out if n is even or odd.
n = length(geyser);
if rem(n,2) == 1
    odd = 1;
else
    odd = 0;
end
if odd
```

```
    q1 = median(geyser(1:(n+1)/2));
    q3 = median(geyser((n+1)/2:end));
else
    q1 = median(geyser(1:n/2));
    q3 = median(geyser(n/2:end));
end
```

The sample quartiles are: 59, 76, 83. The reader is asked in the exercises to verify that these make sense by looking at a histogram and other plots. We provide a function called **quartiles** that implements this code for general use.
❑

Recall from introductory statistics that the sample *interquartile range* (IQR) is the difference between the first and the third sample quartiles. This gives the range of the middle 50% of the data. It is found from the following:

$$IQR = q(0.75) - q(0.25).$$

We need to define two more quantities to determine what observations qualify as potential outliers. These limits are the *lower limit* (LL) and the *upper limit* (UL). They are calculated from the IQR as follows

$$\begin{aligned} LL &= q(0.25) - 1.5 \times IQR \\ UL &= q(0.75) + 1.5 \times IQR. \end{aligned} \tag{9.8}$$

Observations outside these limits are *potential outliers*. In other words, observations smaller than the LL and larger than the UL are flagged as interesting points because they are outlying with respect to the bulk of the data. *Adjacent values* are the most extreme observations in the data set that are within the lower and the upper limits. If there are no potential outliers, then the adjacent values are simply the maximum and the minimum data points.

Just as the definition of quartiles varies with different software packages, so can the definition of outliers. In some cases multipliers other than 1.5 are used in Equation 9.8, again leading to different boxplots. Hoaglin, Iglewicz and Tukey [1986] examine this problem and show how it affects the number of outliers displayed in a boxplot.

The original boxplot, as defined by Tukey, did not include the display of outliers, as it does today. He called the boxplot with outliers the *schematic plot*. It is the default now in most statistics packages (and text books) to construct the boxplot with outliers as we outline below.

To construct a boxplot, we place horizontal lines at each of the three quartiles and draw vertical lines at the edges to create a box. We then extend a line from the first quartile to the smallest adjacent value and do the same for the third quartile and largest adjacent value. These lines are sometimes called the whiskers. Finally, any possible outliers are shown as an asterisk or some other plotting symbol. An example of a boxplot with labels showing the various pieces is shown in Figure 9.6.

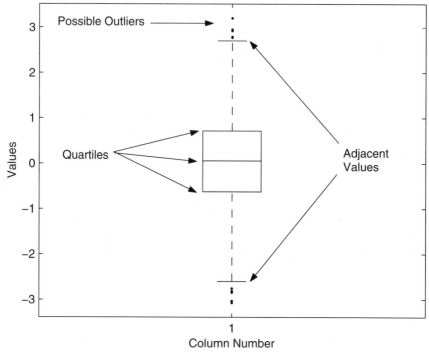

FIGURE 9.6
This is an example of a boxplot with possible outliers.

Boxplots for different univariate samples can be plotted together for visually comparing the corresponding distributions, and they can also be plotted horizontally rather than vertically.

Example 9.5
We now show how to do a boxplot by hand using the **defsloc** variable in the **software** data because it has some potential outliers. First we load the data and transform it using the logarithm.

```
load software
% Take the log of the data.
x = log(sort(defsloc));
n = length(x);
```

The next step is to find the quartiles, the interquartile range, and the upper and lower limits.

```
% First get the quartiles.
q = quartiles(x);
% Find the interquartile range.
iq = q(3) - q(1);
% Find the outer limits.
UL = q(3) + 1.5*iq;
LL = q(1) - 1.5*iq;
```

We can find observations that are outside these limits using the following code:

```
% Find any outliers.
ind = [find(x > UL); find(x < LL)];
outs = x(ind);
% Get the adjacent values. Find the
% points that are NOT outliers.
inds = setdiff(1:n,ind);
% Get their min and max.
adv = [x(inds(1)) x(inds(end))];
```

Now we have all of the quantities necessary to draw the plot.

```
% Now draw the necessary pieces.
% Draw the quartiles.
plot([1 3],[q(1),q(1)])
hold on
plot([1 3],[q(2),q(2)])
plot([1 3],[q(3),q(3)])
% Draw the sides of the box
plot([1 1],[q(1),q(3)])
plot([3 3],[q(1),q(3)])
% Draw the whiskers.
plot([2 2],[q(1),adv(1)],[1.75 2.25],[adv(1) adv(1)])
plot([2 2],[q(3),adv(2)],[1.75 2.25],[adv(2) adv(2)])
% Now draw the outliers with symbols.
plot(2*ones(size(outs)), outs,'o')
hold off
axs = axis;
axis([-1 5 axs(3:4)])
set(gca,'XTickLabel',' ')
ylabel('Defects per SLOC (log)')
```

The boxplot is shown in Figure 9.7, where we see that the distribution is not exactly symmetric.

❑

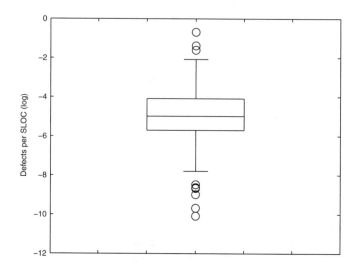

FIGURE 9.7
This shows the boxplot for the number of defects per SLOC (log) from the **software** data set.

We provide a function called **boxp** that will construct boxplots, including some of the variations discussed below. The MATLAB Statistics Toolbox also has a **boxplot** function that the reader is asked to explore in the exercises.

9.2.2 Variations of the Basic Boxplot

We now describe some enhancements and variations of the basic boxplot described above. When we want to understand the significance of the differences between the medians, we can display boxplots with notches [McGill, Tukey and Larsen, 1978]. The notches in the sides of the boxplots represent the uncertainty in the locations of central tendency and provide a rough measure of the significance of the differences between the values. If the intervals represented by the notches do not overlap, then there is evidence that the medians are significantly different. The MATLAB Statistics Toolbox function **boxplot** will produce boxplots with notches, as explained in the exercises.

Vandervieren and Huber [2004] present a robust version of the boxplot for skewed distributions. In this type of distribution, too many observations can be classified as outliers. Their generalization to the boxplot has a robust measure of skewness that is used to find the whiskers. They show that their adjusted boxplot provides a more accurate representation of the data distribution than the basic boxplot. See Appendix B for information on where to download functions for this and other robust analysis methods.

Another enhancement of the boxplot also comes from McGill, Tukey and Larsen [1978]. This is called the *variable-width boxplot*, and it incorporates a measure of the sample size. Instead of having boxplots with equal widths, we could make the widths proportional to some function of n. McGill, Tukey and Larsen recommend using widths proportional to the square root of n and offer it as the standard. They suggest others, such as making the widths directly proportional to sample size or using a logit scale.

Benjamini [1988] uses the width of the boxplot to convey information about the density of the data rather than the sample size. He offers two types of boxplots that incorporate this idea: the *histplot* and the *vaseplot*. In a histplot, the lines at the three quartiles are drawn with a width that is proportional to an estimate of the associated density at these positions. His implementation uses the density histogram, but any other density estimation method can also be used. He then extends this idea by drawing the width of the box at each point proportional to the estimated density at that point. This is called the vaseplot, because it produces a vase-like shape. One can no longer use the notches to show the confidence intervals for the medians in these plots, so Benjamini uses shaded bars instead. All of these extensions adjust the width of the boxes only; the whiskers stay the same.

The *box-percentile plot* [Esty and Banfield, 2003] also uses the sides of the boxplot to convey information about the distribution of the data over the range of data values. They no longer draw the whiskers or the outliers, so there is no ambiguity about how to define these characteristics of the boxplot.

To construct a box-percentile plot, we do the following. From the minimum value up to the 50th percentile, the width of the 'box' is proportional to the percentile of that height. Above the 50th percentile, the width is proportional to 100 minus the percentile. The box-percentile plots are wide in the middle, like boxplots, but narrowing as they get further away from the middle.

We now describe the procedure in more detail. Let w indicate the maximum width of the box-percentile plot, which will be the width at the median. We obtain the order statistics of our observed random sample, $x_{(1)}, ..., x_{(n)}$. Then the sides of the box-percentile plot are obtained as follows:

1. For $x_{(k)}$ less than or equal to the median, we plot the observation at height $x_{(k)}$ at a distance $kw/(n + 1)$ on either side of a vertical axis of symmetry.

2. For $x_{(k)}$ greater than the median, we plot the point at height $x_{(k)}$ at a distance $(n + 1 - k)w/(n + 1)$ on either side of the axis of symmetry.

We illustrate the box-percentile plot and the histplot in the next example. Constructing a variable-width boxplot is left as an exercise to the reader.

Example 9.6

We use some simulated data sets similar to those in Esty and Banfield [2003] to illustrate the histplot and the box-percentile plots. We did not have their exact distributional models, but we tried to reproduce them as much as possible. The first data set is a standard normal distribution. The second one is uniform in the range $[-1.2, 1.2]$, with some outliers close to -3 and 3. The third random sample is trimodal. These three data sets have similar quartiles and ranges, as illustrated in the boxplots in Figure 9.8. The code used to generate the samples follows; we saved the data in a MAT-file for your use.

```
% Generate some standard normal data.
X(:,1) = randn(400,1);
% Generate some uniform data.
tmp = 2.4*rand(398,1) - 1.2;
% Add some outliers to it.
X(:,2) = [tmp; [-2.9 2.9]'];
tmp1 = randn(300,1)*.5;
tmp2 = randn(50,1)*.4-2;
tmp3 = randn(50,1)*.4+2;
X(:,3) = [tmp1; tmp2; tmp3];
save  example96 X
```

We show the side-by-side boxplots in Figure 9.8, where we used the MATLAB Statistics Toolbox **boxplot** function with long whiskers so no outliers will be shown.

```
% This is from the Statistics Toolbox:
figure,boxplot(X,0,[],1,10)
```

We can gain more insights into the distributions by looking at the histplots. The same function called **boxp** mentioned earlier includes this capability; its other functionality will be explored in the exercises.

```
% We can get the histplot. This function is
% included with the text.
boxp(X,'hp')
```

This plot is shown in Figure 9.9, where differences are now apparent. The histplot functionality provided with this text uses a kernel probability density estimate [Scott, 1992; Martinez & Martinez, 2002] for the density at the quartiles. Now, we show the box-percentile plot, which we have coded in the function **boxprct**. This function will construct both constant width boxes, as well as ones with variable width. See the **help** on the function for more information on its capabilities. The plain box-percentile plots are created with this syntax:

```
% Let's see what these look like using the
% box-percentile plot.
boxprct(X)
```

This plot is shown in Figure 9.10. Note that the sides of the plots now give us more information and insights into the distributions, and that the differences between them are more apparent. In a box-percentile plot, outliers like we

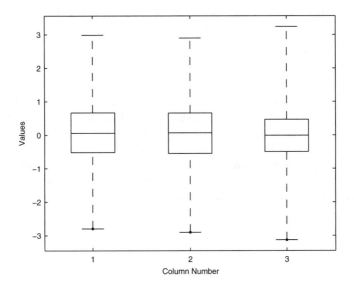

FIGURE 9.8
This shows the boxplots for the generated data in Example 9.6. The first one is for a standard normal distribution. The second is uniform with outliers at the extremes of –3 and 3. The third is from a trimodal distribution. Notice that these distributions do not seem very different based on these boxplots.

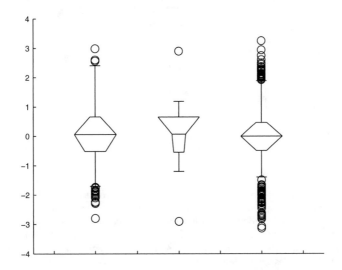

FIGURE 9.9
This is the histplot version of the data in Example 9.6. We now see some of the differences in the distributions.

have in the second distribution will appear as long, skinny lines. To get a variable width box-percentile plot (the maximum width is proportional to the square root of *n*), use the syntax

```
boxprct(X,'vw')
```

❑

In our opinion, one of the problems with the histplot is its dependence on the estimated density. This density estimate is highly dependent on the bin width (or window width in the case of kernel estimation), so the histplots might be very different as these change. The nice thing about the box-percentile plot is that we gain a better understanding of the distribution, and we do not have to arbitrarily set parameters such as the limits to determine outliers or the bin widths to find the density.

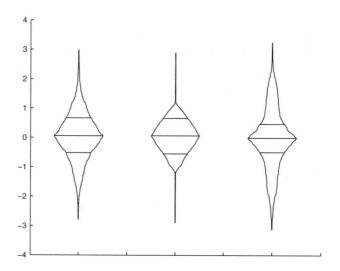

FIGURE 9.10
This is the box-percentile plot of the simulated data in Example 9.6. The plot for the normal distribution shows a single mode at the median, symmetry about the median, and concave sides. The uniform distribution has outliers at both ends, which are shown as long, thin lines. The center part of the plot is a diamond shape, since the percentile plot of a uniform distribution is linear. Although somewhat hard to see, we have several modes in the last one, indicated by valleys (with few observations) and peaks on the sides.

9.3 Quantile Plots

As an alternative to the boxplots, we can use quantile-based plots to visually compare the distributions of two samples. These are also appropriate when we want to compare a known theoretical distribution and a sample. In making the comparisons, we might be interested in knowing how they are shifted relative to each other or to check model assumptions, such as normality.

In this section, we discuss several versions of quantile-based plots. These include *probability plots, quantile-quantile plots* (sometimes called *q-q plots*), and *quantile plots*. The probability plot has historically been used to compare sample quantiles with the quantiles from a known theoretical distribution, such as normal, exponential, etc. Typically, a q-q plot is used to determine whether two random samples were generated by the same distribution. The q-q plot can also be used to compare a random sample with a theoretical distribution by generating a sample from the theoretical distribution as the second sample. Finally, we have the quantile plot that conveys information about the sample quantiles.

9.3.1 Probability Plots

A probability plot is one where the theoretical quantiles are plotted against the ordered data, i.e., the sample quantiles. The main purpose is to visually determine whether or not the data could have been generated from the given theoretical distribution. If the sample distribution is similar to the theoretical one, then we would expect the relationship to follow an approximate straight line. Departures from a linear relationship are an indication that the distributions are different.

To get this display, we plot the $x_{(i)}$ on the vertical axis, and on the other axis we plot

$$F^{-1}\left(\frac{i - 0.5}{n}\right), \tag{9.9}$$

where $F^{-1}(\cdot)$ denotes the inverse of the cumulative distribution function for the hypothesized distribution. If the sample arises from the same distribution represented by Equation 9.9, then the theoretical quantiles and the sample quantiles should fall approximately on a straight line. As we discussed before, the 0.5 in the above argument can be different [Cleveland, 1993]. For example, we could use $i/(n + 1)$. See Kimball [1960] for other options. A well-known example of a probability plot is the *normal probability plot*, where the theoretical quantiles from the normal distribution are used.

The MATLAB Statistics Toolbox has two functions for obtaining probability plots. One is called **normplot** to assess the assumption that a data set comes from a normal distribution. There is also a function for constructing a probability plot that compares a data set to the Weibull distribution. This is called **weibplot**. Probability plots for other theoretical distributions can be obtained using the MATLAB code given below, substituting the appropriate function to get the theoretical quantiles.

Version 5 of the Statistics Toolbox has several new functions that can be used to explore distribution shapes. One is the **probplot** function. The default for this function is to construct a normal probability plot with a reference line. However, this can be changed via an input argument for **'distname'** that specifies the desired distribution. There is also a new GUI tool for fitting distributions. To start the tool, type **dfittool** at the command line. This tool allows you to load data from the workspace, fit distributions to the data, plot the distributions, and manage/evaluate different fits.

Example 9.7

This example illustrates how you can display a probability plot in MATLAB, where we return to the **galaxy** data. From previous plots, these data look approximately normally distributed, so we will check that assumption using the probability plot. First, we get the sorted sample, and then we obtain the corresponding theoretical quantiles for the normal distribution. The resulting quantile plot is shown in Figure 9.11.

```
load galaxy
% We will use the EastWest data again.
x = sort(EastWest);
n = length(x);
% Get the probabilities.
prob = ((1:n)-0.5)/n;
% Now get the theoretical quantiles for
% a normal distribution.
qp = norminv(prob,0,1);
% Now plot theoretical quantiles versus
% the sorted data.
plot(qp,x,'.')
ylabel('Sorted Data')
xlabel('Standard Normal Quantiles')
```

We see in the plot that there is slight curvature at the ends, indicating some departure from normality. However, these plots are exploratory only, and the results are subjective. Data analysts should use statistical inference methods (e.g., goodness-of-fit tests) to assess the significance of any departure from the theoretical distribution.
❑

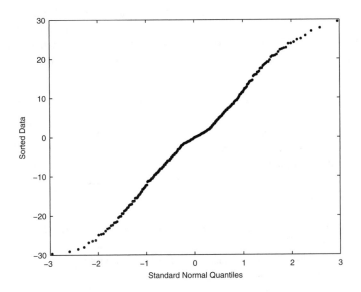

FIGURE 9.11
This is a probability plot for the **EastWest** variable in the **galaxy** data set. The curvature indicates that the data are not exactly normally distributed.

9.3.2 Quantile-quantile Plot

The q-q plot was originally proposed by Wilk and Gnanadesikan [1968] to visually compare two distributions by graphing the quantiles of one versus the quantiles of the other. Either or both of these distributions may be empirical or theoretical. Thus, the probability plot is a special case of the q-q plot.

Say we have two data sets consisting of univariate measurements. We denote the order statistics for the first data set by

$$x_{(1)}, x_{(2)}, \dots, x_{(n)}.$$

Let the order statistics for the second data set be

$$y_{(1)}, y_{(2)}, \dots, y_{(m)},$$

where without loss of generality, $m \leq n$.

In the next example, we show how to construct a q-q plot where the sizes of the data sets are equal, so $m = n$. In this case, we simply plot the sample quantiles of one data set versus the other data set as points.

Example 9.8

We will generate two sets of normal random variables and construct a q-q plot. Constructing a q-q plot for random samples from different distributions and different sample sizes will be covered in the next example. The first simulated data set is standard normal; the second one has a mean of 1 and a standard deviation of 0.75.

```
% Generate the samples - same size.
x = randn(1,300);
% Make the next one a different mean
% and standard deviation.
y = randn(1,300)*.75 + 1;
% Find the order statistics - sort them.
xs = sort(x);
ys = sort(y);
% Construct the q-q plot - do a scatterplot.
plot(xs, ys, '.')
xlabel('Standard Normal - Sorted Data')
ylabel('Normal - Sorted Data')
title('Q-Q Plot')
```

The q-q plot is shown in Figure 9.12. The data appear to be from the same family of distributions, since the relationship between them is approximately linear.
❑

We now look at the case where the sample sizes are not equal and $m < n$. To obtain the q-q plot, we graph the $y_{(i)}$, $i = 1, \ldots, m$ against the $(i - 0.5)/m$ quantile of the other data set. The $(i - 0.5)/m$ quantiles of the x data are usually obtained via interpolation.

Users should be aware that q-q plots provide only a rough idea of how similar the distribution is between two random samples. If the sample sizes are small, then a lot of variation is expected, so comparisons might be suspect. To help aid the visual comparison, some q-q plots include a reference line. These are lines that are estimated using the first and third quartiles of each data set and extending the line to cover the range of the data. The MATLAB Statistics Toolbox provides a function called **qqplot** that displays this type of plot. We show below how to add the reference line.

Example 9.9

This example shows how to do a q-q plot when the samples do not have the same number of points. We use the function provided with this book called **quantileseda**[1] to get the required sample quantiles from the data set that has the larger sample size. We then plot these versus the order statistics of the

---

[1] The Statistics Toolbox has similar functions called **quantile** and **prctile** to calculate the sample quantiles and percentiles.

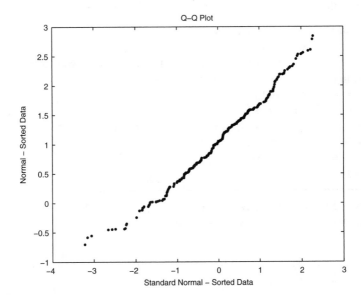

FIGURE 9.12
This is a q-q plot of two generated random samples, each one from a normal distribution. We see that the relationship between them is approximately linear, as expected.

other sample. Note that we add a reference line based on the first and third quartiles of each data set, using the function **polyfit**. We first generate the data sets, both from different distributions.

```
% We will generate some samples - one will be
% from the normal distribution, the other will
% be uniform.
n = 100;
m = 75;
x = randn(1,n);
y = rand(1,m);
```

Next we get the order statistics from the *y* values and the corresponding quantiles from the *x* data set.

```
% Sort y; these are the order statistics.
ys = sort(y);
% Now find the associated quantiles using the x.
% Probabilities for quantiles:
p = ((1:m) - 0.5)/m;
% The next function comes with this text.
xs = quantileseda(x,p);
```

Now we can construct the plot, adding a reference line based on the first and third quartiles to help assess the linearity of the relationship.

```
% Construct the plot.
plot(xs,ys,'.')
% Get the reference line. Use the 1st and 3rd
% quartiles of each set to get a line.
qy = quartiles(y);
qx = quartiles(x);
[pol, s] = polyfit(qx([1,3]),qy([1,3]),1);
% Add the line to the figure.
yhat = polyval(pol,xs);
hold on
plot(xs,yhat,'k')
xlabel('Sample Quantiles - X'),
ylabel('Sorted Y Values')
hold off
```

We see in Figure 9.13 that the data do not come from the same distribution, since the relationship is not linear.
❏

A major benefit of the quantile-based plots discussed so far is that they do not require the two samples (or the sample and theoretical distribution) to have the same location and scale parameter. If the distributions are the same, but differ in location or scale, then we would still expect them to produce a straight line. This will be explored in the exercises.

9.3.3 Quantile Plot

Cleveland [1993] describes another version of a quantile-based plot. He calls this the quantile plot, where we have the u_i values (Equation 9.7) along the horizontal axis and the ordered data $x_{(i)}$ along the vertical. These ordered pairs are plotted as points, joined by straight lines. Of course, this does *not* have the same interpretation or provide the same information as the probability plot or q-q plot described previously. This quantile plot provides an initial look at the distribution of the data, so we can search for unusual structure or behavior. We also obtain some information about the sample quantiles and their relationship to the rest of the data. This type of display is discussed in the next example.

Example 9.10

We use a data set from Cleveland [1993] that contains the heights (in inches) of singers in the New York Choral Society. The following MATLAB code constructs the quantile plot for the **Tenor_2** data.

```
load singer
```

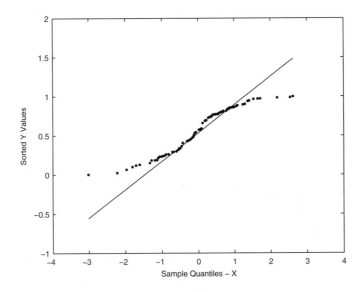

FIGURE 9.13

This shows the q-q plot for Example 9.9. The x values are normally distributed with $n = 100$. The y values are uniformly distributed with $m = 75$. The plot shows that these data do not come from the same distribution.

```
n = length(Tenor_2);
% Sort the data for the y-axis.
ys = sort(Tenor_2);
% Get the associated u values.
u = ((1:n)-0.5)/n;
plot(u,ys,'-o')
xlabel('u_i value')
ylabel('Tenor 2 Height (inches)')
```

We see from the plot shown in Figure 9.14 that the values of the quartiles (and other quantiles of interest) are easy to see.

❑

A word of caution is in order regarding the quantile-based plots discussed in this section. Some of the names given to them are not standard throughout statistics books. For example, the quantile plot is also called the q-q plot in Kotz and Johnson [Vol. 7, p. 233, 1986].

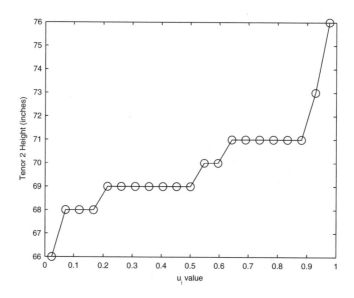

FIGURE 9.14
This is the quantile plot for the **Tenor_2** data, which is part of the **singer** data.

9.4 Bagplots

The *bagplot* is a bivariate box-and-whiskers plot developed by Rousseeuw, Ruts and Tukey [1999]. This is a generalization of the univariate boxplot discussed previously. It uses the idea of the location depth of an observation with respect to a bivariate data set. This extends the idea of ranking or ordering univariate data to the bivariate case, so we can find quantities analogous to the univariate quartiles. The bagplot consists of the following components:

1. A *bag* that contains the inner 50% of the data, similar to the IQR;
2. A cross (or other symbol) indicating the depth median (described shortly);
3. A *fence* to determine potential outliers; and
4. A *loop* that indicates the points that are between the bag and the fence.

We need to define several concepts to construct the bagplot. First, we have the notion of the *halfspace location depth* introduced by Tukey [1975]. The halfspace location depth of a point θ relative to a bivariate data set is given by the smallest number of data points contained in a closed half-plane where the boundary line passes through θ. Next, we have the ***depth region*** D_k, which is the set of all θ with halfspace location depth greater than or equal to k. Donoho and Gasko [1992] define the ***depth median*** of the bivariate point cloud, which in most cases is the center of gravity of the deepest region. Rousseeuw and Ruts [1996] provide a time-efficient algorithm for finding the location depth and depth regions, as well as an algorithm for calculating the depth median [1998].

The bag is constructed in the following manner. We let $\# D_k$ be the number of data points in the depth region D_k. First, we find the value k for which

$$\# D_k \leq \left\lfloor \frac{n}{2} \right\rfloor < \# D_{k-1},$$

where the notation $\lfloor x \rfloor$ denotes the greatest integer less than or equal to x. Then, the bag is obtained by linearly interpolating between D_k and D_{k-1}, relative to the depth median. The fence is constructed by inflating the bag by some factor relative to the depth median. Any points that are outside the fence are flagged as potential outliers. Finally, the loop is the outer boundary of the bagplot; i.e., it is the convex hull of the bag and the nonoutlying points.

Example 9.11

Rousseuw, Ruts and Tukey provide Fortran and MATLAB code to construct the bagplot; see Appendix B for download information. We compiled the code and provide the executable file **bagmat.exe**. This *must* be executed using your data before you call the **bagplot** function from MATLAB. First, your bivariate data has to be saved in the current directory and in ascii row by column format. You can invoke **bagmat.exe** from the MATLAB command window or from a Microsoft CMD window. We show below how to use the MATLAB command line option. We use the **environmental** data set for this example, where the two variables of interest are temperature and ozone.

```
% First we load up some data to be used in the plots.
load environmental
% Now we put the data set together.
data = [Temperature,Ozone];
% Save the data in ascii format.
save data -ascii
```

Note that we have $n = 111$ observations in this data set. Now we invoke the **bagmat.exe** program from the command line using:

```
! bagmat
```

You will see a new blank line at the command window, but no text. Type in the name of the data file and hit return. You will see another blank line; now type in the number of observations and hit return. This is what you see in the command window after following these steps:

```
data
111
PLEASE GIVE THE NAME OF THE DATA FILE
PLEASE GIVE THE NUMBER OF DATA POINTS
```

After the program executes, it will write three text files to your current directory: **interpol.dat**, **datatyp.dat**, and **tukmed.dat**. These three files are required for the **bagplot** function. To get the basic bagplot with all observations plotted, shading for the bag and loop, and no plotting of the fence, just use

```
bagplot
```

at the command line. This plot is shown in Figure 9.15.
❑

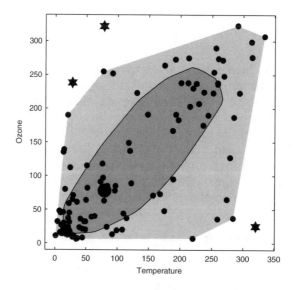

FIGURE 9.15

This is a bagplot showing two variables in the **environmental** data set. Note that the MATLAB version of the bagplot shows the depth median with a large filled circle.

9.5 Summary and Further Reading

We presented several methods for visualizing the shapes of distributions for continuous random variables. The easiest and most intuitive methods to use are the histograms and boxplots. We discussed several variations of these plots. For histograms, we presented frequency, relative frequency, density, and 2-D histograms. Enhancements to the boxplot included histplots, box-percentile plots, variable-width boxplots, and others. We concluded the chapter with quantile-based plots and a generalization of the univariate boxplot to 2-D.

There is an extensive literature on probability density estimation, both univariate and multivariate. The best comprehensive book in this area is Scott [1992]. He discusses histograms, frequency polygons, kernel density estimation, and the average shifted histograms. He covers the theoretical foundation of these methods, how to choose the smoothing parameters (e.g., bin widths), and many practical examples. For a MATLAB perspective on these methods, please see Martinez and Martinez [2002]. The book called *Visualizing Data* by William Cleveland [1993] is an excellent resource on many aspects of data visualization, including the quantile-based plots discussed in this chapter.

We only covered the quantile-based plots for continuous data. However, versions of these are also available for discrete data, such as binomial or Poisson. MATLAB implementations of quantile-based plots for discrete distributions can be found in Martinez and Martinez [2002]. Another resource for the visualization of categorical data is Friendly [2000], where SAS software is used for implementation of the ideas. Finally, we recommend Hoaglin and Tukey [1985] for a nice summary of various methods for checking the shape of discrete distributions.

Exercises

9.1 Repeat Example 9.1 using 5 and 50 bins. Compare these results with what we had in Figure 9.1.

9.2 Repeat Example 9.1 using the **forearm** data.

9.3 Apply the various bin width rules to the **forearm** and to the **galaxy** data. Discuss the results.

9.4 The histogram methods presented in the chapter call the **hist** function with the desired number of bins. Do a **help** on **hist** or **histc** to see how to set the bin centers instead. Use this option to construct histograms with a specified bin width and origin.

9.5   Using the code from problem 9.4, show how changing the bin starting point affects histograms of the **galaxy** data.

9.6   Using the data in Example 9.2, construct a normal curve, using the sample mean and standard deviation. Superimpose this over the histogram and analyze your results.

9.7   Plot the histogram in Example 9.3 using the **surf** plot, using this code:

```
% Plot as a surface plot.
% Get some axes that make sense.
[XX,YY]=...
meshgrid(linspace(min(x(:,1)),max(x(:,1)),nb1),...
linspace(min(x(:,2)),max(x(:,2)),nb2));
% Z is the height of the bins in Example 9.3.
surf(XX,YY,Z)
```

9.8   Use a boxplot and a histogram to verify that the quartiles in Example 9.4 for the **geyser** data make sense.

9.9   Generate some standard normal data, with sample sizes $n = 30$, $n = 50$, and $n = 100$. Use the function **boxp** to first get a set of plain boxplots and then use it to get variable width boxplots. The following code might help:

```
% Generate some standard normal data with
% different sample sizes. Put into a cell array.
X{1} = randn(30,1);
X{2} = randn(50,1);
X{3} = randn(100,1);
% First construct the plain boxplot.
boxp(X)
% Next we get the boxplot with variable
% widths.
boxp(X,'vw')
```

9.10   Generate some random data. Investigate the **histfit** function in the Statistics Toolbox.

9.11   Show that the area represented by the bars in the density histogram sums to one.

9.12   Load the data used in Example 9.6 (**load example96**). Construct histograms (use 12 bins) of the columns of **X**. Compare with the boxplots and box-percentiles plots.

9.13   Type **help boxplot** at the MATLAB command line to learn more about this Statistics Toolbox Function. Do side-by-side boxplots of the **oronsay** data set. The length of the whisker is easily adjusted using optional input arguments to **boxplot**. Try various values for this option.

9.14   Explore the **'notches'** option of the **boxplot** function. Generate bivariate normal data, where the columns have the same means. Construct notched boxplots with this matrix. Now generate bivariate normal data where the columns have very different means. Construct notched boxplots with this matrix. Discuss your results.

9.15 Apply the **boxplot** with notches to the **oronsay** data and discuss the results.

9.16 Generate two data sets, each from a normal distribution with different location and scale parameters. Construct a q-q plot (see Example 9.9) and discuss your results.

9.17 Reconstruct Figure 9.2 using the density histogram. Superimpose the normal curve over this one. Discuss the difference between this and Figure 9.2.

9.18 Construct a bagplot using the BPM data from Chapter 7, Example 7.8. Compare with the polar smoothing.

9.19 A *rootogram* is a histogram where the heights of the bins correspond to the square root of the frequency. Write a MATLAB function that will construct this type of plot. Use it on the **galaxy** data.

9.20 Repeat Example 9.6 using variable width box-percentile plots.

9.21 Use the MATLAB **boxplot** function to get side-by-side boxplots of the following data sets and discuss the results. Construct boxplots with and without notches.

a. **skulls**

b. **sparrow**

c. **pollen**

d. BPM data sets (after using ISOMAP)

e. gene expression data sets

f. **spam**

g. **iris**

h. **software**

9.22 Apply some of the other types of boxplots to the data in problem 9.21.

9.23 Generate uniform random variables (use **rand**) and construct a normal probability plot (or a q-q plot with the other data set generated according to a standard normal). Do the same thing with random variables generated from an exponential distribution (see **exprnd** in the Statistics Toolbox). Discuss your results.

Chapter 10

Multivariate Visualization

In this chapter, we present several methods for visualizing and exploring multivariate data. We have already seen some of these methods in previous chapters. For example, we had grand tours and projection pursuit in Chapter 4, where the dimensionality of the data is first reduced to 2-D and then visualized in scatterplots. The primary focus of this chapter is to look at ways to visualize and explore all of the dimensions in our data at once.

The first thing we cover is glyph plots. Then we present information about scatterplots, both 2-D and 3-D, as well as scatterplot matrices. Next, we talk about some dynamic graphics, such as linking and brushing. These techniques enable us to find connections between points in linked graphs, delete and label points, and highlight subsets of points. We then cover coplots that convey information about conditional dependency between variables. This is followed by dot charts that can be used to visualize summary statistics and other data values. Next, we discuss how to visualize each of our observations as curves via Andrews' plots or as broken line segments in parallel coordinates. We also show how these methods can be combined with the plot matrix concept and the grand tour.

10.1 Glyph Plots

We first briefly discuss some of the multivariate visualization methods that will *not* be covered in detail in this text. Most of these are suitable for small data sets only, so we do not think that they are in keeping with the trend towards analyzing massive, high-dimensional data sets, as seen in most applications of EDA and data mining.

The first method we present is due to Chernoff [1973]. His idea was to represent each observation (with dimensionality $p \leq 18$) by a cartoon face. Each feature of the face, such as length of nose, mouth curvature, eyebrow shape, size of eyes, etc., would correspond to a value of the variable. This technique is useful for understanding the overall regularities and anomalies in the data, but it has several disadvantages. The main disadvantage is the lack of quantitative visualization of the variables; we just get a qualitative understanding of the values and trends. Another problem is the subjective assignment of features to the variables. In other words, assigning a different variable to the eyebrows and other facial features can have a significant effect on the final shape of the face. We show an example of ***Chernoff faces*** for the **cereal** data (described in Appendix C) in Figure 10.1.

Star diagrams [Fienberg 1979] are a similar plot, in that we have one glyph or star for each observation, so they suffer from the same restrictions as the faces regarding sample size and dimensionality. Each observed data point in the sample is plotted as a star, with the value of each measurement shown as a radial line from a common center point. Thus, each measured value for an observation is plotted as a spoke that is proportional to the size of the measured variable with the ends of the spokes connected with line segments to form a star. We show the star plot for the same cereal data in Figure 10.2.

The Statistics Toolbox (Version 5) has a function called **glyphplot** that will construct either Chernoff faces or star diagrams for each observation. The use of this function will be explored in the exercises.

Other glyphs and similar plots have been described in the literature [Kleiner and Hartigan, 1981; du Toit, Steyn and Stumpf, 1986], and most of them suffer from the same drawbacks. These include star-like diagrams, where the rays emanate from a circle, and the end points are not connected. We also have profile plots, where each observation is rendered as a bar chart, with the height of the bar indicating the value of the variable. Another possibility is to represent each observation by a box, where the height, length, and width correspond to the variables.

10.2 Scatterplots

We have already introduced the reader to ***scatterplots*** and ***scatterplot matrices*** in previous chapters, but we now examine these methods in more detail, especially how to construct them in MATLAB. We also present an enhanced scatterplot based on ***hexagonal binning*** that is suitable for massive data sets.

10.2.1 2-D and 3-D Scatterplots

The scatterplot is a visualization technique that enjoys widespread use in data analysis and is a powerful way to convey information about the relationship between two variables. To construct one of these plots in 2-D, we simply plot the individual (x_i, y_i) pairs as points or some other symbol. For 3-D scatterplots, we add the third dimension and plot the (x_i, y_i, z_i) triplets as points.

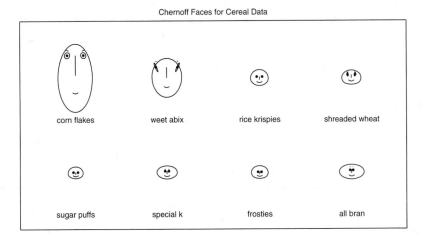

Chernoff Faces for Cereal Data

FIGURE 10.1
This shows the Chernoff faces for the **cereal** data, where we have 8 observations and 11 variables. The shape and size of various facial features (head, eyes, brows, mouth, etc.) correspond to the values of the variables. The variables represent the percent agreement to statements about the cereal. The statements are: comes back to, tastes nice, popular with all the family, very easy to digest, nourishing, natural flavor, reasonably priced, a lot of food value, stays crispy in milk, helps to keep you fit, fun for children to eat.

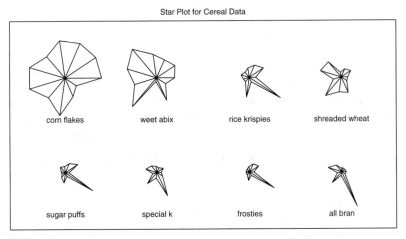

Star Plot for Cereal Data

FIGURE 10.2
This shows the star plots for the same **cereal** data. There is one ray for each variable. The length of the ray indicates the value of the attributed.

The main MATLAB package has several ways we can construct 2-D and 3-D scatterplots, as shown in Example 10.1. The Statistics Toolbox also has a function to create 2-D scatterplots called **gscatter** that will construct a scatterplot, where different plotting symbols are used for each group. However, as we will see in Example 10.1 and the exercises, similar results can be obtained using the **scatter** and **plot** functions.

Example 10.1

In this example, we illustrate the 2-D and 3-D scatterplot functions called **scatter** and **scatter3**. Equivalent scatterplots can also be constructed using the basic **plot** and **plot3** functions, which will be explored in the exercises. Since a scatterplot is generally for 2-D or 3-D, we need to extract a subset of the variables in the **oronsay** data. So we've chosen variables 8, 9, and 10: 0.18-0.25mm, 0.125-0.18mm, and 0.09-0.125mm. We first use variables 8 and 9 to construct a basic 2-D scatterplot.

```
% First load up the data and get the
% variables of interest.
load oronsay
% Use the oronsay data set. Just plot two
% of the variables. Now for the plot:
scatter(oronsay(:,8),oronsay(:,9))
xlabel(labcol{8})
ylabel(labcol{9})
```

This plot is shown in Figure 10.3 (top). The basic syntax for the scatter function is

```
scatter(X,Y,S,C,M)
```

where **X** and **Y** are the data vectors to be plotted, and the other arguments are optional. **S** can be either a scalar or a vector indicating the area (in units of points-squared) of each marker. **M** is an alternative marker (default is the circle), and **C** is a vector of colors. Next we show how to use the color vector to plot the observations in different colors, according to their midden group membership. See color plate Figure 10.3 (top) for the resulting plot.

```
% If we want to use different colors for the groups,
% we can use the following syntax. Note that this
% is not the only way to do the colors.
ind0 = find(midden==0); % Red
ind1 = find(midden==1); % Green
ind2 = find(midden==2); % Blue
% This creates an RGB - 3 column colormap matrix.
C = zeros(length(midden),3);
C(ind0,1) = 1;
C(ind1,2) = 1;
C(ind2,3) = 1;
```

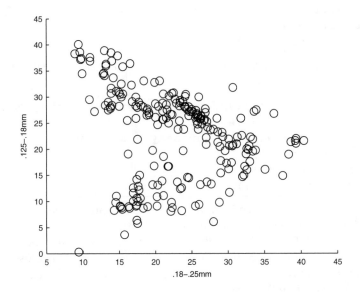

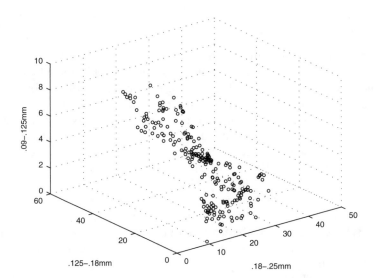

FIGURE 10.3

The top figure shows the 2-D scatterplot for two variables (i.e., columns 8 and 9) of the **oronsay** data set. See the corresponding color plate for the color version, with color indicating the midden class membership. The lower plot shows the 3-D scatterplot for columns 8 through 10. This plot also uses the midden class membership. See the associated color figure following page 144.

```
scatter(oronsay(:,8),oronsay(:,9),5,C)
xlabel(labcol{8})
ylabel(labcol{9})
zlabel(labcol{10})
```

3-D scatterplots can also be very useful, and they are easily created using the **scatter3** function, as shown below.

```
% Now show scatter3 function. Syntax is the same;
% just add third vector.
scatter3(oronsay(:,8),oronsay(:,9),oronsay(:,10),5,C)
xlabel(labcol{8})
ylabel(labcol{9})
zlabel(labcol{10})
```

The 3-D scatterplot is shown in Figure 10.3 (bottom) and in the corresponding color plate. MATLAB has another useful feature in the **Figure Window** toolbar buttons. This is the familiar *rotate* button ⬢ that, when selected, allows the user to click on the 3-D axis and rotate the plot. The user can see the current elevation and azimuth (in degrees) in the lower left corner of the figure window while the axes are being rotated.
❑

10.2.2 Scatterplot Matrices

Scatterplot matrices are suitable for multivariate data, when $p > 2$. They show all possible 2-D scatterplots, where the axis of each plot is given by one of the variables. The scatterplots are then arranged in a matrix-like layout for easy viewing and comprehension. Some implementations of the scatterplot matrix show the plots in the lower triangular portion of the matrix layout only, since showing both is somewhat redundant. However, we feel that showing all of them makes it easier to understand the relationships between the variables. As we see in Example 10.2, the MATLAB functions for scatterplot matrices show all plots.

One of the benefits of a scatterplot matrix is that one can look across a row or column and see the scatterplots of a given variable against all other variables. We can also view the scatterplot matrix as a way of partially linking points in different views, especially when observations of interest are shown with different marker styles (symbols and/or colors).

The main MATLAB package has a function called **plotmatrix** that will produce a scatterplot matrix for a data matrix **X**. The diagonal boxes of the scatterplot matrix contain histograms showing the distribution of each variable (i.e., the column of **X**). Please see the **help** on this function for its other uses. The Statistics Toolbox has an enhanced version called **gplotmatrix**, where one can provide group labels, so observations belonging to different groups are shown with different symbols and colors.

Example 10.2

The **plotmatrix** function will construct a scatterplot matrix when the first argument is a matrix. An alternative syntax allows the user to plot the columns of one matrix against the columns of the other. We use the following commands to construct a scatterplot matrix of the three variables of the **oronsay** data used in the previous example.

```
% Use the same 3 variables in previous example.
X = [oronsay(:,8),oronsay(:,9),oronsay(:,10)];
plotmatrix(X,'.');
% Let's make the symbols slightly smaller.
Hdots = findobj('type','line');
set(Hdots,'markersize',1)
```

We chose these from the scatterplot matrix of the full data set, because they seemed to show some interesting structure. It is also interesting to note the histograms of the variables, as they provide information about their distribution.

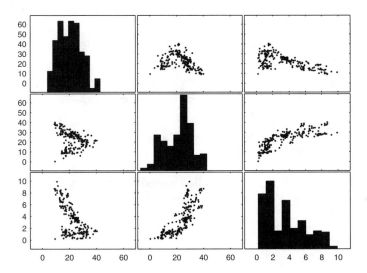

FIGURE 10.4

This is the scatterplot matrix for columns 8 through 10 of the **oronsay** data. The first row of the plots shows us column 8 plotted against column 9 and then against column 10. We have similar plots for the other two rows.

10.2.3 Scatterplots with Hexagonal Binning

Carr, et. al [1987] introduced several scatterplot matrix methods for situations where the size of the data set n is large. In our view, n is large when the scatterplot can have a lot of overplotting, so that individual points are difficult to see. When there is significant overplotting, then it is often more informative to convey an idea of the density of the points, rather than the observations alone. Thus, Carr, et. al suggest displaying bivariate densities represented by gray scale or symbol area instead of individual observations.

To do this, we first need to estimate the density in our data. We have already introduced this issue in Chapter 9, where we looked at estimating the bivariate density using a histogram with bins that are rectangles or squares. Recall, also, that we used vertical bars to represent the value of the density, rather than a scatterplot. In this chapter, we are going to use bins that are hexagons instead of rectangles, and the graphic used to represent data density will be symbol size, as well as color.

Carr, et. al recommended the hexagon as an alternative to using the square as a symbol, because the square tends to create bins that appear stretched out in the vertical or horizontal directions. The procedure we use for hexagonal binning is outlined below.

Procedure - Hexagonal Binning for Scatterplots

1. Find the length r of the side of the hexagon bins based on a given number of bins.
2. Obtain a set of hexagonal bins over the range of the data.
3. Bin the data.
4. Scale the hexagons with *nonzero* bin counts, such that the bin with maximum frequency has sides with length r and the smallest frequency (*nonzero*) has length $0.1r$.
5. Display a hexagon at the center of each bin, where the sides correspond to the lengths found in step 4.

We provide a MATLAB function called **hexplot** and show how it is used in the next example.

Example 10.3
The basic syntax for **hexplot** is:

```
hexplot(X,nbin,flag)
```

The first two arguments *must* be provided. **X** is a matrix with n rows and 2 columns, and **nbin** is the approximate number of bins for the dimension with the larger range. We use the **oronsay** data from the previous examples to illustrate the function.

```
X = [oronsay(:,8),oronsay(:,9)];
% Construct a hexagon scatterplot with
% 15 bins along the longer dimension.
```

```
hexplot(X,15);
```

This plot is shown in Figure 10.5 (top). Since the bins and resulting plot depend on the number of bins, it is useful to construct other scatterplots with different **nbin** values to see if other interesting density structure becomes apparent. The optional input argument **flag** (this can be any value) produces a scatterplot where the color of the hexagon symbols corresponds to the probability density at that bin. The probability density is found in a similar manner to a bivariate histogram with rectangular bins, except that we now normalize using the area of the hexagonal bin. An example of this is shown in Figure 10.5 (bottom), and it was produced with the following MATLAB statements:

```
hexplot(X,15,1)
colormap(gray)
```

❑

10.3 Dynamic Graphics

We now present some methods for dynamic graphics. These allow us to interact with our plots to uncover structure, remove outliers, locate groups, etc. Specifically, we cover labeling observations of interest, deleting points, finding and displaying subsets of our data, linking and brushing. As with the tour methods discussed in Chapter 4, it is difficult to convey these ideas via static graphs on the pages of this book, so the reader is encouraged to try these methods to understand the techniques.

10.3.1 Identification of Data

One can identify points in a plot in several ways. First, we can add labels to certain data points of interest or we can highlight observations by using some other plotting symbol or color [Becker, Cleveland and Wilks, 1987].

We label points in our plot by adding some text that identifies the observation. Showing all labels is often not possible because overplotting occurs, so nothing can be distinguished. Thus, a way to selectively add labels to our plot is a useful capability. One can look at accomplishing this in two ways. The user can click on a point (or select several points) on the plot, and the labels are added. Or, the user can select an observation (or several) from a list, at which point the observations are labeled. We explore both of these methods in the next example.

We might also have the need to interactively delete points because they could be outliers, and they make it difficult to view other observations. For example, we might have an extreme value that pushes all of the remaining

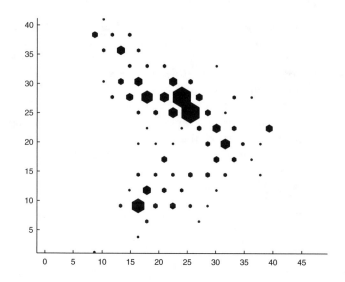

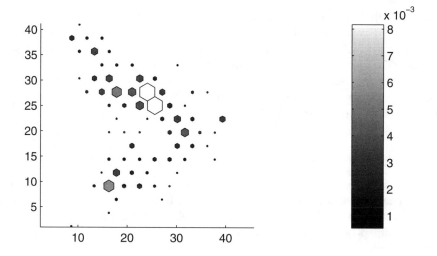

FIGURE 10.5

The top plot is a scatterplot with hexagonal bins. This would be an alternative to the plot shown in Figure 10.3. The bottom plot is for the same data, but the color of the symbols encodes the value of the probability density at that bin. The density is obtained in a manner similar to the bivariate histogram. See the associated color figure following page 144.

data into one small region of the plot, making it difficult to resolve the bulk of the observations.

Instead of labeling individual cases or observations, we might have the need to highlight the points that belong to some subset of our data. For example, with the document clustering data, we could view the scatterplot of each topic separately or highlight the points in one class with different color and symbol type. These options allow direct comparison on the same scale, but overlap and overplotting can hinder our understanding. We can also plot these groups separately in panels, as we do in scatterplot matrices.

A dynamic method for visualizing subsets of the data that is reminiscent of data tours is called *alternagraphics* [Tukey, 1973]. This type of tour cycles through the subsets of the data (e.g., classes), showing each one separately in its own scatterplot. The cycle pauses at each plot for some fixed time step, so it is important that all plots be on a common scale for ease of comparison. Alternatively, one could show all of the data throughout the cycle, and highlight each subset at each step using different color and/or symbol. We leave the implementation of this idea as an exercise to the reader.

Example 10.4

The MATLAB Statistics Toolbox provides a function for labeling data on a plot called **gname**. The user can invoke this function using an input argument containing strings for the case names. This must be in the form of a string matrix. An alternative syntax is to call **gname** without any input argument, in which case observations are labeled with their case number. Once the function is called, the figure window becomes active, and a set of crosshairs appears. The user can click near the observation to be labeled, and the label will appear. This continues until the return key is pressed. Instead of using the crosshairs on individual points, one can use a bounding box (click on the plot, hold the left button down and drag), and enclosed points are identified. We use the animal data to show how to use **gname**. This data set contains the brain weights and body weights of several animal types [Crile and Quiring, 1940].

```
load animal
% Plot BrainWeight against BodyWeight.
scatter(log(BodyWeight),log(BrainWeight))
xlabel('Log Body Weight (log grams)')
ylabel('Log Brain Weight (log grams)')
% Change the axis to provide more room.
axis([0 20 -4 11])
% Need to convert animal names to string matrix.
% Input argument must be string matrix.
cases = char(AnimalName);
gname(cases)
```

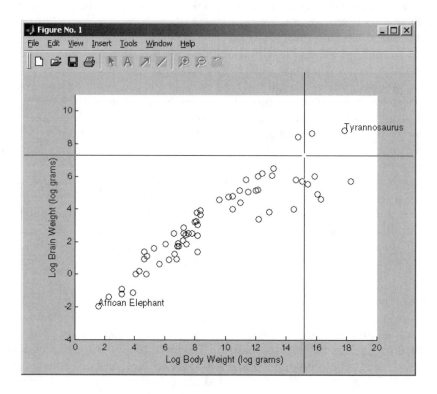

FIGURE 10.6

We constructed a scatterplot of the **animal** data. We then call the **gname** function using the animal names (converted to a string matrix). When **gname** is invoked, the figure window becomes active and a crosshair appears. The user clicks near a point and the observation is labeled. Instead of using a crosshair, one can use a bounding box to label enclosed points. See Example 10.4 for the MATLAB commands.

The scatterplot with two labeled observations is shown in Figure 10.6. We provide an alternative function to **gname** that allows one to perform many of the identification operations discussed previously. The function is called **scattergui**, and it requires two input arguments. The first is the $n \times 2$ matrix to be plotted; the default symbol is blue dots. The user can right click somewhere in the axes (but not on a point) to bring up the shortcut menu. Several options are available, such as selecting subsets of data or cases for identification and deleting points. See the **help** on this function for more information on its use. The MATLAB code given below shows how to call this function.

```
% Now let's look at scattergui using the BPM data.
load L1bpm
% Reduce the dimensionality using Isomap.
options.dims = 1:10;     % These are for ISOMAP.
```

```
options.display = 0;
[Yiso, Riso, Eiso] = isomap(L1bpm, 'k', 7, options);
% Get the data out.
X = Yiso.coords{2}';
scattergui(X,classlab)
% Right click on the axes, and a list box comes up.
% Select one of the classes to highlight.
```

When the user clicks on the **Select Class** menu option, a list box comes up with the various classes available for highlighting. We chose class 6, and the result is shown in Figure 10.7, where we see the class 6 data displayed as red x's.
❏

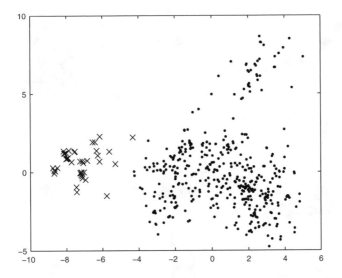

FIGURE 10.7
This shows the BPM data reduced to 2-D using ISOMAP and displayed using **scattergui**. We selected class 6 for highlighting as red x's via the shortcut menu (available by right-clicking inside the axes). See the associated color figure following page 144.

While we illustrated and implemented these ideas using a 2-D scatterplot, they carry over easily into other types of graphs, such as parallel coordinates or Andrews' curves (Section 10.6).

10.3.2 Linking

The idea behind *linking* is to make connections between multiple views of our data with the goal of providing information about the data as a whole. An early idea for linking observations was proposed by Diaconis and

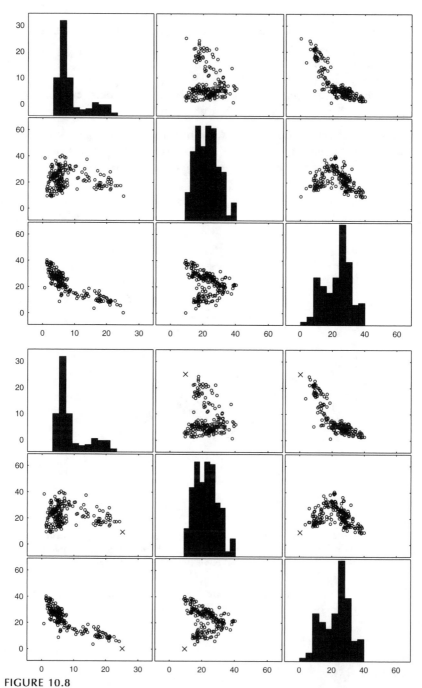

FIGURE 10.8
We have a default scatterplot matrix for three variables of the **oronsay** data (columns 7 through 9) in the top figure. We link observation 71 in all panels and display it using the 'x' symbol. This is shown in the bottom figure.

Friedman [1980]. They proposed drawing a line connecting the same observation in two scatterplots. Another idea is one we've seen before: use different colors and/or symbols for the same observations in all of the scatterplot panels. Finally, we could manually select points by drawing a polygon around the desired subset of points and subsequently highlighting these in all scatterplots.

We have already seen one way of linking views via the grand tour or a partial linking of scatterplot graphs in a scatterplot matrix. While we can apply these ideas to linking observations in all open plots (scatterplots, histograms, dendrograms, etc.), we restrict our attention to linking observations in panels of a scatterplot matrix.

Example 10.5

In this example, we show how to do linking in a brute-force way (i.e., non-interactive) using the **plotmatrix** function. We return to the **oronsay** data, but we use different variables. This creates an initial scatterplot matrix, as we've seen before.

```
load oronsay
X = [oronsay(:,7),oronsay(:,8),oronsay(:,9),];
% Get the initial plot.
% We need some of the handles to the subplots.
[H,AX,BigAx,P,PAx] = plotmatrix(X,'o');
Hdots = findobj('type','line');
set(Hdots,'markersize',3)
```

We called the **plotmatrix** function with output arguments that contain some handle information that we can use next to display linked points with different symbols. The scatterplot matrix is shown in Figure 10.8 (top). The following code shows how to highlight observation 71 in all scatterplots.

```
% The matrix AX contains the handles to the axes.
% Loop through these and change observation 71 to a
% different marker.
% Get the point that will be linked.
linkpt = X(71,:);
% Remove it from the other matrix.
X(71,:) = [];
% Now change in all of the plots.
for i = 1:2
    for j = (i+1):3
        % Change that observation to 'x'.
        axes(AX(i,j))
        cla, axis manual
        line('xdata',linkpt(j),'ydata',linkpt(i),...
            'markersize',5,'marker','x')
        line('xdata',X(:,j),'ydata',X(:,i),...
```

```
                'markersize',3,'marker','o',...
                'linestyle','none')
        axes(AX(j,i))
        cla, axis manual
        line('xdata',linkpt(i),'ydata',linkpt(j),...
            'markersize',5,'marker','x')
        line('xdata',X(:,i),'ydata',X(:,j),...
            'markersize',3,'marker','o',...
            'linestyle','none')
    end
end
```

This plot is given in Figure 10.8 (bottom). This plot would be much easier to do using the **gplotmatrix** function in the Statistics Toolbox, but we thought showing it this way would help motivate the need for interactive graphical techniques. The next example presents a function that allows one to interactively highlight points in a scatterplot panel and link them to the rest of the plots by plotting in a different color.
❑

10.3.3 Brushing

Brushing was first described by Becker and Cleveland [1987] in the context of scatterplots, and it encompassed a set of dynamic graphical methods for visualizing and understanding multivariate data. One of its main uses is to interactively link data between scatterplots of the data. A brush consists of a square or rectangle created in one plot. The brush can be of default size and shape (rectangle or square), or it can be constructed interactively (e.g., creating a bounding box with the mouse). The brush is under the control of the user; the user can click on the brush and drag it within the plot.

Several brushing operations are described by Becker and Cleveland. These include highlight, delete, and label. When the user drags the brush over observations in the plot, then the operation is carried out on corresponding points in all scatterplots. The outcome of the delete and label operations is obvious. In the highlight mode, brushed observations are shown with a different symbol and/or a different color.

Three brushing modes when using the highlighting operation are also available. The first is the *transient* paint mode. In this case, only those points that are in the current brush are highlighted. As observations move outside the scope of the brush, they are no longer highlighted. The *lasting* mode is the opposite; once points are brushed, they stay brushed. Finally, we can use the *undo* mode to remove the highlighting.

Example 10.6

We wrote a function called **brushscatter** that implements the highlighting operation and the three modes discussed above. The basic syntax is shown below, where we use the **oronsay** data as in Example 10.2.

```
% Use the same oronsay columns as in Example 10.2
load oronsay
X = [oronsay(:,8),oronsay(:,9),oronsay(:,10)];
% Get the labels for these.
clabs = labcol(8:10);
% Call the function - the labels are optional.
brushscatter(X,clabs)
```

The scatterplot matrix is shown in Figure 10.9, where we see some of the points have been brushed using the brush in the second panel of row one. The brush is in the transient mode, so only the points inside the brush are highlighted in all scatterplots. Note that the axes labels are not used on the scatterplots to maximize the use of the display space. However, we provide the range of the variables in the corners of the diagonal boxes. This is how they were implemented in the early literature. Several options (e.g., three modes, deleting the brush and resetting the plots to their original form) are available by right-clicking on one of the diagonal plots – the ones with the variable names. Brushes can be constructed in any of the scatterplot panels by creating a bounding box in the usual manner. A default brush is not implemented in this function.
❑

10.4 Coplots

As we have seen in earlier chapters, we sometimes need to understand how a response variable depends on one or more predictor variables. We could explore this by estimating a function that represents the relationship and then visualizing it using lines, surfaces, or contours. We will not delve into this option any further in this text. Instead, we present *coplots* for showing slices of relationships for given values of another variable. We look only at the three variable cases (one is conditional). The reader is referred to Cleveland [1993] for an extension to coplots with two conditional variables.

The idea behind coplots is to arrange subplots of one dependent variable against the independent variable. These subplots can be scatterplots, with or without smooths, or some other graphic indicating the relationship between them. Each subplot displays the relationship for a range of data over a given interval of the second variable.

The subplots are called *dependence panels*, and they are arranged in a matrix-like layout. The *given panel* is at the top, and this shows the interval

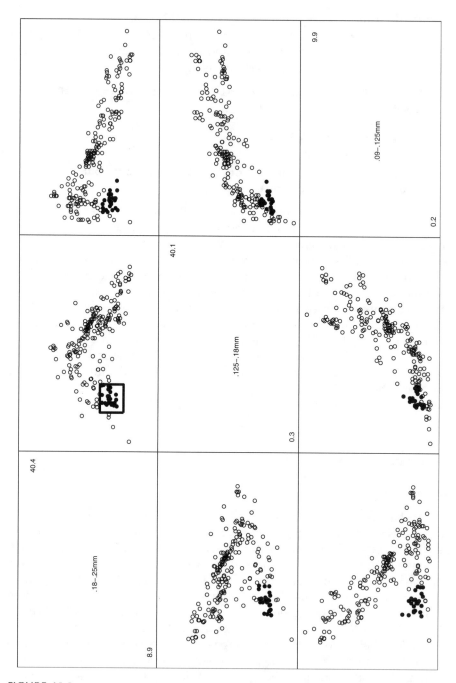

FIGURE 10.9

This is the scatterplot matrix with brushing and linking. This mode is transient, where only points inside the brush are highlighted. Corresponding points are highlighted in all scatterplots. See the associated color figure following page 144.

of values for each subplot. The usual arrangement of the coplots is left to right and bottom to top.[1] Note that the intervals of the given variable have two main properties [Becker and Cleveland, 1991]. First, we want to have approximately the same number of observations in each interval. Second, we want the overlap to be the same in successive intervals. We require the dependence intervals to be large enough so there are enough points for effects to be seen and relationships to be estimated (via smoothing or some other procedure). On the other hand, if the length of the interval is too big, then we might get a false view of the relationship. Cleveland [1993] presents an equal-count algorithm for selecting the intervals, which is used in the **coplot** function illustrated in the next example.

Example 10.7

We turn to Cleveland [1993] and the Data Visualization Toolbox function[2] called **coplot**. We updated the function to make it compatible with later versions of MATLAB. These data contain three variables: abrasion loss, tensile strength, and hardness. Abrasion loss is a response variable, and the others are predictors, and we would like to understand how abrasion loss depends on the factors. The following MATLAB code constructs the coplot in Figure 10.10. Note that the conditioning variable must be in the first column of the input matrix.

```
load abrasion
% Get the data into one matrix.
% We are conditioning on hardness.
X = [hardness(:) tensile(:) abrasion(:)];
labels = {'Hardness'; 'Tensile Strength';...
          'Abrasion Loss'};
% Set up the parameters for the coplot.
% These are the parameters for the intervals.
np = 6;       % Number of given intervals.
overlap = 3/4;  % Amount of interval overlap.
intervalParams = [np overlap];
%  Parameters for loess curve:
alpha = 3/4;
lambda = 1;
robustFlag = 0;
fitParams = [alpha lambda robustFlag];
% Call the function.
coplot(X,labels,intervalParams,fitParams)
```

The coplot is shown in Figure 10.10. A loess smooth is fit to each of the subsets of data, based on the conditioning variable. We see by the curves that

---

[1] This is backwards from MATLAB's usual way of numbering subplots: left to right and top to bottom.

[2] Also see the Data Visualization Toolbox M-file **book_4_3.m**.

most of them have a similar general shape – decreasing left to right with a possible slight increase at the end. However, the loess curve in the top row, third panel shows a different pattern – a slight increase at first, followed by a downward trend. The reader is asked to explore abrasion loss against hardness, with the tensile strength serving as the conditioning variable. We note that other plots could be used in the panels, such as scatterplots alone, histograms, line plots, etc., but these are not implemented at this time.
❑

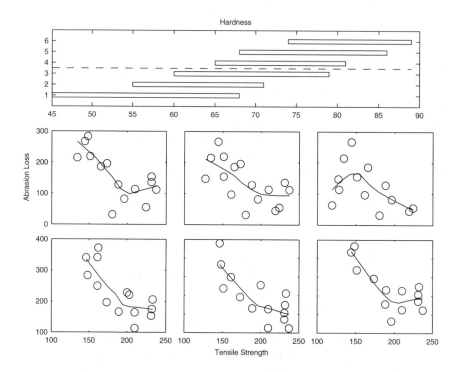

FIGURE 10.10
This is a coplot of abrasion loss against tensile strength, with the hardness as the given variable. The loess curves follow a similar pattern, except for the one in the upper row, third column.

10.5 Dot Charts

A *dot chart* is a visualization method that is somewhat different than others presented in this chapter in that it is typically used with smaller data sets that have labels. Further, it is not used for multivariate data in the sense that we have been looking at so far. However, we think it is a useful way to graphically summarize interesting statistics describing our data, and thus, it is a part of EDA. We first describe the basic dot chart and several variations, such as using scale breaks, error bars, and adding labels for the sample cumulative distribution function. This is followed by a multiway dot chart, where individual dot charts are laid out in panels according to some categorical variable.

10.5.1 Basic Dot Chart

A data analyst could use a dot chart as a replacement for bar charts [Cleveland, 1984]. They are sometimes called *dot plots* [Cleveland, 1993], but should not be confused with the other type of dot plot sometimes seen in introductory statistics books.[3] An example of a dot chart is shown in Figure 10.11. The labels are shown along the left vertical axis, and the value of the datum is given by the placement of a dot (or circle) along the horizontal axis. Dotted lines connecting the dots with the labels help the viewer connect the observation with the label. If there are just a few observations, then the lines can be omitted.

If the data are ordered, then the dots are a visual summary of the sample cumulative distribution function. Cleveland [1984] suggests that this be made explicit by specifying the sample cumulative distribution function on the right vertical axis of the graph. Thus, we can use the following label at the *i*-th order statistic:

$$\frac{(i - 0.5)}{n}.$$

Another useful addition to the dot chart is to convey a sense of the variation via error bars. This is something that is easily shown on dot charts, but is difficult to show on bar charts. As described earlier in dynamic graphics, we might have an outlying observation(s) that makes the remaining values difficult to understand. Before, we suggested just deleting the point, but in some cases, we need to keep all points. So, we can use a break in the scale. Scale breaks are sometimes highlighted via hash marks along the axis, but these are not very noticeable and false impressions of the data can result. Cleveland recommends a full scale break, which is illustrated in the exercises.

Example 10.8

We provide a function for constructing dot charts. In this example, we show how to use some of the options in the basic **dotchart** function. First, we load the **oronsay** data and find the means and the standard deviations. We will plot the means as dots in the dot chart.

---

[3] This type of dot plot has a horizontal line covering the range of the data. A dot is placed above each observation, in a vertical stack. This dot plot is reminiscent of a histogram or a bar chart in its final appearance.

```
load oronsay
% Find the means and standard deviations of
% each column.
mus = mean(oronsay);
stds = std(oronsay);
% Construct the dotchart using lines to the dots.
dotchart(mus,labcol)
% Change the axes limits.
axis([-1 25 0 13])
```

The dot chart is shown in Figure 10.11 (top), where we see the dotted lines extending only to the solid dot representing the mean for that variable. We can add error bars using the following code:

```
% Now try error bar option.
dotchart(mus,labcol,'e',stds)
```

The resulting dot chart is shown in Figure 10.11 (bottom), where we have the error bars extending from +/– one standard deviation either side of the mean. Note that this provides an immediate visual impression of some summary statistics for the variables.
❏

10.5.2 Multiway Dot Chart

With *multiway data*, we have observations that have more than one categorical variable, and for which we have at least one quantitative variable that we would like to explore. The representation of the data can be laid out into panels, where each panel contains a dot chart and each row of the dot chart represents a level. To illustrate this visualization technique, we use an example from Cleveland [1993].

A census of farm animals in 26 countries was conducted in 1987 to study air pollution arising from the feces and urine of livestock [Buijsman, Maas and Asman, 1987]. These countries include those in Europe and the former Soviet Union. A log (base 10) transformation was applied to the data to improve the resolution on the graph.

The multiway dot chart is shown in Figure 10.12, where each of the panels corresponds to livestock type. The quantitative variable associated with these is the number of livestock, for each of the different levels (categorical variable for country). We could also plot these using the countries for each panel and animal type for the levels. This will be explored in the exercises.

Note that the order of the countries (or levels) in Figure 10.12 is not in alphabetical order. It is sometimes more informative to order the data based on some summary statistic. In this case, the median of the five counts was used, and they increase from bottom to top. That is, Albania has the smallest median and Russia, et al. has the largest median. We can also order the panels in a similar manner. The median for horses is the smallest over the five

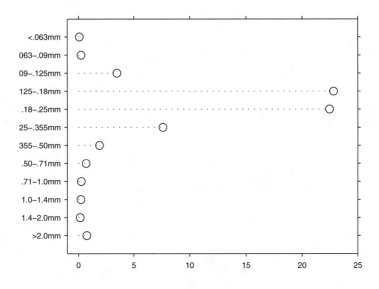

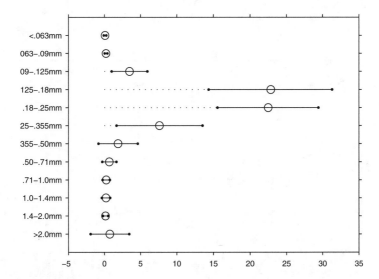

FIGURE 10.11
A regular dot chart is shown in the top panel. The dots represented the average weight for each of the sieve sizes. The dot chart at the bottom shows the same information with error bars added. These bars extend from +/− 1 standard deviation either side of the mean.

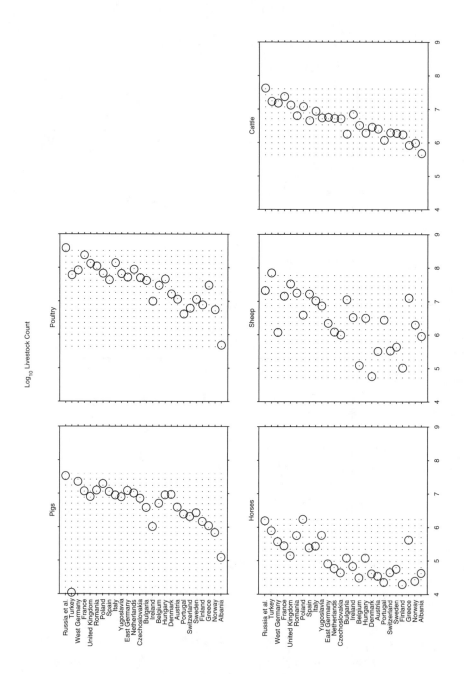

FIGURE 10.12

This is the multiway dot charts for livestock counts for various animals (shown in the panels) and countries (rows of the dot chart). Note that the dotted lines now indicate the range of the data rather than connecting values to the label, as before.

animal types, and the median for poultry is the largest. From these charts, we can obtain an idea of the differences of these values in the various countries. For example, Turkey has very few pigs; poultry seems to be the more numerous type across most of the countries; and the number of cattle seems to be fairly constant among the levels. On the other hand, there is more variability in the counts of horses and sheep, with horses being the least common animal type.

Example 10.9

We provide a function to construct a multiway dot chart called **multiwayplot**. We apply it below to the **oronsay** data, where we use the beach/dune/midden classification for the categorical variable. Thus, we will have three panels showing the average particle weight for each classification and sieve size.

```
load oronsay
% Get the means according to each midden class.
% The beachdune variable contains class
% labels for midden (0), beach (1), and
% dune (2). Get the means for each group.
ind = find(beachdune==0);
middenmus = mean(oronsay(ind,:));
ind = find(beachdune==1);
beachmus = mean(oronsay(ind,:));
ind = find(beachdune==2);
dunemus = mean(oronsay(ind,:));
X = [middenmus(:), beachmus(:), dunemus(:)];
% Get the labels for the groups and axes.
bdlabs = {'Midden'; 'Beach'; 'Dune'};
labx = 'Average Particle Weight';
% Get the location information for the plots.
sublocs{1} = [1,3];
sublocs{2} = [1 2 3];
multiwayplot(X,labcol,labx,bdlabs,sublocs)
```

The plot given in Figure 10.13 shows that the larger sieve sizes (and the two smallest) have approximately the same average weight, while the average weight in the two sieves from 0.125mm to 0.25mm are different among the classes. Note that the horizontal axes have the same limits for easier comparison.

❑

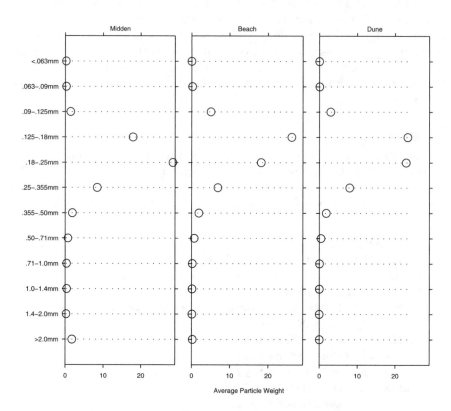

FIGURE 10.13
This is the multiway plot described in Example 10.9. Here we have the dot charts for the average particle weight, given that they come from the beach, dune, or midden.

10.6 Plotting Points as Curves

In this section, we present two methods for visualizing high-dimensional data: parallel coordinate plots and Andrews' curves. These methods are not without their problems, as we discuss shortly, but they are an efficient way of visually representing multi-dimensional relationships.

10.6.1 Parallel Coordinate Plots

In the Cartesian coordinate system, the axes are orthogonal, so the most we can view is three dimensions projected onto a computer screen or paper. If instead we draw the axes parallel to each other, then we can view many axes on the same 2-D display. This technique was developed by Wegman [1986] as a way of viewing and analyzing multi-dimensional data and was introduced by Inselberg [1985] in the context of computational geometry and computer vision.

A parallel coordinate plot for p-dimensional data is constructed by drawing p lines parallel to each other. We draw p copies of the real line representing the coordinate axes for $x_1, x_2,..., x_p$. The lines are the same distance apart and are perpendicular to the Cartesian y axis. Additionally, they all have the same positive orientation as the Cartesian x axis, as illustrated in Figure 10.14. Some versions of parallel coordinates draw the parallel axes perpendicular to the Cartesian x axis.

Let's look at the following 4-D point:

$$\mathbf{c} = \begin{bmatrix} 1 \\ 3 \\ 7 \\ 2 \end{bmatrix}$$

This is shown in Figure 10.14, where we see that the point is a polygonal line with vertices at $(c_i, i - 1), i = 1,..., p$ in Cartesian coordinates on the x_i parallel axis. Thus, a point in Cartesian coordinates is represented in parallel coordinates as a series of connected line segments.

We can plot observations in parallel coordinates with colors designating what class they belong to or use some other line style to indicate group membership. The parallel coordinate display can also be used to determine the following: a) class separation in a given coordinate, b) correlation between pairs of variables (explored in the exercises), and c) clustering or groups. We could also include a categorical variable indicating the class or group label as one of the parallel axes. This helps identify groups, when color is used for each category and n is large.

Example 10.10

We are going to use a subset of the BPM data from Example 10.4 to show how to use the parallel coordinate function **csparallel**. We plot topics 6 and 9 in parallel coordinates, using different colors and linestyles.

```
load example104
% This loads up the reduced BPM features using ISOMAP.
% Use the 3-D data.
X = Yiso.coords{3}';
% Find the observations for the two topics: 6 and 9.
ind6 = find(classlab == 6);
ind9 = find(classlab == 9);
% Put the data into one matrix.
```

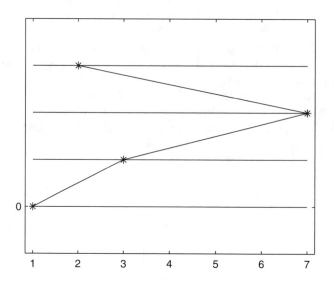

FIGURE 10.14
This shows the parallel coordinate representation for the 4-D point $c^T = (1,3,7,2)$. The reader should note that the parallel axes in future plots will have the parallel axes ordered from top to bottom: $x_1, x_2, ..., x_p$.

```
x = [X(ind6,:);X(ind9,:)];
% Use the csparallel function from the Computational
% Statistics Toolbox.⁴
% Construct the plot.
csparallel(x)
```

The parallel coordinates plot is illustrated in Figure 10.15. Several features should be noted regarding this plot. First, there is evidence of two groups in dimensions one and three. These two variables should be good features to use in classification and clustering. Second, the topics seem to overlap in dimension two, so this is not such a useful feature. Finally, although we did not use different colors or line styles to make this more apparent, it appears that observations within a topic have similar line shapes, and they are different from those in the other topic. Example 10.13 should make this point clearer.
❑

   In parallel coordinate plots, the order of the variables makes a difference. Adjacent parallel axes provide some insights about the relationship between consecutive variables. To see other pairwise relationships, we must permute the order of the parallel axes. Wegman [1990] provides a systematic way of

---

⁴This is available free. For download information, see Appendix B.

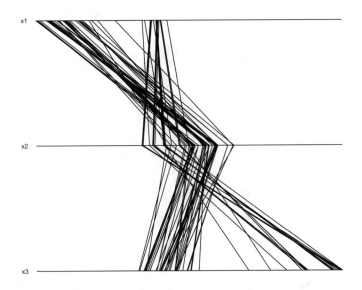

FIGURE 10.15
In this figure we have the parallel coordinates plot for the BPM data in Example 10.10. There is evidence for two groups in dimensions one and three. We see considerable overlap in dimension two.

finding all permutations such that all adjacencies in the parallel coordinate display will be visited. These tours will be covered at the end of the chapter.

10.6.2 Andrews' Curves

Andrews' curves [Andrews, 1972] were developed as a method for visualizing multi-dimensional data by mapping each observation onto a function. This is similar to star plots in that each observation or sample point is represented by a glyph, except that in this case the glyph is a curve. The Andrews' function is defined as

$$f_{\mathbf{x}}(t) = x_1/\sqrt{2} + x_2\sin t + x_3\cos t + x_4\sin 2t + x_5\cos 2t + \ldots, \qquad (10.1)$$

where the range of t is given by $-\pi \le t \le \pi$. We see by Equation 10.1 that each observation is projected onto a set of orthogonal basis functions represented by sines and cosines, and that each sample point is now represented by a curve. Because of this definition, the Andrews' functions produce infinitely many projections onto the basis vectors over the range of t. We now illustrate the MATLAB code to obtain Andrews' curves.

Example 10.11

We use a small data set to show how to get Andrews' curves. The data we have are the following observations:

$$\mathbf{x}_1 = (2, \ 6, \ 4)$$
$$\mathbf{x}_2 = (5, \ 7, \ 3)$$
$$\mathbf{x}_3 = (1, \ 8, \ 9).$$

Using Equation 10.1, we construct three curves, one corresponding to each data point. The Andrews' curves for the data are:

$$f_{\mathbf{x}_1}(t) = 2 \div \sqrt{2} + 6\sin t + 4\cos t$$
$$f_{\mathbf{x}_2}(t) = 5 \div \sqrt{2} + 7\sin t + 3\cos t$$
$$f_{\mathbf{x}_3}(t) = 1 \div \sqrt{2} + 8\sin t + 9\cos t.$$

We can plot these three functions in MATLAB using the following commands.

```
% Get the domain.
t = linspace(-pi,pi);
% Evaluate function values for each observation.
f1 = 2/sqrt(2)+6*sin(t)+4*cos(t);
f2 = 5/sqrt(2)+7*sin(t)+3*cos(t);
f3 = 1/sqrt(2)+8*sin(t)+9*cos(t);
plot(t,f1,'-.',t,f2,':',t,f3,'--')
legend('F1','F2','F3')
xlabel('t')
```

The Andrews' curves for these data are shown in Figure 10.16.
❑

It has been shown [Andrews, 1972; Embrechts and Herzberg, 1991] that because of the mathematical properties of the trigonometric functions, the Andrews' functions preserve means, distances (up to a constant), and variances. One consequence of this is that observations that are close together should produce Andrews' curves that are also closer together. Thus, one use of these curves is to look for clustering of the data points.

Embrechts and Herzberg [1991] discuss how other projections could be constructed and used with Andrews' curves. One possibility is to set one of the variables to zero and re-plot the curve. If the resulting plots remain nearly the same, then that variable has *low discriminating power*. If the curves are greatly affected by setting the variable to zero, then it carries a lot of information. Of course, other types of projections (such as nonlinear

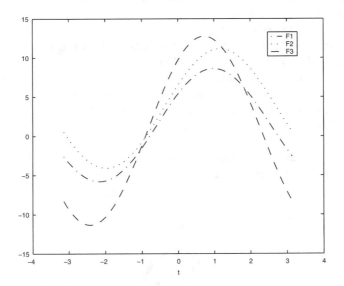

FIGURE 10.16
Andrews' curves for the three data points in Example 10.11.

dimensionality reduction, PCA, SVD, etc.) can be constructed and the results viewed using Andrews' curves.

Andrews' curves are dependent on the order of the variables. Lower frequency terms exert more influence on the shape of the curves, so re-ordering the variables and viewing the resulting plot might provide insights about the data. By lower frequency terms, we mean those that are first in the sum given in Equation 10.1. Embrechts and Herzberg [1991] also suggest that the data be rescaled so they are centered at the origin and have covariance equal to the identity matrix. Andrews' curves can be extended by using orthogonal bases other than sines and cosines. For example, Embrechts and Herzberg illustrate Andrews' curves using Legendre polynomials and Chebychev polynomials.

Example 10.12

We now construct the Andrews' curves for the data from Example 10.10 using a function called **csandrews**. The following code yields the plot in Figure 10.17. Not surprisingly, we see features similar to what we saw in parallel coordinates. The curves for each group have similar shapes, although topic 6 is less coherent. Also, the curves for topic 6 are quite a bit different than those for topic 9. One of the disadvantages of the Andrews' curves over the parallel coordinates is that we do not see information about the separate variables as we do in parallel coordinates.

```
load example104
```

```
% This loads up the reduced BPM features using ISOMAP.
% Use the 3-D data.
X = Yiso.coords{3}';
% Find the observations for the two topics: 6 and 9.
ind6 = find(classlab == 6);
ind9 = find(classlab == 9);
% This function is from the Comp Statistics Toolbox.
% Construct the plot for topic 6.
csandrews(X(ind6,:),'-','r')
hold on
% Construct the plot for topic 9.
csandrews(X(ind9,:),':','g')
```

❑

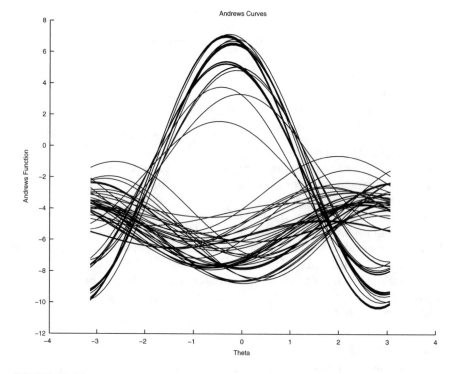

FIGURE 10.17
This is the Andrews' curves version of Figure 10.15. Similar curve shapes for each topic and different overall shapes indicate two groups in the data. The solid line is topic 6, and the dashed line is topic 9.

The Statistics Toolbox, Version 5, has functions for parallel coordinate plots and Andrews' curves. The **parallelcoords** function constructs horizontal parallel coordinate plots. Options include grouping (plotting lines using different colors based on class membership), standardizing (PCA or z-score), and plotting only the median and/or other quantiles. The function to construct Andrews' curves is called **andrewsplot**. It has the same options found in **parallelcoords**.

10.6.3 More Plot Matrices

So far, we have discussed the use of Andrews' functions and parallel coordinate plots for locating groups and understanding structure in multivariate data. When we know the true groups or categories in our data set, then we can use different line styles and color to visually separate them on the plots. However, we might have a large sample size and/or many groups, making it difficult to explore the data set.

We borrow from the scatterplot matrix and panel graphs (e.g., multiway dot charts, coplots) concepts and apply these to Andrews' curves and parallel coordinate plots. We simply plot each group separately in its own subplot, using either Andrews' curves or parallel coordinates. Common scales are used for all plots to allow direct comparison of the groups.

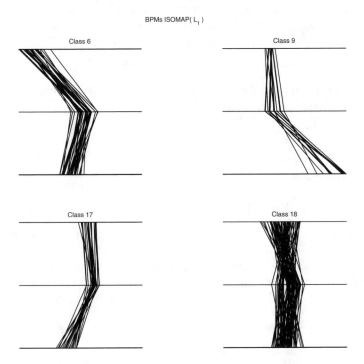

FIGURE 10.18
This shows the plot matrix of parallel coordinate plots for the BPM data in Example 10.13.

Example 10.13

Continuing with the same BPM data, we now add two more topics (17 and 18) and plot each topic separately in its own subplot.

```
load example104
% This loads up the reduced BPM features using ISOMAP.
% Use the 3-D data.
X = Yiso.coords{3}';
% Find the observations for the topics.
inds = ismember(classlab,[6 9 17 18]);
% This function comes with the text:
plotmatrixpara(X(inds,:),classlab(inds),[],...
    'BPMs ISOMAP( L_1 )')
```

See Figure 10.18 for this plot. Using these three features, we have some confidence that we would be able to discriminate between the three topics: 6, 9, and 17. Each has differently shaped line segments, and the line segments are similar within each group. However, topic 18 is a different story. The lines in this topic indicate that we would have some trouble distinguishing topic 17 and 18. Also, the topic 18 lines are not coherent; they seem to have different shapes within the topic. We can do something similar with Andrews' curves. The code for this is given below, and the plot is given in Figure 10.19. Analysis of this plot is left as an exercise to the reader.

```
plotmatrixandr(X(inds,:),classlab(inds))
```

The functions **plotmatrixandr** and **plotmatrixpara** are provided with the EDA Toolbox.
❑

10.7 Data Tours Revisited

In Chapter 4, we discussed the basic ideas behind tours and motion graphics, but we only used 2-D scatterplots for displaying the data. Some of these ideas are easily extended to higher dimensional representations of the data. In this section, we discuss how the grand tour can be used with scatterplot matrices and parallel coordinates, as well as a new type of animated graphical technique called permutation tours.

10.7.1 Grand Tour

Wegman [1991] and Wegman and Solka [2002] describe the grand tour in k dimensions, where $k \leq p$. The basic procedure outlined in Chapter 4 remains

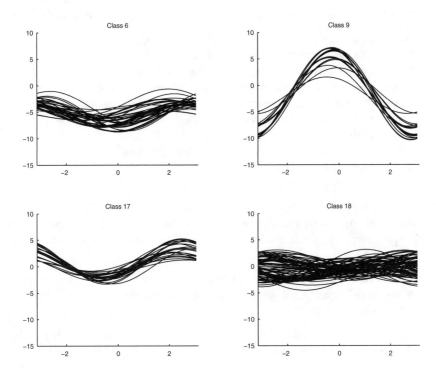

FIGURE 10.19
Here we have the plot matrix with Andrews' curves for the data in Example 10.13.

the same, but we replace the manifold of two-planes with a manifold of k-planes. Thus, we would use

$$\mathbf{A}_K = \mathbf{Q}_K \mathbf{E}_{1 \ldots k}$$

to project the data, where the columns of $\mathbf{E}_{1 \ldots k}$ contain the first k basis vectors.

The other change we must make is in how we display the data to the user; 2-D scatterplots can no longer be used. Now that we have some ways to visualize multivariate data, we can combine these with the grand tour. For example, we could use a 3-D scatterplot if $k = 3$. For this or higher values of k, we might use the scatterplot matrix display, parallel coordinates, or Andrews' curves. We illustrate this in the next example.

Example 10.14

To go on a grand tour in k dimensions, use the function called **kdimtour**. This implements the torus grand tour, as described in Chapter 4, but now the display can either be parallel coordinates or Andrews' curves. The user can specify the maximum number of iterations, the type of display, and the number of dimensions $k \le p$. We tour the topic 6 data from the previous examples and show the tour after a few iterations in Figure 10.20 (top).

```
% We show the tour at several iterations.
% Parallel coordinates will be used here.
% Andrews' curves are left as an exercise.
% We see that this shows some grouping of
% the lines - all seem to follow the same 'structure'.
ind6 = find(classlab==6);
x = X(ind6,:);
% Default tour is parallel coordinates.
% We have 10 iterations and k = 3.
kdimtour(x,10,3)
```

If the lines stay as a cohesive group for most of the tour, then that is an indication of true groups in the data, because the grouping is evident under various rotations/projections. Let's see what happens if we go further along in the tour.

```
% Now at the 90th iteration.
% We see that the grouping falls apart.
kdimtour(x,90,3)
```

The lines stay together until the end of this tour (90 iterations), at which time, they become rather incoherent. This plot is given in Figure 10.20 (bottom).
❑

10.7.2 Permutation Tour

One of the criticisms of the parallel coordinate plots and Andrews' curves is the dependency on the order of the variables. In the case of parallel coordinate displays, the position of the axes is important in that the relationships between pairwise axes are readily visible. In other words, the relationship between variables on nonadjacent axes is difficult to understand and compare. With Andrews' curves, the variables that are placed first carry more weight in the resulting curve. Thus, it would be useful to have a type of tour we call a *permutation tour,* where we look at replotting the points based on reordering the variables or axes.

A permutation tour can be one of two types: either a full tour of all possible permutations or a partial one. In the first case, we have $p!$ permutations or possible steps in our tour, but this yields many duplicate adjacencies in the case of parallel coordinates. We describe a much shorter permutation tour first described in Wegman [1990] and later in Wegman and Solka [2002]. This could also be used with Andrews' curves, but a full permutation tour might be more useful in this case, since knowing what variables (or axes) are adjacent is not an issue.

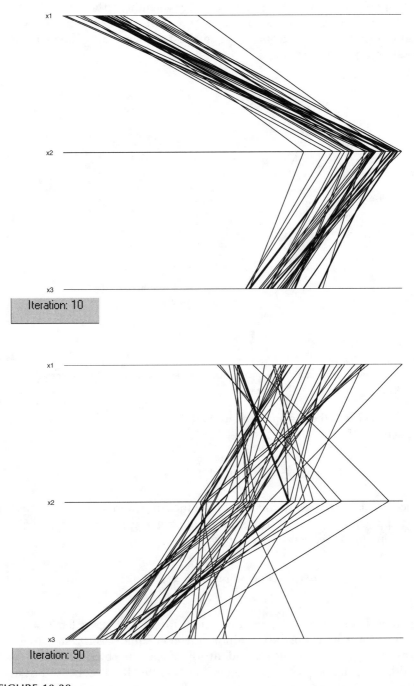

FIGURE 10.20
The top parallel coordinates plot is the 10-*th* iteration in our grand tour. The bottom plot is later on in the tour.

To illustrate this procedure, we first draw a graph, where each vertex corresponds to an axis. We place edges between the vertices to indicate that the two axes (vertices) are adjacent. We use the zig-zag pattern shown in Figure 10.21 (for $p = 6$) to obtain the smallest set of orderings such that every possible adjacency is present.

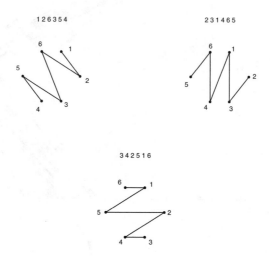

FIGURE 10.21
This figure shows the minimal number of permutations needed to obtain all the adjacencies for $p = 6$.

We can obtain these sequences in the following manner. We start with the sequence

$$p_{k+1} = (p_k + (-1)^{k+1}k) \bmod p \qquad k = 1, 2, \dots, p-1, \qquad (10.2)$$

where we have $p_1 = 1$. In this use of the mod function, it is understood that $0 \bmod p = p \bmod p = p$. To get all of the zig-zag patterns, we apply the following to the sequence given above

$$p_k^{(j+1)} = (p_k^j + 1) \bmod p \qquad j = 1, 2, \dots, \left\lfloor \frac{p-1}{2} \right\rfloor, \qquad (10.3)$$

where $\lfloor \bullet \rfloor$ is the greatest integer function. To get started, we let $p_k^1 = p_k$.

For the case of p even, the definitions given in Equations 10.2 and 10.3 yield an extra sequence, so some redundant adjacencies are obtained. In the case of p odd, the extra sequence is needed to generate all adjacencies, but again some redundant adjacencies will occur. To illustrate, we show in Example 10.15 how the sequences in Figure 10.21 are obtained using this formulation.

Example 10.15

We start by getting the first sequence given in Equation 10.2. We must alter the results from the MATLAB function **mod** to include the correct result for the case of 0 mod p and p mod p.

```
p = 6;
N = ceil((p-1)/2);
% Get the first sequence.
P(1) = 1;
for k = 1:(p-1)
    tmp(k) = (P(k) + (-1)^(k+1)*k);
    P(k+1) = mod(tmp(k),p);
end
% To match our definition of 'mod':
P(find(P==0)) = p;
```

We find the rest of the permutations by applying Equation 10.3 to the previous sequence.

```
for j = 1:N;
    P(j+1,:) = mod(P(j,:)+1,p);
    ind = find(P(j+1,:)==0);
    P(j+1,ind) = p;
end
```

We now apply this idea to parallel coordinates and Andrews' curves. Use the function we wrote called **permtourparallel** for a parallel coordinate permutation tour, based on Wegman's minimum permutation scheme. The syntax is

```
permtourparallel(X)
```

Note that this is very different from the grand tour in Example 10.14. Here we are just swapping adjacent axes; we are not rotating the data as in the grand tour. The plot freezes at each iteration of the tour; the user must hit any key to continue. This allows one to examine the plot for structure before moving on. We also include a function for Andrews' curves permutation tours. The user can either do Wegman's minimal tour or the full permutation tour (i.e., all possible permutations). The syntax for this function is

```
permtourandrews(X)
```

for a full permutation tour. To run the Wegman minimal permutation tour, use

```
permtourandrews(X,flag)
```

The input argument **flag** can be anything. As stated before, the full tour with Andrews' curves is more informative, because we are not concerned about adjacencies, as we are with parallel coordinates.
□

10.8 Summary and Further Reading

We start off by recommending some books that describe scientific and statistical visualization in general. One of the earliest ones in this area is *Semiology of Graphics: Diagrams, Networks, Maps* by Jacques Bertin [1983], originally published in French in 1967. This book discusses rules and properties of a graphic system and provides many examples. Edward Tufte wrote several books on visualization and graphics. His first book, *The Visual Display of Quantitative Information* [Tufte, 1983], shows how to depict numbers. The second in the series is called *Envisioning Information* [Tufte, 1990], and illustrates how to deal with pictures of nouns (e.g., maps, aerial photographs, weather data). The third book is entitled *Visual Explanations* [Tufte, 1997], and it discusses how to illustrate pictures of verbs (e.g., dynamic data, information that changes over time). These three books also provide many examples of good graphics and bad graphics. For more examples of graphical mistakes, we highly recommend the book by Wainer [1997]. Wainer discusses this subject in a way that is accessible to the general reader. We referenced this book several times already, but the reader should consult Cleveland's *Visualizing Data* [1993] for more information and examples on univariate and multivariate data. He includes extensive discussions on visualization tools, the relationship of visualization to classical statistical methods, and even some of the cognitive aspects of data visualization and perception. Another excellent resource on graphics for data analysis is Chambers, et al. [1983]. For books on visualizing categorical data, see Blasius and Greenacre [1998] and Friendly [2000]. Finally, we mention Wilkinson's *The Grammar of Graphics* [1999] for those who are interested in applying statistical and scientific visualization in a distributed computing environment. This book provides a foundation for quantitative graphics and is based on a Java graphics library.

Many papers have been written on visualization for the purposes of exploratory data analysis and data mining. One recent one by Wegman [2003] discusses techniques and strategies for visual data mining on high-dimensional and large data sets. Wegman and Carr [1993] present many visualization techniques, such as stereo plots, mesh plots, parallel coordinates, and more. Other survey articles include Anscombe [1973], Weihs and Schmidli [1990], Young, et al. [1993], and McLeod and Provost [2001]. Carr, et al. [1987] include other scatterplot methods besides hexagonal binning, such as sunflower plots. For an entertaining discussion of various plotting methods, such as scatterplots, pie charts, line charts, etc., and their inventors, see Friendly and Wainer [2004].

An extensive discussion of brushing scatterplots can be found in the paper by Becker and Cleveland [1987]. They show several brushing techniques for linking single points and clusters, conditioning on a single variable (a type of coplot), conditioning on two variables, and subsetting with categorical variables. Papers that provide nice reviews of these methods and others for dynamic graphics are Becker and Cleveland [1991], Buja, et al. [1991], Stuetzle [1987], Swayne, Cook and Buja [1991], and Becker, Cleveland and Wilks [1987].

For more information on dot charts and scaling, see Cleveland [1984]. He compares dot charts with bar charts, shows why the dot charts are a useful graphical method in EDA, and presents grouped dot charts. He also has an excellent discussion of the importance of a meaningful baseline with dot charts. If there is a zero on the scale or another baseline value, then the dotted lines should end at the dot. If this is not the case, then the dotted lines should continue across the graph.

Parallel coordinate techniques were expanded upon and described in a statistical setting by Wegman [1990]. Wegman [1990] also gave a rigorous explanation of the properties of parallel coordinates as a projective transformation and illustrated the duality properties between the parallel coordinate representation and the Cartesian orthogonal coordinate representation. Extensions to parallel coordinates that address the problem of over-plotting include saturation and brushing [Wegman and Luo, 1997; Wegman, 2003] and conveying aggregated information [Fua, et al., 1999].

In the previous pages, we mentioned the papers by Andrews [1972] and Embrechts and Herzberg [1991] that describe Andrews' curves plots and their extensions. Additional information on these plots can be found in Jackson [1991] and Jolliffe [1986]

Exercises

10.1 Write a MATLAB function that will construct a profile plot, where each observation is a bar chart with the height of the bar corresponding to the value of the variable for that data point. The functions **subplot** and **bar** might be useful. Try your function on the **cereal** data. Compare to the other glyph plots.

10.2 The MATLAB Statistics Toolbox, Version 5 has a function that will create glyph plots called **glyphplot**. Do a **help** on this function and recreate Figures 10.1 and 10.2.

10.3 Do a help on **gscatter** and repeat Example 10.1.

10.4 Repeat Example 10.1 using **plot** and **plot3**, where you specify the plot symbol to get a scatterplot.

10.5 Construct a scatterplot matrix using all variables of the **oronsay** data set. Analyze the results and compare to Figure 10.4. Construct a grouped scatterplot matrix using just a subset of the most interesting variables and both classifications. Do you see evidence of groups?

10.6   Repeat Example 10.3 and vary the number of bins. Compare your results.

10.7   Write a MATLAB function that will construct a scatterplot matrix using hexagonal bins.

10.8   Load the **hamster** data and construct a scatterplot matrix. There is an unusual observation in the scatterplot of spleen (variable 5) weight against liver (variable 3) weight. Use linking to find and highlight that observation in all plots. By looking at the plots, decide whether this is because the spleen is enlarged or the liver is underdeveloped. [Becker and Cleveland, 1991]

10.9   Construct a dot chart using the **brainweight** data. Analyze your results.

10.10   Implement Tukey's alternagraphics in MATLAB. This function should cycle through given subsets of data and display them in a scatterplot. Use it on the **oronsay** data.

10.11   Randomly generate a set of 2-D ($n = 20$) data that have a correlation coefficient of 1, and generate another set of 2-D data that have a correlation coefficient of -1. Construct parallel coordinate plots of each and discuss your results.

10.12   Generate a set of bivariate data ($n = 20$), with one group in the first dimension (first column of **X**) and two clusters in the second dimension (i.e., second column of **X**). Construct a parallel coordinates plot. Now generate a set of bivariate data that has two groups in both dimensions and graph them in parallel coordinates. Comment on the results.

10.13   Use the **playfair** data and construct a dot chart with the cdf (cumulative distribution) option. Do a **help** and try some of the other options in the **dotchart** function.

10.14   Using the **livestock** data, construct a multiway dot chart as in Figure 10.12. Use the countries for the panels and animal type for the levels. Analyze the results.

10.15   Repeat Example 10.7 for the **abrasion** loss data using tensile strength as the conditioning variable. Discuss your results.

10.16   Construct coplots using the **ethanol** and **software** data sets. Analyze your results.

10.17   Embrechts and Herzberg [1991] present the idea of projections with the Andrews' curves, as discussed in Section 10.6.2. Implement this in MATLAB and apply it to the **iris** data to find any variables that have low discriminating power.

10.18   Try the **scattergui** function with the following data sets. Note that you will have to either reduce the dimensionality or pick two dimensions to use.

        a. **skulls**

        b. **sparrow**

        c. **oronsay** (both classifications)

d. BPM data sets

e. **spam**

f. gene expression data sets

10.19   Repeat Example 10.5 using **gplotmatrix**.

10.20   Use the **gplotmatrix** function with the following data sets.

a. **iris**

b. **oronsay** (both classifications)

c. BPM data sets

10.21   Run the permutation tour with Andrews' curves and parallel coordinates. Use the data in Example 10.14.

10.22   Find the adjacencies as outlined in Example 10.15 for a permutation tour with $p = 7$. Is the last sequence needed to obtain all adjacencies? Are some of them redundant?

10.23   Analyze the Andrews' curves shown in Figure 10.19.

10.24   Apply the **permtourandrews**, **kdimtour** and **permtourparallel** to the following data sets. Reduce the dimensionality first, if needed.

a. **environmental**

b. **oronsay**

c. **iris**

d. **posse** data sets

e. **skulls**

f. BPM data sets.

g. **pollen**

h. gene expression data sets.

10.25   The MATLAB Statistics Toolbox, Version 5 has functions for Andrews' curves and parallel coordinate plots. Do a **help** on these for information on how to use them and the available options. Explore their capabilities using the **oronsay** data. The functions are called **parallelcoords** and **andrewsplot**.

10.26   Do a **help** on the **scatter** function. Construct a scatterplot of the data in Example 10.1, using the argument that controls the symbol size to make the circles smaller. Compare your plot with Figure 10.1 (top) and discuss the issue of overplotting.

Appendix A

Proximity Measures

We provide some information on proximity measures in this appendix. *Proximity measures* are important in clustering, multi-dimensional scaling, and nonlinear dimensionality reduction methods, such as ISOMAP. The word *proximity* indicates how close objects or measurements are in space, time, or in some other way (e.g., taste, character, etc.). How we define the nearness of objects is important in our data analysis efforts and results can depend on what measure is used. There are two types of proximity measures: similarity and dissimilarity. We now describe several examples of these and include ways to transform from one to the other.

A.1 Definitions

Similarity measures indicate how alike objects are to each other. A high value means they are similar, while a small value means they are not. We denote the similarity between objects (or observations) i and j as s_{ij}. Similarities are often scaled so the maximum similarity is one (e.g., the similarity of an observation with itself is one, $s_{ii} = 1$). *Dissimilarity measures* are just the opposite. Small values mean the observations are close together and thus are alike. We denote the dissimilarity by δ_{ij}, and we have $\delta_{ii} = 0$. In both cases, we have $s_{ij} \geq 0$ and $\delta_{ij} \geq 0$.

Hartigan [1967] and Cormack [1971] provide a taxonomy of twelve proximity structures; we list just the top four of these below. As we move down in the taxonomy, constraints on the measures are relaxed. Let the observations in our data set be denoted by the set O, then the proximity measure is a real function defined on O × O. The four structures S that we consider here are

S1   S defined on O × O is Euclidean distance;

S2   S defined on O × O is a metric;

S3   S defined on O × O is symmetric real-valued;

S4   S defined on O × O is real-valued.

The first structure S1 is very strict with the proximity defined as the familiar *Euclidean distance* given by

$$\delta_{ij} = \sqrt{\sum_k (x_{ik} - x_{jk})^2}, \tag{A.1}$$

where x_{ik} is the k-th element of the i-th observation.

If we relax this and require our measure to be a metric only, then we have structure S2. A dissimilarity is a metric if the following holds:

1. $\delta_{ij} = 0$, if and only if $i = j$.
2. $\delta_{ij} = \delta_{ji}$.
3. $\delta_{ij} \le \delta_{it} + \delta_{tj}$.

The second requirement given above means the dissimilarity is symmetric, and the third one is the triangle inequality. Removing the metric requirement produces S3, and allowing nonsymmetry yields S4. Some of the later structures in the taxonomy correspond to similarity measures.

We often represent the interpoint proximity between all objects i and j in an $n \times n$ matrix, where the ij-th element is the proximity between the i-th and j-th observation. If the proximity measure is symmetric, then we only need to provide the $n(n - 1)/2$ unique values (i.e., the upper or lower part of the *interpoint proximity matrix*. We now give examples of some commonly used proximity measures.

A.1.1 Dissimilarities

We have already defined Euclidean distance (Equation A.1), which is probably used most often. One of the problems with the Euclidean distance is its sensitivity to the scale of the variables. If one of the variables is more dispersed than the rest, then it will dominate the calculations. Thus, we recommend transforming or scaling the data as discussed in Chapter 1.

The Mahalanobis distance (squared) takes the covariance into account and is given by

$$\delta_{ij}^2 = (\mathbf{x}_i - \mathbf{x}_j)^T \Sigma^{-1} (\mathbf{x}_i - \mathbf{x}_j),$$

where Σ is the covariance matrix. Often Σ must be estimated using the data, so it can be affected by outliers.

The *Minkowski metric* defines a whole family of dissimilarities, and it is used extensively in nonlinear multidimensional scaling (see Chapter 3). It is defined by

$$\delta_{ij} = \left\{ \sum_k |x_{ik} - x_{jk}|^\lambda \right\}^{\frac{1}{\lambda}} \qquad \lambda \geq 1 . \tag{A.2}$$

When the parameter $\lambda = 1$, we have the *city block metric* (sometimes called *Manhattan distance*) given by

$$\delta_{ij} = \sum_k |x_{ik} - x_{jk}| .$$

When $\lambda = 2$, then Equation A.2 becomes the Euclidean distance (Equation A.1).

The MATLAB Statistics Toolbox provides a function called **pdist** that calculates the interpoint distances between the n data points. It returns the values in a vector, but they can be converted into a square matrix using the function **squareform**. The basic syntax for the **pdist** function is

```
Y = pdist(X, distance)
```

The **distance** argument can be one of the following choices:

```
'euclidean'    - Euclidean distance
'seuclidean'   - Standardized Euclidean distance, each
   coordinate in the sum of squares is inverse weighted
   by the sample variance of that coordinate
'cityblock'    - City Block distance
'mahalanobis'  - Mahalanobis distance
'minkowski'    - Minkowski distance with exponent 2
'cosine'       - One minus the cosine of the included
   angle between observations (treated as vectors)
'correlation'  - One minus the sample correlation
   between observations (treated as sequences of
   values).
'hamming'      - Hamming distance, percentage of
   coordinates that differ
'jaccard'      - One minus the Jaccard coefficient, the
   percentage of nonzero coordinates that differ
```

The **cosine, correlation,** and **jaccard** measures specified above are originally similarity measures, which have been converted to dissimilarities, two of which are covered below. It should be noted that the **minkowski** option can be used with an additional argument designating other values of the exponent.

There is a new distance available with Version 5 of the Statistics Toolbox. This is the Chebychev distance, which is the maximum coordinate difference between two vectors. Use the flag `'chebychev'` to get this distance.

A.1.2 Similarity Measures

A common similarity measure that can be used with real-valued vectors is the *cosine similarity measure*. This is the cosine of the angle between the two vectors and is given by

$$s_{ij} = \frac{\mathbf{x}_i^T \mathbf{x}_j}{\sqrt{\mathbf{x}_i^T \mathbf{x}_i} \sqrt{\mathbf{x}_j^T \mathbf{x}_j}}.$$

The *correlation similarity measure* between two real-valued vectors is similar to the one above, and is given by the expression

$$s_{ij} = \frac{(\mathbf{x}_i - \bar{\mathbf{x}})^T (\mathbf{x}_j - \bar{\mathbf{x}})}{\sqrt{(\mathbf{x}_i - \bar{\mathbf{x}})^T (\mathbf{x}_i - \bar{\mathbf{x}})} \sqrt{(\mathbf{x}_j - \bar{\mathbf{x}})^T (\mathbf{x}_j - \bar{\mathbf{x}})}}.$$

A.1.3 Similarity Measures for Binary Data

The proximity measures defined in the previous sections can be used with quantitative data that are continuous or discrete, but not binary. We now describe some measures for binary data.

When we have binary data, we can calculate the following frequencies:

		j-th Observation		
		1	0	
i-th Observation	1	a	b	$a + b$
	0	c	d	$c + d$
		$a + c$	$b + d$	$a + b + c + d$

This table shows the number of elements where the i-th and j-th observations both have a 1 (a), both have a 0 (d), etc. We use these quantities to define the following similarity measures for binary data.

First is the *Jaccard coefficient*, given by

$$s_{ij} = \frac{a}{a + b + c}.$$

Next we have the *Ochiai measure*:

$$s_{ij} = \frac{a}{\sqrt{(a+b)(a+c)}}.$$

Finally, we have the **simple matching coefficient**, that calculates the proportion of matching elements:

$$s_{ij} = \frac{a+d}{a+b+c+d}.$$

A.1.4 Dissimilarities for Probability Density Functions

In some applications, our observations might be in the form of relative frequencies or probability distributions. For example, this often happens in the case of measuring semantic similarity between two documents. Or, we might be interested in calculating a distance between two probability density functions, where one is estimated and one is the true density function. We discuss three types of measures here: Kullback-Leibler information, L_1 norm, and information radius.

Say we have two probability density functions f and g (or any function in the general case). **Kullback-Leibler** (KL) **information** measures how well function g approximates function f. The KL information is defined as

$$KL(f, g) = \int f(\mathbf{x}) \log\left(\frac{f(\mathbf{x})}{g(\mathbf{x})}\right) d\mathbf{x} \,,$$

for the continuous case. A summation is used in the discrete case, where f and g are probability mass functions. The KL information measure is sometimes called the **discrimination information**, and it measures the divergence between two functions. The higher the measure, the greater the difference between the functions.

There are two problems with the KL information measure. First, there could be cases where we get a value of infinity, which can happen often in natural language understanding applications [Manning and Schütze, 2000]. The other potential problem is that the measure is not necessarily symmetric.

The second measure we consider is the **information radius** (IRad) that overcomes these problems. It is based on the KL information and is given by

$$IRad(f, g) = KL\left(f, \frac{f+g}{2}\right) + KL\left(g, \frac{f+g}{2}\right).$$

The IRad measure tries to quantify how much information is lost if we describe two random variables that are distributed according to f and g with their average distribution. The information radius ranges from 0 for identical distributions to 2log2 for distributions that are maximally different. Here we assume that 0log0 = 0. Note that the IRad measure is symmetric and is not subject to infinite values [Manning and Schütze, 2000].

The third measure of this type that we cover is the L_1 norm. It is symmetric and well-defined for arbitrary probability density functions f and g (or any function in the general case). We can think of it as a measure of the expected proportion of events that are going to be different between distributions f and g. This norm is defined as

$$L_1(f, g) = \int |f(\mathbf{x}) - g(\mathbf{x})|\, dx \ .$$

The L_1 norm has a nice property. It is bounded below by zero and bounded above by two:

$$0 \le L_1(f, g) \le 2,$$

when f and g are valid probability density functions.

A.2 Transformations

In many instances, we start with a similarity measure and then convert to a dissimilarity measure for further processing. There are several transformations that one can use, such as

$$\delta_{ij} = 1 - s_{ij}$$
$$\delta_{ij} = c - s_{ij} \qquad \text{for some constant } c$$
$$\delta_{ij} = \sqrt{2(1 - s_{ij})} \ .$$

The last one is valid only when the similarity has been scaled such that $s_{ii} = 1$. For the general case, one can use

$$\delta_{ij} = \sqrt{s_{ii} - 2s_{ij} + s_{jj}} \ .$$

In some cases, we might want to transform from a dissimilarity to a similarity. One can use the following to accomplish this:

$$s_{ij} = (1 + \delta_{ij})^{-1} \ .$$

A.3 Further Reading

Most books on clustering and multidimensional scaling have extensive discussions on proximity measures, since the measure used is fundamental to these methods. We describe a few of these here.

The clustering book by Everitt, Landau and Leese [2001] has a chapter on this topic that includes measures for data containing both continuous and categorical variables, weighting variables, standardization, and the choice of proximity measure. For additional information from a classification point of view, we recommend Gordon [1999].

Cox and Cox [2001] has an excellent discussion of proximity measures as they pertain to multidimensional scaling. Their book also includes the complete taxonomy for proximity structures mentioned previously. Finally, Manning and Schütze [2000] provide an excellent discussion of measures of similarity that can be used when measuring the semantic distance between documents, words, etc.

There are also several survey papers on dissimilarity measures. Some useful ones are Gower [1966] and Gower and Legendre [1986]. They review properties of dissimilarity coefficients, and they emphasize their metric and Euclidean nature. For a summary of distance measures that can be used when one wants to measure the distance between two functions or statistical models, we recommend Basseville [1989]. The work described in Basseville takes the signal processing point of view. Jones and Furnas [1987] study proximity measures using a geometric analysis and apply it to several measures used in information retrieval applications. Their goal is to demonstrate the relationship between a measure and its performance.

Appendix B

Software Resources for EDA

The purpose of this appendix is to provide information on internet resources for EDA. Most of these were discussed in the body of the text, but some were not. Also, most of them are for MATLAB code, but some of them are stand-alone packages.

B.1 MATLAB Programs

In this section of Appendix B, we provide websites and references to MATLAB code that can be used for EDA. Some of the code is included with the EDA Toolbox (see Appendix E). However, we recommend that users periodically look for the most recent versions.

Bagplots (included with the EDA Toolbox)

The code for the bagplot can be found at

www.agoras.ua.ac.be/Public99.htm

The authors include software for FORTRAN, S-Plus, and MATLAB. Papers are also available for download from this same website.

Computational Statistics Toolbox

This toolbox was written for the book *Computational Statistics Handbook with MATLAB*. It contains many useful functions, some of which were used in this book. It is available for download from

www.stat.unipg.it/stat/statlib/

Some of the functions from this toolbox are included with the EDA Toolbox (see Appendix E).

Data Visualization Toolbox

The authors of this MATLAB toolbox provide functions that implement the graphical techniques described in *Visualizing Data* by William Cleveland. Software, data sets, documentation, and a tutorial are available from

www.datatool.com/Dataviz_home.htm

Some of these functions are included with the EDA Toolbox (see Appendix E).

Generative Topographic Mapping

The GTM Toolbox is available for download from

www.ncrg.aston.ac.uk/GTM/

This website includes links to documentation, papers on the GTM, and data sets. This software is available for noncommercial use under the GNU license. This toolbox is included when you download the EDA Toolbox.

Hessian Eigenmaps

The home page for Hessian eigenmaps or HLLE is

http://basis.stanford.edu/WWW/HLLE/frontdoc.htm

This website also has code that runs the Swiss roll demo. The HLLE code is included with the EDA Toolbox.

ISOMAP

The main website for ISOMAP is found at

http://isomap.stanford.edu/

This also has links to related papers, data sets, convergence proofs, and supplemental figures. The ISOMAP functions are in the EDA Toolbox.

Locally Linear Embedding – LLE

The main website for LLE is

www.cs.toronto.edu/~roweis/lle/

From here you can download the MATLAB code, papers, and data sets. The LLE function is part of the EDA Toolbox.

MATLAB Central

You can find lots of user-contributed code at this website:

www.mathworks.com/matlabcentral/

Always look here first before you write your own functions. What you need might be here already.

Microarray Analysis - MatArray Toolbox

The following link takes you to a webpage where you can download a toolbox that implements techniques for the analysis of microarray data. This toolbox includes functions for k-means, hierarchical clustering, and others.

www.ulb.ac.be/medecine/iribhm/microarray/toolbox/

Model-Based Clustering Toolbox

The code from the Model-Based Clustering Toolbox is included with the EDA Toolbox. However, for those who are interested in more information on MBC, please see

www.stat.washington.edu/mclust/

This website also contains links for model-based clustering functions that will work in S-Plus and R.

Robust Analysis

A toolbox for robust statistical analysis is available for download at

http://www.wis.kuleuven.ac.be/stat/robust/LIBRA.html

The toolbox has functions for univariate location and scale, multivariate location and covariance, regression, PCA, principal component regression, partial least squares and robust classification. Graphical tools are also included for model checking and outlier detection.

Self-Organizing Map

The website for the SOM Toolbox is

www.cis.hut.fi/projects/somtoolbox/links/

This website also contains links to documentation, theory, research, and related information. The software may be used for noncommercial purposes and is governed by the terms of the GNU General Public License. Some functions in the SOM Toolbox are contributed by users and might be governed by their own copyright. Some of the functions in the SOM Toolbox are included in the EDA Toolbox.

SiZer

This software allows one to explore data through smoothing. It contains various functions and GUIs for MATLAB. It helps answer questions about what features are really there or what might be just noise. In particular, it provides information about smooths at various levels of the smoothing parameter. For code, see:

> `www.stat.unc.edu/faculty/marron/marron_software.html`

For notes on a short course in SiZer and how it can be used in EDA, go to:

> `www.galaxy.gmu.edu/interface/I02/I2002Proceedings/...`
> `MarronSteve/MarronSiZerShortCourse.pdf`

Statistics Toolboxes - No Cost

The following are links to statistics toolboxes:

> `www.statsci.org/matlab/statbox.html`
> `www.maths.lth.se/matstat/stixbox/`

Here is a link to a toolbox on statistical pattern recognition:

> `cmp.felk.cvut.cz/~xfrancv/stprtool/index.html`

Text to Matrix Generator (TMG) Toolbox

The TMG Toolbox creates new term-document matrices from documents. It also updates existing term-document matrices. It includes various term-weighting methods, normalization, and stemming. The website to download this toolbox is

> `scgroup.hpclab.ceid.upatras.gr/scgroup/Projects/TMG/`

B.2 Other Programs for EDA

The following non-MATLAB programs are available for download (at no charge).

Crystal Vision

CrystalVision, copyright (c) 2000 by Crystal Data Technologies (Qiang Luo, Edward J. Wegman, and Xiaodong Fu), is a Windows 95/98/NT package for Wintel computers. A demonstration version of CrystalVision is available at

> `ftp://www.galaxy.gmu.edu/pub/software/CrystalVisionDemo.exe`

This software includes parallel coordinates, brushing, linking, plot matrices, and others.

Finite Mixtures - EMMIX

A FORTRAN program for fitting finite mixtures with normal and t-distribution components can be downloaded from

`http://www.maths.uq.edu.au/~gjm/emmix/emmix.html`

GGobi

GGobi is a data visualization system for viewing high-dimensional data, and it includes brushing, linking, plot matrices, multi-dimensional scaling and many other capabilities. The home page for GGobi is

`www.ggobi.org/`

Versions are available for Windows and Linux.

MANET

For Macintosh users, one can use MANET (missings are now equally treated) found at:

`www1.math.uni-augsburg.de/MANET/`

MANET provides various graphical tools that are designed for studying multivariate features.

Model-Based Clustering

Software for model-based clustering can be downloaded at

`http://www.stat.washington.edu/mclust/`

Versions are available for S-Plus and R. One can also obtain some of the papers and technical reports at this website, in addition to joining a mailing list.

R for Statistical Computing

R is a language and environment for statistical computing. It is distributed as free software under the GNU license. It operates under Windows, UNIX systems (such as Linux), and the Macintosh. One of the benefits of the R language is the availability of many user-contributed packages. Please see the following website for more information.

`www.r-project.org/`

B.3 EDA Toolbox

The Exploratory Data Analysis Toolbox is available for download from two sites:

http://lib.stat.cmu.edu

and

http://www.infinityassociates.com

Please review the **readme** file for installation instructions and information on any changes.

The functions in the EDA Toolbox are available via a graphical user interface called **eda**, as well as the command line. The EDA Toolbox will be available when this book is published, but the **eda** GUI will not be available in the toolbox until February, 2005.

Most of the functions and capabilities in the toolbox have been discussed in this text, but several of them have not. Please see Appendix E for a list of functions that are included in the EDA Toolbox, along with a brief description of what they do. To view a *current* list of available functions (i.e., including those written after this book was published), type **help eda** at the command line. For complete functionality, you must have the Statistics Toolbox, version 4 or higher. For more information on the use of the EDA Toolbox, please see the accompanying documentation.

Appendix C

Description of Data Sets

In this appendix, we describe the various data sets used in the book. All data sets are downloaded with the accompanying software. We provide the data sets in MATLAB binary format (MAT-files).

abrasion

These data are the results from an experiment measuring three variables for 30 rubber specimens [Davies, 1957; Cleveland and Devlin, 1988]: tensile strength, hardness and abrasion loss. The abrasion loss is the amount of material abraded per unit of energy. Tensile strength measures the force required to break a specimen (per unit area). Hardness is the rebound height of an indenter dropped onto the specimen. Abrasion loss is measured in g/hp-hour; tensile strength is measured in kg/cm²; and the unit for hardness is ° Shore. The intent of the experiment was to determine the relationship between abrasion loss with respect to hardness and tensile strength.

animal

This data set contains the brain weights and body weights of several types of animals [Crile and Quiring, 1940]. According to biologists, the relationship between these two variables is interesting, since the ratio of brain weight to body weight is a measure of intelligence [Becker, Cleveland and Wilks, 1987]. The MAT-file contains variables: **AnimalName**, **BodyWeight**, and **BrainWeight**.

BPM data sets

There are several BPM data sets included with the text. We have **iradbpm**, **ochiaibpm**, **matchbpm** and **L1bpm**. Each data file contains the interpoint distance matrix for 503 documents and an array of class labels, as described in Chapter 1. These data can be reduced using ISOMAP before applying other analyses.

calibrat

This data set reflects the relationship between radioactivity counts (**counts**) to hormone level for 14 immunoassay calibration values (**tsh**). The original source of the data is Tiede and Pagano [1979], and we downloaded them from the website for Simonoff [1996]: **www.stern.nyu.edu/SOR/SmoothMeth**.

cereal

These data were obtained from ratings of eight brands of cereal [Chakrapani and Ehrenberg, 1981; Venables and Ripley, 1994]. The **cereal** file contains a matrix where each row corresponds to an observation and each column represents one of the variables or the percent agreement to statements about the cereal. The statements are: comes back to, tastes nice, popular with all the family, very easy to digest, nourishing, natural flavor, reasonably priced, a lot of food value, stays crispy in milk, helps to keep you fit, fun for children to eat. It also contains a cell array of strings (**labs**) for the type of cereal.

environmental

This file contains data comprising 111 measurements of four variables. These include **Ozone** (PPB), **SolarRadiation** (Langleys), **Temperature** (Fahrenheit), and **WindSpeed** (MPH). These were initially examined in Bruntz, et al. [1974], where the intent was to study the mechanisms that might lead to air pollution.

ethanol

A single-cylinder engine was run with either ethanol or indolene. This data set contains 110 measurements of compression ratio (**Compression**), equivalence ratio (**Equivalence**), and NO_x in the exhaust (**NOx**). The goal was to understand how NO_x depends on the compression and equivalence ratios [Brinkman, 1981; Cleveland and Devlin, 1988].

example96

This is the data used in Example 9.6 to illustrate the box-percentile plots.

example104

This loads the data used in Examples 10.10 and 10.12 to illustrate parallel coordinate plots. It is a subset of the BPM data that was reduced using ISOMAP as outlined in Example 10.4.

forearm

These data consist of 140 measurements of the length in inches of the forearm of adult males [Hand, et al., 1994; Pearson and Lee, 1903].

galaxy

The **galaxy** data set contains measurements of the velocities of the spiral galaxy NGC 7531. The array **EastWest** contains the velocities in the east-west direction, covering around 135 arc sec. The array **NorthSouth** contains the velocities in the north-south direction, covering approximately 200 arc sec. The measurements were taken at the Cerro Tololo Inter-American Observatory in July and October of 1981 [Buta, 1987].

geyser
These data represent the waiting times (in minutes) between eruptions of the Old Faithful geyser at Yellowstone National Park [Hand, et al., 1994; Scott, 1992].

hamster
This data set contains measurements of organ weights for hamsters with congenital heart failure [Becker and Cleveland, 1991]. The organs are heart, kidney, liver, lung, spleen and testes.

iris
The **iris** data were collected by Anderson [1935] and were analyzed by Fisher [1936] (and many statisticians since then!). The data set consists of 150 observations containing four measurements based on the petals and sepals of three species of iris. The three species are: *Iris setosa*, *Iris virginica*, and *Iris versicolor*. When the **iris** data file is loaded, you get three 50×4 matrices, one corresponding to each species.

leukemia
The **leukemia** data set is described in detail in Chapter 1. It measures the gene expression levels of patients with acute leukemia.

lsiex
This file contains the term-document matrix used in Example 2.3.

lungA, lungB
The **lung** data set is another one that measures gene expression levels. Here the classes correspond to various types of lung cancer.

oronsay
The **oronsay** data set consists of particle size measurements. It is described in Chapter 1. The data can be classified according to the sampling site as well as the type (beach, dune, midden).

playfair
This data set is described in Cleveland [1993] and Tufte [1983], and it is based on William Playfair's (1801) published displays of demographic and economic data. The **playfair** data set consists of the 22 observations representing the populations (thousands) of cities at the end of the 1700s and the diameters of the circles Playfair used to encode the population information. This MAT-file also includes a cell array containing the names of the cities.

pollen
This data set was generated for a data analysis competition at the 1986 Joint Meetings of the American Statistical Association. It contains 3848

observations, each with five fictitious variables: ridge, nub, crack, weight, and density. The data contain several interesting features and structures. See Becker, et al. [1986] and Slomka [1986] for information and results on the analysis of these artificial data.

posse

The **posse** file contains several data sets generated for simulation studies in Posse [1995b]. These data sets are called **croix** (a cross), **struct2** (an L-shape), **boite** (a donut), **groupe** (four clusters), **curve** (two curved groups), and **spiral** (a spiral). Each data set has 400 observations in 8-D. These data can be used in PPEDA and other data tours.

salmon

The **salmon** data set was downloaded from the website for the book by Simonoff [1996]: **www.stern.nyu.edu/SOR/SmoothMeth**. The MAT-file contains 28 observations in a 2-D matrix. The first column represents the size (in thousands of fish) of the annual spawning stock of Sockeye salmon along the Skeena River from 1940 to 1967. The second column represents the number of new catchable-size fish or recruits, again in thousands of fish.

scurve

This file contains data randomly generated from an S-curve manifold. See Example 3.5 for more information.

singer

This file contains several variables representing the height in inches of singers in the New York Choral Society [Cleveland, 1993; Chambers, et al., 1983]. There are four voice parts: sopranos, altos, tenors, and basses. The sopranos and altos are women, and the tenors and basses are men.

skulls

These data were taken from Cox and Cox [2000]. The data originally came from a paper by Fawcett [1901], where they detailed measurements and statistics of skulls belonging to the Naqada race in Upper Egypt. The **skulls** file contains an array called **skullsdata** for forty observations, 18 of which are female and 22 are male. The variables are greatest length, breadth, height, auricular height, circumference above the superciliary ridges, sagittal circumference, cross-circumference, upper face height, nasal breadth, nasal height, cephalic index, and ratio of height to length.

software

This file contains data collected on software inspections. The variables are normalized by the size of the inspection (the number of pages or SLOC – single lines of code). The file **software.mat** contains the preparation time in minutes (**prepage, prepsloc**), the total work hours in minutes for the meeting (**mtgsloc**), and the number of defects found (**defpage**, **defsloc**). A more detailed description can be found in Chapter 1.

spam

These data were downloaded from the UCI Machine Learning Repository:

http://www.ics.uci.edu/~mlearn/MLRepository/.html

Anyone who uses email understands the problem of spam, which is unsolicited email, commercial or otherwise. For example, spam can be chain letters, pornography, advertisements, foreign money-making schemes, etc. This data set came from Hewlett-Packard Labs and was generated in 1999. The **spam** data set consists of 58 variables: 57 continuous and one class label. If an observation is labeled class 1, then it is considered to be spam. If it is of class 0, then it is not considered spam. The first 48 attributes represent the percentage of words in the email that match some specified word corresponding to spam or not spam. There are an additional six variables that specify the percentage of characters in the email that match a specified character. Others refer to attributes relating to uninterrupted sequences of capital letters. More information on the attributes is available at the above internet link. One can use these data to build classifiers that will discriminate between spam and non-spam emails. In this application, a low false positive rate (classifying an email as spam when it is not) is very important.

sparrow

These data are taken from Manly [1994]. They represent some measurements taken on sparrows collected after a storm on February 1, 1898. Eight morphological characteristics and the weight were measured for each bird, five of which are provided in this data set. These are on female sparrows only. The variables are total length, alar extent, length of beak and head, length of humerus, and length of keel of sternum. All lengths are in millimeters. The first 21 of these birds survived, and the rest died.

swissroll

This data file contains a set of variables randomly generated from the Swiss roll manifold with a hole in it. It also has the data in the reduced space (from ISOMAP and HLLE) that was used in Example 3.5.

votfraud

These data represent the Democratic over Republican pluralities of voting machine and absentee votes for 22 Philadelphia County elections. The variable **machine** is the Democratic plurality in machine votes, and the variable **absentee** is the Democratic plurality in absentee votes [Simonoff, 1996].

yeast

The **yeast** data set is described in Chapter 1. It contains the gene expression levels over two cell cycles and five phases.

Appendix D

Introduction to MATLAB[1]

D.1 What Is MATLAB?

MATLAB is a technical computing environment developed by The MathWorks, Inc. for computation and data visualization. It is both an interactive system and a programming language, whose basic data element is an array: scalar, vector, matrix, or multi-dimensional array. Besides basic array operations, it offers programming features similar to those of other computing languages (e.g., functions, control flow, etc.).

In this appendix, we provide a brief summary of MATLAB to help the reader understand the algorithms in the text. We do not claim that this introduction is complete, and we urge the reader to learn more about MATLAB from other sources. The documentation that comes with MATLAB is excellent, and the reader should find the tutorials contained in there to be helpful. For a comprehensive overview of MATLAB, we also recommend Hanselman and Littlefield [1998, 2001]. A new book that introduces numerical computing with MATLAB is by Moler [2004]. This book is available in print version (see references); an electronic version can be downloaded from

http://www.mathworks.com/moler

If the reader needs to understand more about the graphics and GUI capabilities in MATLAB, Marchand and Holland [2003] is the reference to use.

MATLAB will execute on Windows, UNIX/Linux, and Macintosh systems. Here we focus on the Windows version, but most of the information applies to all systems. The main MATLAB software package contains many functions for analyzing data. There are also specialty toolboxes that extend the capabilities of MATLAB. These are available from The MathWorks and

---

[1] Much of this appendix was taken and updated from the corresponding appendix in *Computational Statistics Handbook with MATLAB* [2002].

third party vendors for purchase. Some toolboxes are also on the internet and may be downloaded at no charge. See Appendix B for a partial list.

We assume that readers know how to start MATLAB for their particular platform. When MATLAB is started, you will have a command window with a prompt where you can enter commands. Other windows are available (help window, history window, etc.), but we do not cover those here.

D.2 Getting Help in MATLAB

One useful and important aspect of MATLAB is the **help** feature. There are many ways to get information about a MATLAB function. Not only does the **help** provide information about the function, but it also gives references for other related functions. We discuss below the various ways to get help in MATLAB.

- *Command Line:* Typing **help** and then the function name at the command line will, in most cases, tell you everything you need to know about the function. In this text, we do not write about all the capabilities or uses of a function. The reader is strongly encouraged to use command line **help** to find out more. As an example, typing **help plot** at the command line provides lots of useful information about the basic **plot** function. Note that the command line **help** works with the EDA Toolbox (and others) as well.

- *Help Menu:* The **help** files can also be accessed via the usual **Help** menu. This opens up a separate **help** window. Information can be obtained by clicking on links or searching the index.

D.3 File and Workspace Management

We can enter commands interactively at the command line or save them in an M-file. So, it is important to know some commands for file management. The commands shown in Table D.1 can be used to list, view, and delete files.

Variables created in a session (and not deleted) live in the MATLAB *workspace*. You can recall the variable at any time by typing in the variable name with no punctuation at the end. Note that MATLAB is case sensitive, so **Temp, temp,** and **TEMP** represent different variables.

MATLAB remembers the commands that you enter in the command history. There is a separate command history window available via the **View** menu and certain desktop layouts. One can use this to re-execute old

TABLE D.1

File Management Commands

Command	Usage
`dir, ls`	Shows the files in the present directory.
`delete` *`filename`*	Deletes *`filename`*.
`cd, pwd`	Show the present directory.
`cd dir, chdir`	Changes the directory. There is a pop-up menu on the toolbar that allows the user to change directory.
`type` *`filename`*	Lists the contents of *`filename`*.
`edit` *`filename`*	Brings up *`filename`* in the editor.
`which` *`filename`*	Displays the path to *`filename`*. This can help determine whether a file is part of the standard MATLAB package.
`what`	Lists the **.m** files and **.mat** files that are in the current directory.

TABLE D.2

Commands for Workspace Management

Command	Usage
`who`	Lists all variables in the workspace.
`whos`	Lists all variables in the workspace along with the size in bytes, array dimensions, and object type.
`clear`	Removes all variables from the workspace.
`clear x y`	Removes variables **x** and **y** from the workspace.

commands. One way is to highlight the desired commands (hold the **ctrl** key while clicking), right click for a shortcut menu on the highlighted section, and select **Evaluate Selection**. While in the command window, the arrow keys can be used to recall and edit commands. The up-arrow and down-arrow keys scroll through the commands. The left and right arrows move through the present command. By using these keys, the user can recall commands and edit them using common editing keystrokes.

We can view the contents of the current workspace using the **Workspace Browser**. This is accessed through the **File** menu or the toolbar. All variables in the workspace are listed in the window. The variables can be viewed and edited in a spreadsheet-like window format by double-clicking on the variable name.

The commands contained in Table D.2 help manage the workspace. It is important to be able to get data into MATLAB and to save it. We outline below some of the ways to get data in and out of MATLAB. These are not the only options for file I/O. For example, see **help** on **fprintf**, **fscanf**, and **textread** for more possibilities.

- *Command Line:* The **save** and **load** commands are the main way to perform file I/O in MATLAB. We give some examples of how to use the **save** command. The **load** command works similarly.

Command	Usage
save *filename*	Saves all variables in *filename*.mat.
save *filename* var1 var2	Saves only variables **var1** **var2** in *filename*.mat.
save *filename* var1 -ascii	Saves **var1** in ASCII format in *filename*.

- *File Menu:* There are commands in the **File** menu for saving and loading the workspace.
- *Import Wizard:* There is a spreadsheet-like window for inputting data. To execute the wizard, type **uiimport** at the command line.

D.4 Punctuation in MATLAB

Table D.3 contains some of the common punctuation characters in MATLAB, and how they are used.

TABLE D.3

List of Punctuation

Punctuation	Usage
%	A percent sign denotes a comment line. Information after the % is ignored.
,	When used to separate commands on a single line, a comma tells MATLAB to display the results of the preceding command. When used in concatenation (between brackets [ ]), a comma or a blank space concatenates elements along a row. A comma also has other uses, including separating function arguments and array subscripts.
;	When used after a line of input or between commands on a single line, a semicolon tells MATLAB not to display the results of the preceding command. When used in concatenation (between brackets [ ]), a semicolon begins a new row.
. . .	Three periods denote the continuation of a statement. Comment statements and variable names cannot be continued with this punctuation.
!	An exclamation tells MATLAB to execute the following as an operating system command.
:	The colon specifies a range of numbers. For example, 1:10 means the numbers 1 through 10. A colon in an array dimension accesses all elements in that dimension.
.	The period before an operator tells MATLAB to perform the corresponding operation on each element in the array.

D.5 Arithmetic Operators

Arithmetic operators (*, /, +, −, ^) in MATLAB follow the conventions in linear algebra. If we are multiplying two matrices, **A** and **B**, they must be dimensionally correct. In other words, the number of columns of **A** must be equal to the number of rows of **B**. To multiply two matrices in this manner, we simply use **A*B**. It is important to remember that the default interpretation of an operation is to perform the corresponding array operation.

MATLAB follows the usual order of operations. The precedence can be changed by using parentheses, as in other programming languages.

It is often useful to operate on an array element-by-element. For instance, we might want to square each element of an array. To accomplish this, we add a period before the operator. As an example, to square each element of array A, we use **A.^2**. These operators are summarized in Table D.4.

TABLE D.4

List of Element-by-Element Operators

Operator	Usage
.*	Multiply element-by-element.
./	Divide element-by-element.
.^	Raise elements to powers.

D.6 Data Constructs in MATLAB

Basic Data Constructs

We do not cover the object-oriented aspects of MATLAB here. Thus, we are concerned mostly with data that are floating point (type **double**) or strings (type **char**). The elements in the arrays will be of these two data types.

The fundamental data element in MATLAB is an array. Arrays can be:

- The 0×0 empty array created using [ ].
- A 1×1 scalar array.
- A row vector, which is a $1 \times n$ array.
- A column vector, which is an $n \times 1$ array.
- A matrix with two dimensions, say $m \times n$ or $n \times n$.
- A multi-dimensional array, say $m \times ... \times n$.

Arrays must always be dimensionally conformal and all elements must be of the same data type. In other words, a 2×3 matrix must have 3 elements (e.g., numbers) on each of its 2 rows. Table D.5 gives examples of how to access elements of arrays.

Building Arrays

In most cases, the statistician or engineer will need to import data into MATLAB using **load** or some other method described previously. However, sometimes we might need to type in simple arrays for testing code or entering parameters, etc. Here we cover some of the ways to build small arrays. Note that this can also be used to combine separate arrays into one large array.

Commas or spaces concatenate elements (which can be arrays) as columns. Thus, we get a row vector from the following

```
temp = [1, 4, 5];
```

or we can concatenate two column vectors **a** and **b** into one matrix, as follows

```
temp = [a b];
```

The semi-colon tells MATLAB to concatenate elements as rows. So, we would get a column vector from this command:

```
temp = [1; 4; 5];
```

We note that when concatenating arrays, the sizes of each array element must be conformal. The ideas presented here also apply to cell arrays, discussed below.

Before we continue with cell arrays, we cover some of the other useful functions in MATLAB for building arrays. These are summarized here.

Function	Usage
zeros, ones	These build arrays containing all 0's or all 1's, respectively.
rand, randn	These build arrays containing uniform (0,1) random variables or standard normal random variables, respectively.
eye	This creates an identity matrix.

Cell Arrays

Cell arrays and structures allow for more flexibility. Cell arrays can have elements that contain any data type (even other cell arrays), and they can be of different sizes. The cell array has an overall structure that is similar to the basic data arrays. For instance, the cells are arranged in dimensions (rows, columns, etc.). If we have a 2×3 cell array, then each of its 2 rows has to have 3 cells. However, the *content* of the cells can be different sizes and can contain different types of data. One cell might contain **char** data, another **double**, and some can be empty. Mathematical operations are not defined on cell arrays.

TABLE D.5

Examples of Accessing Elements of Arrays

Notation	Usage
a(i)	Denotes the i-th element (cell) of a row or column vector array (cell array).
A(:,i)	Accesses the i-th column of a matrix or cell array. In this case, the colon in the row dimension tells MATLAB to access all rows.
A(i,:)	Accesses the i-th row of a matrix or cell array. The colon tells MATLAB to gather all of the columns.
A(1,3,4)	This accesses the element in the first row, third column on the fourth entry of dimension 3 (sometimes called the page).

In Table D.5, we show some of the common ways to access elements of arrays, which can be cell arrays or basic arrays. With cell arrays, this accesses the cell element, but not the contents of the cells. Curly braces, **{ }**, are used to get to the elements inside the cell. For example, **A{1,1}** would give us the contents of the cell (type **double** or **char**). Whereas, **A(1,1)** is the cell itself and has data type **cell**. The two notations can be combined to access part of the contents of a cell. To get the first two elements of the contents of **A{1,1}**, assuming it contains a vector, we can use

$$A\{1,1\} \ (1:2).$$

Cell arrays are very useful when using strings in plotting functions such as **text**.

Structures

Structures are similar to cell arrays in that they allow one to combine collections of dissimilar data into a single variable. Individual structure elements are addressed by names called *fields*. We use the *dot notation* to access the fields. Each element of a structure is called a *record*.

For example, suppose we had a structure called **data** that had fields called **name**, **dob**, **test**. Then we could obtain the information in the tenth record using

```
data(10).name
data(10).dob
data(10).test
```

Here **name** and **dob** are vectors of type **char**, and **test** is a numeric. Note that fields may also be structures with their own fields. We may access the elements of the fields using the element accessing syntax discussed earlier.

D.7 Script Files and Functions

MATLAB programs are saved in M-files. These are text files that contain MATLAB commands, and they are saved with the **.m** extension. Any text editor can be used to create them, but the one that comes with MATLAB is recommended. This editor can be activated using the **File** menu or the toolbar.

When script files are executed, the commands are implemented just as if you typed them in interactively. The commands have access to the workspace and any variables created by the script file are in the workspace when the script finishes executing. To execute a script file, simply type the name of the file at the command line or use the option in the **File** menu.

Script files and functions both have the same **.m** extension. However, a function has a special syntax for the first line. In the general case, this syntax is

```
function [out1,...,outM] = func_name(in1,...,inN)
```

A function does not have to be written with input or output arguments. Whether you have these or not depends on the application and the purpose of the function. The function corresponding to the above syntax would be saved in a file called **func_name.m**. These functions are used in the same way any other MATLAB function is used.

It is important to keep in mind that functions in MATLAB are similar to those in other programming languages. The function has its own workspace. So, communicating information between the function workspace and the main workspace is done via input and output variables.

It is always a good idea to put several comment lines at the beginning of your function. These are returned by the **help** command.

D.8 Control Flow

Most computer languages provide features that allow one to control the flow of execution depending on certain conditions. MATLAB has similar constructs:

- **for** loops
- **while** loops
- **if-else** statements
- **switch** statement

These should be used sparingly. In most cases, it is more efficient in MATLAB to operate on an entire array rather than looping through it.

for Loop

The basic syntax for a **for** loop is

```
for i = array
    commands
end
```

Each time through the loop, the loop variable **i** assumes the next value in **array**. The colon notation is usually used to generate a sequence of numbers that **i** will take on. For example,

```
for i = 1:10
```

The commands between the **for** and the **end** statements are executed once for every value in the array. Several **for** loops can be nested, where each loop is closed by **end**.

while Loop

A **while** loop executes an indefinite number of times. The general syntax is:

```
while expression
    commands
end
```

The commands between the **while** and the **end** are executed as long as **expression** is true. Note that in MATLAB a scalar that is nonzero evaluates to true. Usually a scalar entry is used in the **expression**, but an array can be used also. In the case of arrays, all elements of the resulting array must be true for the commands to execute.

if-else Statements

Sometimes, commands must be executed based on a relational test. The **if-else** statement is suitable here. The basic syntax is

```
if expression
    commands
elseif expression
    commands
else
    commands
end
```

Only one **end** is required at the end of the sequence of **if, elseif** and **else** statements. Commands are executed only if the corresponding **expression** is true.

switch Statement

The **switch** statement is useful if one needs a lot of **if, elseif** statements to execute the program. This construct is very similar to that in the C language. The basic syntax is:

```
switch expression
case value1
    commands execute if expression is value1
case value2
    commands execute if expression is value2
...
otherwise
    commands
end
```

The **expression** must be either a scalar or a character string.

D.9 Simple Plotting

For more information on some of the plotting capabilities of MATLAB, the reader is referred to the MATLAB documentation *Using MATLAB Graphics* and *Graphics and GUIs with MATLAB* [Marchand and Holland, 2003]. In this appendix, we briefly describe some of the basic uses of **plot** for plotting 2-D graphics and **plot3** for plotting 3-D graphics. The reader is strongly urged to view the **help** file for more information and options for these functions.

When the function **plot** is called, it opens a **Figure** window (if one is not already there) scales the axes to fit the data, and plots the points. The default is to plot the points and connect them using straight lines. For example,

> **plot(x,y)**

plots the values in vector **x** on the horizontal axis and the values in vector **y** on the vertical axis, connected by straight lines. These vectors must be the same size or you will get an error.

Any number of pairs can be used as arguments to **plot**. For instance, the following command plots two curves,

> **plot(x,y1,x,y2)**

on the same axes. If only one argument is supplied to **plot**, then MATLAB plots the vector versus the index of its values.

The default is a solid line, but MATLAB allows other choices. These are given in Table D.6.

TABLE D.6

Line Styles for Plots

Notation	Line Type
-	Solid Line
:	Dotted Line
-.	Dash-dot Line
--	Dashed Line

If several lines are plotted on one set of axes, then MATLAB plots them in different colors. The predefined colors are listed in Table D.7.

TABLE D.7

Line Colors for Plots

Notation	Color
b	blue
g	green
r	red
c	cyan
m	magenta
y	yellow
k	black
w	white

Plotting symbols (e.g., *, x, o, etc.) can be used for the points. Since the list of plotting symbols is rather long, we refer the reader to the online **help** for **plot** for more information. To plot a curve where both points and a connected curve are displayed, use

```
plot(x, y, x, y, 'b*')
```

or

```
plot(x,y,'b*-')
```

This command first plots the points in **x** and **y**, connecting them with straight lines. It then plots the points in **x** and **y** using the symbol * and the color blue. See the **help** on **graph2d** for more 2-D plots.

The **plot3** function works the same as **plot**, except that it takes three vectors for plotting:

```
plot3(x, y, z)
```

All of the line styles, colors, and plotting symbols apply to **plot3**. Other forms of 3-D plotting (e.g., **surf** and **mesh**) are available. See the **help** on **graph3d** for more capabilities. Titles and axes labels can be created for all plots using **title**, **xlabel**, **ylabel**, and **zlabel**.

Before we finish this discussion on simple plotting techniques in MATLAB, we present a way to put several axes or plots in one **figure** window. This is through the use of the **subplot** function. This creates an $m \times n$ matrix of plots (or axes) in the current **figure** window. We provide an example below, where we show how to create two plots side-by-side.

```
% Create the left-most plot.
subplot(1,2,1)
plot(x,y)
% Create the right-most plot
subplot(1,2,2)
plot(x,z)
```

The first two arguments to **subplot** tell MATLAB about the layout of the plots within the **figure** window. The third argument tells MATLAB which plot to work with. The plots are numbered from top to bottom and left to right. The most recent plot that was created or worked on is the one affected by any subsequent plotting commands. To access a previous plot, simply use the **subplot** function again with the proper value for the third argument. You can think of the **subplot** function as a *pointer* that tells MATLAB what set of axes to work with.

Through the use of MATLAB's low-level Handle Graphics functions, the data analyst has complete control over graphical output. We do not present any of that here, because we make limited use of these capabilities. However, we urge the reader to look at the online **help** for **propedit**. This graphical user interface allows the user to change many aspects or properties of the plots without resorting to Handle Graphics.

D.10 Where to get MATLAB Information

For MATLAB product information, please contact:

> The MathWorks, Inc.
> 3 Apple Hill Drive
> Natick, MA, 01760-2098 USA
> Tel: 508-647-7000
> Fax: 508-647-7101
> E-mail: info@mathworks.com
> Web: www.mathworks.com

There are two useful resources that describe new products, programming tips, algorithm development, upcoming events, etc. One is the monthly electronic newsletter called the *MATLAB Digest*. Another is called *MATLAB News & Notes*, published quarterly. You can subscribe to both of these at **www.mathworks.com** or send an email request to

subscribe@mathworks.com

Back issues of these documents are available on-line.

The MathWorks has an educational area that contains many resources for professors and students. You can find contributed course materials, product recommendations by curriculum, tutorials on MATLAB, and more at:

www.mathworks.com/products/education/

The MathWorks also provides free live (and archived) webinars that demonstrate their products. See

www.mathworks.com/webinars

for a schedule and registration information.

There is an active news group to discuss MATLAB. Sign up at

comp.soft-sys.matlab

This is a good place to ask questions and share ideas. Another useful source for MATLAB users is MATLAB Central. This is on the main website at The MathWorks:

www.mathworks.com/matlabcentral/

It has an area for sharing MATLAB files, a web interface to the news group, and other features. It is always a good idea to check here first before writing your own functions, since what you need might be there already.

Appendix E

MATLAB Functions

For the convenience of the reader, we provide a list of functions and commands that were mentioned in the text and others that we think might be of interest to the exploratory data analyst. Be sure to check the **help** files on the Statistics Toolbox and EDA Toolbox for a list of all functions that are available.

E.1 MATLAB

abs	Absolute value of a vector or matrix.
axes	Low-level function to set axes properties.
axis	Command or function to change axes.
bar, bar3	Construct a 2-D or 3-D bar graph.
cart2pol	Change from Cartesian to polar coordinates.
ceil	Round up to the next integer.
char	Creates character array.
cla	Clears the current axes.
colorbar	Puts a color scale on the figure.
colormap	Changes the color map for the figure.
contour	Constructs a contour plot.
convhull	Finds the convex hull.
cos	Finds the cosine of the angle (radians).
cov	Returns the sample covariance matrix.
cumsum	Calculates the cumulative sum of vector elements.
diag	Diagonal matrices and diagonals of a matrix.
diff	Finds the difference between elements.
drawnow	Force MATLAB to update the graphics screen.
eig, eigs	Eigenvalues/eigenvectors for full and sparse matrices.
exp	Find the exponential of the argument.
eye	Provides an identity matrix.
figure	Creates a blank figure window.
find	Find indices of nonzero elements.
findobj	Find objects with certain properties.
flipud	Flip matrix in up/down direction.

`floor`	Round down to the nearest integer.
`gammaln`	Logarithm of the gamma function.
`gray`	A pre-specified gray-scale color map.
`help` *funcname*	Command line way to get help on a function.
`hist`	Create a histogram.
`hold`	Freeze the current plot, graphics are added.
`imagesc`	Scaled image of the data.
`inpolygon`	True if the point is inside the polygon, zero otherwise.
`int2str`	Convert integer to char.
`intersect`	Elements in the set intersection.
`legend`	Add a legend to the axes.
`length`	Returns the length of a vector.
`line`	Low-level function to create a line or curve.
`linspace`	Provide linearly-spaced points in a given 1-D range.
`load`	Load some data – MAT-file or ASCII.
`log`	Take logarithm of elements.
`mean`	Find the sample mean.
`median`	Find the median.
`meshgrid`	Provide linearly-spaced points (X, Y) in a 2-D domain.
`min, max`	Find the minimum or maximum element in a vector.
`mod`	Returns modulus after division.
`norm`	Calculates the vector or matrix norm.
`ones`	Provides an array with all ones.
`pause`	Pause all activity in MATLAB.
`pcolor`	Pseudo-color plot – elements specify colors.
`pinv`	Pseudo-inverse of a matrix.
`plot, plot3`	2-D or 3-D plots.
`plotmatrix`	Matrix of scatterplots.
`pol2cart`	Convert polar to Cartesian coordinates.
`polyfit`	Fit data to a polynomial.
`polyval`	Evaluate a polynomial.
`rand`	Generate uniform (0, 1) random variables.
`randn`	Generate standard normal random variables.
`repmat`	Construct a new array with replicated/tiled elements.
`reshape`	Reshape or change the size of a matrix.
`round`	Round towards the nearest integer.
`scatter, scatter3`	Construct 2-D and 3-D scatterplots.
`semilogy`	Construct a 2-D plot, with a vertical log scale.
`set`	Set properties of graphics objects.
`setdiff`	Find the set difference between vectors.
`sin`	Calculate the sine (radians.)
`size`	Find the size of an array.
`sort`	Sort in ascending order.
`sqrt`	Find the square root.
`std`	Find the sample standard deviation.
`strmatch`	Find matches for a string.
`sum`	Add up the elements in a vector.
`surf`	Construct a surface plot.
`svd`	Singular value decomposition.
`tan`	Find the tangent of an angle (radians).

`title`	Put a title on a plot.
`vander`	Vandermonde matrix.
`xlabel,ylabel`	Add labels to the axes in a plot.
`zeros`	Returns an array with all zeros.

E.2 Statistics Toolbox - Versions 4 and 5

Unless otherwise noted, the following functions are included in version 4 of the Statistics Toolbox.

`biplot` (V5)	Biplot of variable/factor coefficients and scores.
`boxplot`	Displays boxplots of the columns of a matrix.
`cluster`	Get clusters from hierarchical clustering.
`clusterdata`	Construct clusters from data (whole process).
`cmdscale`	Classical multi-dimensional scaling.
`cophenet`	Calculates the cophenetic coefficient.
`dendrogram`	Constructs the dendrogram.
`dfittool` (V5)	GUI for fitting distributions to data.
`exprnd`	Generate random variables from exponential.
`factoran`	Factor analysis.
`gname`	Interactive labeling of scatterplot.
`gplotmatrix`	Grouped scatterplot matrix.
`gscatter`	Grouped scatter plot.
`hist3` (V5)	Histogram of bivariate data.
`iqr`	Interquartile range.
`kmeans`	k-means clustering.
`linkage`	Agglomerative hierarchical clustering.
`mdscale` (V5)	Metric and nonmetric multidimensional scaling.
`mvnpdf`	Multivariate normal probability density function.
`mvnrnd`	Generate multivariate normal random variables.
`norminv`	Inverse of the normal cumulative distribution function.
`normpdf`	1-D normal probability density function.
`normplot`	Normal probability plot.
`normrnd`	Generate 1-D normal random variables.
`pcacov`	PCA using the covariance matrix.
`pdist`	Find interpoint distances.
`probplot` (V5)	Probability plots for specified distributions.
`princomp`	Principal component analysis.
`qqplot`	Empirical q-q plot
`quantile` (V5)	Find the quantiles of a sample.
`silhouette`	Construct silhouette plot and return silhouette values.
`squareform`	Convert output of `pdist` to matrix.
`unifrnd`	Generate uniform random variables over range (a, b).
`weibplot`	Weibull probability plot.

E.3 Exploratory Data Analysis Toolbox

adjrand	Adjusted Rand index to compare groupings.
agmclust	Model-based agglomerative clustering.
bagmat.exe	Executable file to get arrays needed for bagplot.
bagplot	M-file to construct actual bagplot.
boxp	Boxplot - regular.
boxprct	Box-percentile plot.
brushscatter	Scatterplot brushing and linking.
coplot	Coplot from Data Visualization Toolbox.
csandrews	Andrews' curves plot.
csparallel	Parallel coordinates plot.
csppstrtrem	Remove structure in PPEDA.
dotchart	Dot chart plot.
genmix	GUI to generate random variables from finite mixture.
gtm_pmd	Calculates posterior mode projection (GTM Toolbox).
gtm_pmn	Calculates posterior mean projection (GTM Toolbox).
gtm_stp2	Generates components of a GTM (GTM Toolbox).
gtm_trn	Train the GTM using EM (GTM Toolbox).
hexplot	Hexagonal binning for scatterplot.
hlle	Hessian eigenmaps.
idpettis	Intrinsic dimensionality estimate.
intour	Interpolation tour of the data.
isomap	ISOMAP nonlinear dimensionality reduction.
kdimtour	k-dimensional grand tour.
lle	Locally linear embedding.
loess	1-D loess scatterplot smoothing.
loess2	2-D loess smoothing from Data Visualization Toolbox.
loessenv	Loess upper and lower envelopes.
loessr	Robust loess scatterplot smoothing.
mbcfinmix	Model-based finite mixture estimation - EM.
mbclust	Model-based clustering.
mixclass	Classification using mixture model.
multiwayplot	Multiway dot charts.
nmmds	Nonmetric multidimensional scaling.
permtourandrews	Permutation tour using Andrews' curves.
permtourparallel	Permutation tour using parallel coordinate plots.
plotbic	Plot the BIC values from model-based clustering.
plotmatrixandr	Plot matrix of Andrews' curves.
plotmatrixpara	Plot matrix of parallel coordinates.
polarloess	Bivariate smoothing using loess.
ppeda	Projection pursuit EDA.
pseudotour	Pseudo grand tour.
quantileseda	Sample quantiles.
quartiles	Sample quartiles using Tukey's fourths.
randind	Rand index to compare groupings.
reclus	ReClus plot to visualize cluster output.

rectplot	Rectangle plot to visualize hierarchical clustering.
scattergui	Scatterplot with interactive labeling.
som_autolabel	Automatic labeling (SOM Toolbox).
som_data_struct	Create a data structure (SOM Toolbox).
som_make	Create, initialize and train SOM (SOM Toolbox).
som_normalize	Normalize data (SOM Toolbox).
som_set	Set up SOM structures (SOM Toolbox).
som_show	Basic SOM visualization (SOM Toolbox).
som_show_add	Shows hits, labels and trajectories (SOM Toolbox).
torustour	Asimov grand tour
treemap	Treemap display for hierarchical clustering.

References

Alter, O., P. O. Brown, and D. Botstein. 2000. "Singular value decomposition for genome-wide expression data processing and modeling," *Proceedings of the National Academy of Science*, **97**:10101-10106.

Anderberg, Michael R. 1973. *Cluster Analysis for Applications*, New York: Academic Press.

Anderson, E. 1935. "The irises of the Gaspe Peninsula," *Bulletin of the American Iris Society*, **59**:2-5.

Andrews, D. F. 1972. "Plots of high-dimensional data," *Biometrics*, **28**:125-136.

Andrews, D. F. 1974. "A robust method of multiple linear regression," *Technometrics*, **16**:523-531.

Andrews, D. F. and A. M. Herzberg. 1985. *Data: A Collection of Problems from Many Fields for the Student and Research Worker*, New York: Springer-Verlag.

Anscombe, F. J. 1973. "Graphs in statistical analysis," *The American Statistician*, **27**: 17-21.

Asimov, Daniel. 1985. "The grand tour: A tool for viewing multidimensional data," *SIAM Journal of Scientific and Statistical Computing*, **6**:128-143.

Asimov, D. and A. Buja. 1994. "The grand tour via geodesic interpolation of 2-frames," in *Visual Data Exploration and Analysis, Symposium on Electronic Imaging Science and Technology, IS&T/SPIE*.

Baeza-Yates, Ricardo and Berthier Ribero-Neto. 1999. *Modern Information Retrieval*, New York, NY: ACM Press.

Bailey, T. A. and R. Dubes. 1982. "Cluster validity profiles," *Pattern Recognition*, **15**:61-83.

Balasubramanian, M. and E. L. Schwartz. 2002. "The isomap algorithm and topological stability (with rejoinder)," *Science*, **295**:7.

Banfield, A. D. and A. E. Raftery. 1993. "Model-based Gaussian and non-Gaussian clustering," *Biometrics*, **49**:803-821.

Basseville, Michele. 1989. "Distance measures for signal processing and pattern recognition," *Signal Processing*, **18**:349-369.

Becker, R. A., and W. S. Cleveland. 1987. "Brushing scatterplots," *Technometrics*, **29**:127-142.

Becker, R. A., and W. S. Cleveland. 1991. "Viewing multivariate scattered data," *Pixel*, July/August, 36-41.

Becker, R. A., W. S. Cleveland, and A. R. Wilks. 1987. "Dynamic graphics for data analysis," *Statistical Science*, **2**:355-395.

Becker, R. A., L. Denby, R. McGill, and A. Wilks. 1986. "Datacryptanalysis: A case study," *Proceedings of the Section on Statistical Graphics*, 92-91.

Benjamini, Yoav. 1988. "Opening the box of a boxplot," *The American Statistician*, **42**: 257-262.

Bennett, G. W. 1988. "Determination of anaerobic threshold," *Canadian Journal of Statistics*, **16**:307-310.

Bensmail, H., G. Celeux, A. E. Raftery, and C. P. Robert. 1997. "Inference in model-based cluster analysis," *Statistics and Computing*, 7:1-10.

Berry, Michael W., and Murray Browne. 1999. *Understanding Search Engines: Mathematical Modeling and Text Retrieval*, Philadelphia, PA: SIAM.

Berry, M. W., S. T. Dumais, and G. W. O'Brien. 1995. "Using linear algebra for intelligent information retrieval," *SIAM Review*, 37:573-595.

Berry, M. W., Z. Drmac, and E. R. Jessup. 1999. "Matrices, vector spaces, and information retrieval," *SIAM Review*, 41:335-362.

Bertin, Jacques. 1983. *Semiology of Graphics: Diagrams, Networks, Maps*. Madison, WI: The University of Wisconsin Press.

Bhattacharjee, A., W. G. Richards, J. Staunton, C. Li, S. Monti, P. Vasa, C. Ladd, J. Beheshti, R. Bueno, M. Gillette, M. Loda, G. Weber, E. J. Mark, E. S. Lander, W. Wong, B. E. Johnson, T. R. Bolub, D. J. Sugarbaker, & M. Meyerson. 2001. "Classification of human lung carcinomas by mRNA expression profiling reveals distinct adenocarcinoma subclasses," *Proceedings of the National Academy of Science*, **98**:13790-13795.

Biernacki, C. and G. Govaert. 1997. "Using the classification likelihood to choose the number of clusters," *Computing Science and Statistics*, **29**(2):451-457.

Biernacki, C., G. Celeux, and G. Govaert. 1999. "An improvement of the NEC criterion for assessing the number of clusters in a mixture model," *Pattern Recognition Letters*, **20**:267-272.

Binder, D. A. 1978. "Bayesian cluster analysis," *Biometrika*, **65**:31-38.

Bishop, Christopher M., G. E. Hinton, and I. G. D. Strachan. 1997. "GTM through time," *Proceedings IEEE 5th International Conference on Artificial Neural Networks*, Cambridge, UK, 111-116.

Bishop, Christopher M., Markus Svensén, and Christopher K. I. Williams. 1996. "GTM: The generative topographic mapping," Neural Computing Research Group, Technical Report NCRG/96/030.

Bishop, Christopher M., Markus Svensén, and Christopher K. I. Williams. 1997a. "Magnification factors for the SOM and GTM algorithms, *Proceedings 1997 Workshop on Self-Organizing Maps*, Helsinki University of Technology, 333-338.

Bishop, Christopher M., Markus Svensén, and Christopher K. I. Williams. 1997b. "Magnification factors for the GTM algorithm," *Proceedings IEEE 5th International Conference on Artificial Neural Networks*, Cambridge, UK, 64-69.

Bishop, Christopher M., Markus Svensén, and Christopher K. I. Williams. 1998a. "The generative topographic mapping," *Neural Computation*, **10**:215-234.

Bishop, Christopher M., Markus Svensén, and Christopher K. I. Williams. 1998b. "Developments of the generative topographic mapping," *Neurocomputing*, **21**:203-224.

Bishop, Christopher M. and M. E. Tipping. 1998. "A hierarchical latent variable model for data visualization," *IEEE Transactions on Pattern Analysis and Machine Intelligence*, **20**:281-293.

Blasius, J. and M. Greenacre. 1998. *Visualizing of Categorical Data*, New York: Academic Press.

Bock, H. 1985. "On some significance tests in cluster analysis," *Journal of Classification*, **2**:77-108.

Bock, H. 1996. "Probabilistic models in cluster analysis," *Computational Statistics and Data Analysis*, **23**:5-28.

Bolton, R. J. and W. J. Krzanowski. 1999. "A characterization of principal components for projection pursuit," *The American Statistician*, **53**:108-109.

Bonner, R. 1964. "On some clustering techniques," *IBM Journal of Research and Development*, **8**:22-32.

Borg, Ingwer and Patrick Groenen. 1997. *Modern Multidimensional Scaling: Theory and Applications*, New York: Springer.

Bowman, A. W. and A. Azzalini. 1997. *Applied Smoothing Techniques for Data Analysis: The Kernel Approach with S-Plus Illustrations*, Oxford: Oxford University Press.

Brinkman, N. D. 1981. "Ethanol fuel - a single-cylinder engine study of efficiency and exhaust emissions," *SAE Transactions*, **90**:1410-1424.

Bruls, D. M., C. Huizing, J. J. van Wijk. 2000. "Squarified Treemaps," in: W. de Leeuw, R. van Liere (eds.), *Data visualization 2000, Proceedings of the Joint Eurographics and IEEE TCVG Symposium on Visualization*, 33-42, New York: Springer.

Bruntz, S. M., W. S. Cleveland, B. Kleiner, and J. L. Warner. 1974. "The dependence of ambient ozone on solar radiation, wind, temperature and mixing weight," *Symposium on Atmospheric Diffusion and Air Pollution*, Boston: American Meteorological Society, 125-128.

Bruske, J. and G. Sommer. 1998. "Intrinsic dimensionality estimation with optimally topology preserving maps," *IEEE Transactions on Pattern Analysis and Machine Intelligence*, **20**:572-575.

Buijsman, E., H. F. M. Maas, and W. A. H. Asman. 1987. "Anthropogenic NH_3 Emissions in Europe," *Atmospheric Environment*, **21**:1009-1022.

Buja, A. and D. Asimov. 1986. "Grand tour methods: An outline," *Computer Science and Statistics*, **17**:63-67.

Buja, A., J. A. McDonald, J. Michalak, and W. Stuetzle. 1991. "Interactive data visualization using focusing and linking," *IEEE Visualization, Proceedings of the 2nd Conference on Visualization '91*, 156-163.

Buta, R. 1987. "The structure and dynamics of ringed galaxies, III: Surface photometry and kinematics of the ringed nonbarred spiral NGC 7531," *The Astrophysical Journal Supplement Series*, **64**:1-37.

Calinski, R. and J. Harabasz. 1974. "A dendrite method for cluster analysis," *Communications in Statistics*, **3**:1-27.

Campbell, J. G., C. Fraley, F. Murtagh, and A. E. Raftery. 1997. "Linear flaw detection in woven textiles using model-based clustering," *Pattern Recognition Letters*, **18**:1539-1549.

Campbell, J. G., C. Fraley, D. Stanford, F. Murtagh, and A. E. Raftery. 1999. "Model-based methods for real-time textile fault detection," *International Journal of Imaging Systems and Technology*, **10**:339-346.

Carr, D., R. Littlefield, W. Nicholson, and J. Littlefield. 1987. "Scatterplot matrix techniques for large N," *Journal of the American Statistical Association*, **82**:424-436.

Cattell, R. B. 1966. "The scree test for the number of factors," *Journal of Multivariate Behavioral Research*, **1**:245-276.

Cattell, R. B. 1978. *The Scientific Use of Factor Analysis in Behavioral and Life Sciences*, New York: Plenum Press.

Celeux, G. and G. Govaert. 1995. "Gaussian parsimonious clustering models," *Pattern Recognition*, **28**:781-793.

Chakrapani, T. K. and A. S. C. Ehrenberg. 1981. "An alternative to factor analysis in marketing research - Part 2: Between group analysis," *Professional Marketing Research Society Journal*, **1**:32-38.

Chambers, John. 1999. "Computing with data: Concepts and challenges," *The American Statistician*, **53**:73-84.

Chambers, J. M., W. S. Cleveland, B. Kleiner, and P. A. Tukey. 1983. *Graphical Methods for Data Analysis*, Boca Raton: CRC/Chapman and Hall.

Charniak, Eugene. 1996. *Statistical Language Learning*, Cambridge, MA: The MIT Press.

Chatfield, C. 1985. "The initial examination of data," *Journal of the Royal Statistical Society, A*, **148**:214-253.

Chee, M., R. Yang, E. Hubbell, A. Berno, X. C. Huang, D. Stern, J. Winkler, D. J. Lockhart, M. S. Morris, and S. P. A. Fodor. 1996. "Accessing genetic information with high-density DNA arrays," *Science*, **274**:610-614.

Chernoff, Herman. 19 73. "The use of faces to represent points in *k*-dimensional space graphically," *Journal of the American Statistical Association*, **68**:361-368.

Cho, R. J., M. J. Campbell, E. A. Winzeler, L. Steinmetz, A. Conway, L. Wodicka, T. G. Wolfsberg, A. E. Gabrielian, D. Landsman, D. J. Lockhart, and R. W. Davis. 1998. "A genome-wide transcriptional analysis of the mitotic cell cycle," *Molecular Cell*, **2**:65-73.

Cleveland, W. S. 1979. "Robust locally weighted regression and smoothing scatterplots," *Journal of the American Statistical Association*, **74**:829-836.

Cleveland, W. S. 1984. "Graphical methods for data presentation: Full scale breaks, dot charts, and multibased logging," *The American Statistician*, **38**:270-280.

Cleveland, W. S. 1993. *Visualizing Data*, New York: Hobart Press.

Cleveland, W. S. and Susan J. Devlin. 1988. "Locally weighted regression: An approach to regression analysis by local fitting," *Journal of the American Statistical Association*, **83**:596-610.

Cleveland, W. S., Susan J. Devlin, and Eric Grosse. 1988. "Regression by local fitting: Methods, properties, and computational algorithms," *Journal of Econometrics*, **37**:87-114.

Cleveland, W. S. and C. Loader. 1996. "Smoothing by local regression: Principles and methods, in Härdle and Schimek (eds.), *Statistical Theory and Computational Aspects of Smoothing*, Heidelberg: Phsyica-Verlag, 10-49.

Cleveland, W. S. and Robert McGill. 1984. "The many faces of a scatterplot," *Journal of the American Statistical Association*, **79**:807-822.

Cohen, A., R. Gnanadesikan, J. R. Kettenring, and J. M. Landwehr. 1977. "Methodological developments in some applications of clustering," in: Applications of Statistics, P. R. Krishnaiah (ed.), Amsterdam: North-Holland.

Cook, D., A. Buja, and J. Cabrera. 1993. "Projection pursuit indexes based on orthonormal function expansions," *Journal of Computational and Graphical Statistics*, 2:225-250.

Cook, D., A. Buja, J. Cabrera, and C. Hurley. 1995. "Grand tour and projection pursuit," *Journal of Computational and Graphical Statistics*, 4:155-172.

Cook, W. J., W. H. Cunningham, W. R. Pulleyblank, and A. Schrijver. 1998. *Combinatorial Optimization*, New York: John Wiley & Sons.

Cormack, R. M. 1971. "A review of classification," *Journal of the Royal Statistical Society, Series A*, 134:321-367.

Costa, Jose A. and Alfred O. Hero. 2004. "Geodesic entropic graphs for dimension and entropy estimation in manifold learning," *IEEE Transactions on Signal Processing*, 52:2210-2221.

Cottrell, M., J. C. Fort, and G. Pages. 1998. "Theoretical aspects of the SOM algorithm," *Neurocomputing*, 21:119-138.

Cox, Trevor F. and Michael A. A. Cox. 2001. *Multidimensional Scaling, 2nd Edition*, Boca Raton: Chapman & Hall/CRC.

Crawford, Stuart. 1991. "Genetic optimization for exploratory projection pursuit," *Proceedings of the 23rd Symposium on the Interface*, 23:318-321.

Crile, G. and D. P. Quiring. 1940. "A record of the body weight and certain organ and gland weights of 3690 animals," *Ohio Journal of Science*, 15:219-259.

Dasgupta, A. and A. E. Raftery. 1998. "Detecting features in spatial point processes with clutter via model-based clustering," *Journal of the American Statistical Association*, 93:294-302.

Davies, O. L. 1957. *Statistical Methods in Research and Production*, New York: Hafner Press.

Davies, David L. and Donald W. Bouldin. 1979. "A cluster separation measure," *IEEE Transactions on Pattern Analysis and Machine Intelligence*, 1:224-227.

Day, N. E. 1969. "Estimating the components of a mixture of normal distributions," *Biometrika*, 56:463-474.

de Leeuw, J. 1977. "Applications of convex analysis to multidimensional scaling," in *Recent Developments in Statistics*, J. R. Barra, R. Brodeau, G. Romier & B. van Cutsem (ed.), Amsterdam, The Netherlands: North-Holland, 133-145.

Deboecek, G. and T. Kohonen. 1998. *Visual Explorations in Finance using Self-Organizing Maps*, London: Springer-Verlag.

Deewester, S., Susan T. Dumais, George W. Furnas, Thomas K. Landauer, and Richard Harshman. 1990. "Indexing by Latent Semantic Analysis," *Journal of the American Society for Information Science*, 41:391-407.

Dempster, A. P., Laird, N. M., and Rubin, D. B. 1977. "Maximum likelihood from incomplete data via the EM algorithm (with discussion)," *Journal of the Royal Statistical Society: B*, 39:1-38.

Diaconis, Persi. 1985. "Theories of data analysis: From magical thinking through classical statistics," in *Exploring Data Tables, Trends, and Shapes*, D. Hoaglin, F. Mosteller, and J. W. Tukey (eds.), New York: John Wiley and Sons.

Diaconis, P. and J. H. Friedman. 1980. "*M* and *N* plots," Technical Report 15, Department of Statistics, Stanford University.

Donoho, D. L. and M. Gasko. 1992. "Breakdown properties of location estimates based on halfspace depth and projected outlyingness," *The Annals of Statistics*, **20**:1803-1827.

Donoho, D. L. and C. Grimes. 2002. Technical Report 2002-27, Department of Statistics, Stanford University.

Donoho, D. L. and C. Grimes. 2003. "Hessian eigenmaps: Locally linear embedding techniques for high-dimensional data," *Proceedings of the National Academy of Science*, **100**:5591-5596.

Draper, N. R. and H. Smith. 1981. *Applied Regression Analysis, 2nd Edition*, New York: John Wiley & Sons.

du Toit, S. H. C., A. G. W. Steyn, and R. H. Stumpf. 1986. *Graphical Exploratory Data Analysis*, New York: Springer-Verlag.

Dubes, R. and A. K. Jain. 1980. "Clustering methodologies in exploratory data analysis," *Advances in Computers, Vol. 19*, New York: Academic Press.

Duda, Richard O. and Peter E. Hart. 1973. *Pattern Classification and Scene Analysis*, New York: John Wiley & Sons.

Duda, Richard O., Peter E. Hart, and David G. Stork. 2001. *Pattern Classification, Second Edition*, New York: John Wiley & Sons.

Edwards, A. W. F. and L. L. Cavalli-Sforza. 1965. "A method for cluster analysis," *Biometrics*, **21**:362-375.

Efromovich, Sam. 1999. *Nonparametric Curve Estimation: Methods, Theory, and Applications*, New York: Springer-Verlag.

Efron, B. and R. J. Tibshirani. 1993. *An Introduction to the Bootstrap*, London: Chapman and Hall.

Embrechts, P. and A. Herzberg. 1991. "Variations of Andrews' plots," *International Statistical Review*, **59**:175-194.

Emerson, John D. and Michael A. Stoto. 1983. "Transforming Data," in *Understanding Robust and Exploratory Data Analysis*, Hoaglin, Mosteller, and Tukey (eds.), New York: John Wiley & Sons.

Estivill-Castro, Vladimir. 2002. "Why so many clustering algorithms - A position paper," *SIGKDD Explorations*, **4**:65-75.

Esty, Warren W. and Jeffrey D. Banfield. 2003. "The box-percentile plot," *Journal of Statistical Software*, **8**, **http://www.jstatsoft.org/v08/i17**.

Everitt, Brian S. 1993. *Cluster Analysis, Third Edition*, New York: Edward Arnold Publishing.

Everitt, B. S. and D. J. Hand. 1981. *Finite Mixture Distributions*, London: Chapman and Hall.

Everitt, B. S., S. Landau, and M. Leese. 2001. *Cluster Analysis, Fourth Edition*, New York: Edward Arnold Publishing.

Fawcett, C. D. 1901. "A second study of the variation and correlation of the human skull, with special reference to the Naqada crania," *Biometrika*, **1**:408-467.

Fieller, N. R. J., E. C. Flenley, and W. Olbricht. 1992. "Statistics of particle size data," *Applied Statistics*, **41**:127-146.

Fieller, N. R. J., D. D. Gilbertson, and W. Olbricht. 1984. "A new method for environmental analysis of particle size distribution data from shoreline sediments," *Nature*, **311**:648-651.

Fienberg, S. 1979. "Graphical methods in statistics," *The American Statistician*, **33**:165-178.

Fisher, R. A. 1936. "The use of multiple measurements in taxonomic problems," *Annals of Eugenics*, **7**:179-188.

Flick, T., L. Jones, R. Priest, and C. Herman. 1990. "Pattern classification using projection pursuit," *Pattern Recognition*, **23**:1367-1376.

Flury, B. and H. Riedwyl. 1988. *Multivariate Statistics: A Practical Approach*, London: Chapman and Hall.

Fowlkes, E. B. and C. L. Mallows. 1983. "A method for comparing two hierarchical clusterings," *Journal of the American Statistical Association*, **78**:553-584.

Frakes, W. B. and Ricardo Baeza-Yates. 1992. *Information Retrieval: Data Structures & Algorithms*, Prentice Hall, New Jersey.

Fraley, C. 1998. "Algorithms for model-based Gaussian hierarchical clustering," *SIAM Journal on Scientific Computing*, **20**:270-281.

Fraley, C. and A. E. Raftery. 1998. "How many clusters? Which clustering method? Answers via model-based cluster analysis," *The Computer Journal*, **41**:578-588.

Fraley, C. and A. E. Raftery. 2002. "Model-based clustering, discriminant analysis, and density estimation: MCLUST," *Journal of the American Statistical Association*, **97**:611-631.

Fraley, C. and A. E. Raftery. 2003. "Enhanced software for model-based clustering, discriminant analysis, and density estimation: MCLUST," *Journal of Classification*, **20**:263-286.

Fraley, C., A. E. Raftery, and R. Wehrens. 2003. "Incremental model-based clustering for large datasets with small clusters," Technical Report 439, Department of Statistics, University of Washington (to appear in *Journal of Computational and Graphical Statistics*).

Freedman, D. and P. Diaconis. 1981. "On the histogram as a density estimator: L_2 theory," *Zeitschrift fur Wahrscheinlichkeitstheorie und verwandte Gebiete*, **57**:453-476.

Fridlyand, Jane and Sandrine Dudoit. 2001. "Applications of resampling methods to estimate the number of clusters and to improve the accuracy of a clustering method," Technical Report #600, Division of Biostatistics, University of California, Berkeley.

Friedman, J. 1987. "Exploratory projection pursuit," *Journal of the American Statistical Association*, **82**:249-266.

Friedman, J. and W. Stuetzle. 1981. "Projection pursuit regression," *Journal of the American Statistical Association*, **76**:817-823.

Friedman, J. and John Tukey. 1974. "A projection pursuit algorithm for exploratory data analysis," *IEEE Transactions on Computers*, **23**:881-889.

Friedman, J., W. Stuetzle, and A. Schroeder. 1984. "Projection pursuit density estimation," *Journal of the American Statistical Association*, **79**:599-608.

Friendly, Michael. 2000. *Visualizing Categorical Data*, Cary, NC: SAS Institute, Inc.

Friendly, M. and E. Kwan. 2003. "Effect ordering for data displays," *Computational Statistics and Data Analysis*, **43**:509-539.

Friendly, Michael and Howard Wainer. 2004. "Nobody's perfect," *Chance*, **17**:48-51.

Frigge, M., D. C. Hoaglin, and B. Iglewicz. 1989. "Some implementations of the boxplot," *The American Statistician*, **43**:50-54.

Fua, Ying-Huey, Matthew O. Ward, and Elke A. Rundensteiner. 1999. "Hierarchical parallel coordinates for exploration of large datasets," *IEEE Visualization, Proceedings of the Conference on Visualization '99*, 43 - 50.

Fukunaga, Keinosuke. 1990. *Introduction to Statistical Pattern Recognition, Second Edition*, New York: Academic Press.

Fukunaka, K. and D. R. Olsen. 1971. "An algorithm for finding intrinsic dimensionality of data," *IEEE Transactions on Computers*, **20**:176-183.

Gentle, James E. 2002. *Elements of Computational Statistics*, New York: Springer-Verlag.

Golub, G. and C. Van Loan. 1996. *Matrix Computations*, Baltimore: Johns Hopkins University Press.

Golub, T. R., Slonim, D. K., P. Tamayo, C. Huard, M. Gaasenbeek, J. P. Mesirov, H. Coller, M. L. Loh, J. R. Downing, M. A. Caligiuri, C. D. Bloomfield, and E. S. Lander. 1999. "Molecular classification of cancer: Class discovery and class prediction by gene expression monitoring," *Science*, **286**:531-537.

Good, I. J. 1983. "The philosophy of exploratory data analysis," *Philosophy of Science*, **50**:283-295.

Gordon, A. D. 1999. *Classification*, London: Chapman and Hall.

Gower, J. C. 1966. "Some distance properties of latent root and vector methods in multivariate analysis," *Biometrika*, **53**:325-338.

Gower, J. C. and Legendre, P. 1986. "Metric and Euclidean properties of dissimilarity coefficients," *Journal of Classification*, **3**:5-48.

Green P. J. and B. W. Silverman. 1994. *Nonparametric Regression and Generalized Linear Models: A Roughness Penalty Approach*, London: Chapman and Hall.

Griffiths, A. J. F., J. H. Miller, D. T. Suzuki, R. X. Lewontin, and W. M. Gelbart. 2000. *An Introduction to Genetic Analysis*, 7th ed., New York: Freeman.

Groenen, P. 1993. *The Majorization Approach to Multidimensional Scaling: Some Problems and Extensions*, Leiden, The Netherlands: DSWO Press.

Guttman, L. 1968. "A general nonmetric technique for finding the smallest coordinate space for a configuration of points," *Psychometrika*, **33**:469-506.

Hand, D., F. Daly, A. D. Lunn, K. J. McConway and E. Ostrowski. 1994. *A Handbook of Small Data Sets*, London: Chapman and Hall.

Hand, David, Heikki Mannila, and Padhraic Smyth. 2001. *Principles of Data Mining*, Cambridge, MA: The MIT Press.

Hanselman, D. and B. Littlefield. 1998. *Mastering MATLAB 5: A Comprehensive Tutorial and Reference*, New Jersey: Prentice Hall.

Hanselman, D. and B. Littlefield. 2001. *Mastering MATLAB 6: A Comprehensive Tutorial and Reference*, New Jersey: Prentice Hall.

Hansen, Pierre, Brigitte Jaumard, and Bruno Simeone. 1996. "Espaliers: A generalization of dendrograms," *Journal of Classification*, **13**:107-127.

Hartigan, J. A. 1967. "Representation of similarity measures by trees," *Journal of the American Statistical Association*, **62**:1140-1158.

Hartigan, J. A. 1975. *Clustering Algorithms*, New York: John Wiley & Sons.

Hartigan, J. A. 1985. "Statistical theory in clustering," *Journal of Classification*, **2**:63-76.

Hartwig, F. and Brian E. Dearing. 1979. *Exploratory Data Analysis*, Newbury Park, CA: Sage University Press.

Hastie, T. J. and C. Loader. 1993. "Local regression: Automatic kernel carpentry (with discussion)," *Statistical Science*, **8**:120-143.

Hastie, T. J. and R. J. Tibshirani. 1986. "Generalized additive models," *Statistical Science*, **1**:297-318.

Hastie, T. J., and R. J.Tibshirani. 1990. *Generalized Additive Models*, London: Chapman and Hall.

Hastie, T. J., R. J. Tibshirani, and J. Friedman. 2001. *The Elements of Statistical Learning: Data Mining, Inference and Prediction*, New York: Springer.

Hastie, T., R. Tibshirani, M. B. Eisen, A. Alizadeh, R. Levy, L. Staudt, W. C. Chan, D. Botstein and P. Brown. 2002. "Gene shaving as a method for identifying distinct sets of genes with similar expression patterns," *Genome Biology*, **1**.

Hoaglin, D. C. 1982. "Exploratory data analysis," in *Encyclopedia of Statistical Sciences, Volume 2*, Kotz, S. and N. L. Johnson, eds., New York: John Wiley & Sons.

Hoaglin, D. C., B. Iglewicz, and J. W. Tukey. 1986. "Performance of some resistant rules for outlier labeling," *Journal of the American Statistical Association*, **81**:991-999.

Hoaglin, D. C. and John Tukey. 1985. "Checking the shape of discrete distributions," in *Exploring Data Tables, Trends and Shapes*, D. Hoaglin, F. Mosteller, J. W. Tukey, eds., New York: John Wiley & Sons.

Hoaglin, D. C., F. Mosteller, and J. W. Tukey (eds.). 1983. *Understanding Robust and Exploratory Data Analysis*, New York: John Wiley & Sons.

Hogg, Robert. 1974. "Adaptive robust procedures: A partial review and some suggestions for future applications and theory (with discussion)," *Journal of the American Statistical Association*, **69**:909-927.

Huber, P. J. 1973. "Robust regression: Asymptotics, conjectures, and Monte Carlo," *Annals of Statistics*, **1**:799-821.

Huber, P. J. 1981. *Robust Statistics*, New York: John Wiley & Sons.

Huber, P. J. 1985. "Projection pursuit (with discussion)," *Annals of Statistics*, **13**:435-525.

Hubert, L. J. and P. Arabie. 1985. "Comparing partitions," *Journal of Classification*, **2**:193-218.

Hurley, Catherine and Andreas Buja. 1990. "Analyzing high-dimensional data with motion graphics," *SIAM Journal of Scientific and Statistical Computing*, **11**:1193-1211.

Inselberg, Alfred. 1985. "The plane with parallel coordinates," *The Visual Computer*, **1**:69-91.

Jackson, J. Edward. 1981. "Principal components and factor analysis: Part III - What is factor analysis?" *Journal of Quality Technology*, **13**:125-130.

Jackson, J. Edward. 1991. *A User's Guide to Principal Components*, New York: John Wiley & Sons.

Jain, Anil K. and Richard C. Dubes. 1988. *Algorithms for Clustering Data*, New York: Prentice Hall.

Jain, Anil K., M. N. Murty, and P. J. Flynn. 1999. "Data clustering: A review," *ACM Computing Surveys*, **31**:264-323.

Jeffreys, H. 1935. "Some tests of significance, treated by the theory of probability," *Proceedings of the Cambridge Philosophy Society*, **31**:203-222.

Jeffreys, H. 1961. *Theory of Probability, Third Edition*, Oxford, U. K.: Oxford University Press.

Johnson, B. and B. Shneiderman. 1991. "Treemaps: A space-filling approach to the visualization of hierarchical information structures," *Proceedings of the 2nd International IEEE Visualization Conference*, 284-291.

Jolliffe, I. T. 1972. "Discarding variables in a principal component analysis I: Artificial data," *Applied Statistics*, **21**:160-173.

Jolliffe, I. T. 1986. *Principal Component Analysis*, New York: Springer-Verlag.

Jones, W. P. and George W. Furnas. 1987. "Pictures of relevance: A geometric analysis of similarity measures," *Journal of the American Society for Information Science*, **38**:420-442.

Jones, M. C. and R. Sibson. 1987. "What is projection pursuit" (with discussion), *Journal of the Royal Statistical Society, Series A*, **150**:1–36.

Kaiser, H. F. 1960. "The application of electronic computers to factor analysis, *Educational and Psychological Measurement*, **20**:141-151.

Kangas, J. and S. Kaski. 1998. "3043 works that have been based on the self-organizing map (SOM) method developed by Kohonen," *Technical Report A50*, Helsinki University of Technology, Laboratory of Computer and Information Science.

Kaski, Samuel. 1997. *Data Exploration Using Self-Organizing Maps*, Ph.D. dissertation, Helsinki University of Technology.

Kaski, S., T. Honkela, K. Lagus, and T. Kohonen. 1998. "WEBSOM - Self-organizing maps of document collections," *Neurocomputing*, **21**:101-117.

Kass, R. E. and A. E. Raftery. 1995. "Bayes factors," *Journal of the American Statistical Association*, **90**:773-795.

Kaufman, Leonard and Peter J. Rousseeuw. 1990. *Finding Groups in Data: An Introduction to Cluster Analysis*, New York: John Wiley & Sons.

Kiang, Melody Y. 2001. "Extending the Kohonen self-organizing map networks for clustering analysis," *Computational Statistics and Data Analysis*, **38**:161-180.

Kimbrell, Roy E. 1988. "Searching for text? Send an N-Gram!," *Byte*, May, 297 - 312.

Kimball, B. F. 1960. "On the choice of plotting positions on probability paper," *Journal of the American Statistical Association*, **55**:546-560.

Kirby, M. 2001. *Geometric Data Analysis: An Empirical Approach to Dimensionality Reduction and the Study of Patterns*, New York: John Wiley & Sons.

Kleiner, B. and J. A. Hartigan. 1981. "Representing points in many dimensions by trees and castles," *Journal of the American Statistical Association*, **76**:260-276

Kohonen, Tuevo. 1998. "The self-organizing map," *Neurocomputing*, **21**:1-6.

Kohonen, Tuevo. 2001. *Self-Organizing Maps, Third Edition*, Berlin: Springer.

Kohonen, T., S. Kaski, K. Lagus, J. Salojarvi, T. Honkela, V. Paatero, and A. Saarela. 2000. "Self organization of a massive document collection," *IEEE Transactions on Neural Networks*, **11**:574-585.

Kotz, Samuel and Norman L. Johnson (eds.). 1986. *Encyclopedia of Statistical Sciences*, New York: John Wiley & Sons.

Kruskal, Joseph B. 1964a. "Multidimensional scaling by optimizing goodness of fit to a nonmetric hypothesis," *Psychometrika*, **29**:1-27.

Kruskal, Joseph B. 1964b. "Nonmetric multidimensional scaling: A numerical method," *Psychometrika*, **29**:115-129.

Kruskal, Joseph B. and Myron Wish. 1978. *Multidimensional Scaling*, Newbury Park, CA: Sage Publications, Inc.

Krzanowski, W. J. and Y. T. Lai. 1988. "A criterion for determining the number of groups in a data set using sum-of-squares clustering," *Biometrics*, **44**:23-34.

Lander, Eric S. 1999. "Array of hope." *Nature Genetics Supplement*, **21**:3-4.

Launer, R., and G. Wilkinson (eds.). 1979. *Robustness in Statistics*, New York: Academic Press.

Lawley, D. N. and A. E. Maxwell. 1971. *Factor Analysis as a Statistical Method, 2nd Edition*, London: Butterworth.

Levina, Elizaveta and Peter Bickel. 2004. "Maximum likelihood estimation of intrinsic dimension," *NIPS 2004* (to appear).

Li, G. and Z. Chen. 1985. "Projection-pursuit approach to robust dispersion matrices and principal components: Primary theory and Monte Carlo," *Journal of the American Statistical Association*, **80**:759-766.

Lindsey, J. C., A. M. Herzberg, and D. G. Watts. 1987. "A method for cluster analysis based on projections and quantile-quantile plots," *Biometrics*, **43**:327-341.

Ling, R. F. 1973. "A computer generated aid for cluster analysis," *Communications of the ACM*, **16**:355-361.

Loader, C. 1999. *Local Regression and Likelihood*, New York: Springer-Verlag.

Manly, B. F. J. 1994. *Multivariate Statistical Methods - A Primer, Second Edition*, London: Chapman & Hall.

Manning, Christopher D. and Hinrich Schütze. 2000. *Foundations of Statistical Natural Language Processing*, Cambridge, MA: The MIT Press.

Mao, J. and A. K. Jain. 1995. "Artificial neural networks for feature extraction and multivariate data projection," *IEEE Transaction on Neural Networks*, **6**:296-317.

Marchand, Patrick and O. Thomas Holland. 2003. *Graphics and GUIs with MATLAB, Third Edition*, Boca Raton: CRC Press.

Marchette, D. J. and J. L. Solka. 2003. "Using data images for outlier detection," *Computational Statistics and Data Analysis*, **43**:541-552.

Martinez, A. R. 2002. *A Framework for the Representation of Semantics*, Ph.D. Dissertation, Fairfax, Va: George Mason University.

Martinez, A. R. and Edward J. Wegman. 2002a. "A text stream transformation for semantic-based clustering," *Computing Science and Statistics*, **34**:184-203.

Martinez, A. R. and Edward J. Wegman. 2002b. "Encoding of text to preserve "meaning"," *Proceedings of the Eighth Annual U. S. Army Conference on Applied Statistics*, 27-39.

Martinez, W. L. and A. R. Martinez. 2002. *Computational Statistics Handbook with MATLAB*, Boca Raton: CRC Press.

McGill, Robert, John Tukey, and Wayne Larsen. 1978. "Variations of box plots," *The American Statistician*, **32**:12-16.

McLachlan, G. J. and K. E. Basford. 1988. *Mixture Models: Inference and Applications to Clustering*, New York: Marcel Dekker.

McLachlan, G. J. and T. Krishnan. 1997. *The EM Algorithm and Extensions*, New York: John Wiley & Sons.

McLachlan, G. J. and D. Peel. 2000. *Finite Mixture Models*, New York: John Wiley & Sons.

McLachlan, G. J., D. Peel, K. E. Basford, and P. Adams. 1999. "The EMMIX software for the fitting of mixtures of normal and t-components," *Journal of Statistical Software*, **4**, `http://www.jstatsoft.org/index.php?vol=4`

McLeod, A. I. and S. B. Provost. 2001. "Multivariate Data Visualization," `www.stats.uwo.ca/faculty/aim/mviz`.

Mead, A. 1992. "Review of the development of multidimensional scaling methods," *The Statistician*, **41**:27-39.

Milligan, G. W. and M. C. Cooper. 1985. "An examination of procedures for determining the number of clusters in a data set," *Psychometrika*, **50**:159-179.

Milligan, G. W. and M. C. Cooper. 1988. "A study of standardization of variables in cluster analysis," *Journal of Classification*, **5**:181-204.

Minnotte, M. and R. West. 1998. "The data image: A tool for exploring high dimensional data sets," *Proceedings of the ASA Section on Statistical Graphics*, Dallas, Texas, 25-33.

Mojena, R. 1977. "Hierarchical grouping methods and stopping rules: An evaluation," *Computer Journal*, **20**:359-363.

Moler, Cleve. 2004. *Numerical Computing with MATLAB*, New York: SIAM.

Montanari, Angela and Laura Lizzani. 2001. "A projection pursuit approach to variable selection," *Computational Statistics and Data Analysis*, **35**:463-473.

Morgan, Byron J. T. and Andrew P. G. Ray. 1995. "Non-uniqueness and inversions in cluster analysis," *Applied Statistics*, **44**:114-134.

Mosteller, F. and J. W. Tukey. 1977. *Data Analysis and Regression: A Second Course in Statistics*, New York: Addison-Wesley.

Mosteller, F. and D. L. Wallace. *Inference and Disputed Authorship: The Federalist Papers*, New York: Addison-Wesley.

Mukherjee, S., E. Feigelson, G. Babu, F. Murtagh, C. Fraley, and A. E. Raftery. 1998. "Three types of gamma ray bursts," *Astrophysical Journal*, **508**:314-327.

Murtagh, F. and A. E. Raftery. 1984. "Fitting straight lines to point patterns," *Pattern Recognition*, **17**:479-483.

Nason, Guy. 1995. "Three-dimensional projection pursuit," *Applied Statistics*, **44**:411–430.

Olbricht, W. 1982. "Modern statistical analysis of ancient sand," MSc Thesis, University of Sheffield, Sheffield, UK.

Panel on Discriminant Analysis, Classification, and Clustering. 1989. "Discriminant Analysis and Clustering," *Statistical Science*, **4**:34-69.

Pearson, K. and A. Lee. 1903. "On the laws of inheritance in man. I. Inheritance of physical characters," *Biometrika*, **2**:357-462.

Pettis, K. W., T. A. Bailey, A. K. Jain, and R. C. Dubes. 1979. "An intrinsic dimensionality estimator from near-neighbor information," *IEEE Transactions on Pattern Analysis and Machine Intelligence*, **1**:25-37.

Porter, M. F. 1980. 'An algorithm for suffix stripping,' *Program*, **14**:130 - 137.

Posse, Christian. 1995a. "Projection pursuit exploratory data analysis," *Computational Statistics and Data Analysis*, **29**:669–687.

Posse, Christian. 1995b. "Tools for two-dimensional exploratory projection pursuit," *Journal of Computational and Graphical Statistics*, **4**:83–100.

Posse, Christian. 2001. "Hierarchical model-based clustering for large data sets," *Journal of Computational and Graphical Statistics*," **10**:464-486.

Rand, William M. 1971. "Objective criteria for the evaluation of clustering methods," *Journal of the American Statistical Association*, **66**:846-850.

Rao, C. R. 1993. *Computational Statistics*, The Netherlands: Elsevier Science Publishers.

Redner, A. R. and H. F. Walker. 1984. "Mixture densities, maximum likelihood and the EM algorithm," *SIAM Review*, **26**:195-239.

Ripley, Brian D. 1996. *Pattern Recognition and Neural Networks*, Cambridge: Cambridge University Press.

Roeder, Kathryn. 1994. "A graphical technique for determining the number of components in a mixture of normals," *Journal of the American Statistical Association*, **89**:487-495.

Rousseeuw, P. J. and A. M. Leroy. 1987. *Robust Regression and Outlier Detection*, New York: John Wiley & Sons.

Rousseeuw, P. J. and I. Ruts. 1996. "Algorithm AS 307: Bivariate location depth," *Applied Statistics (JRSS-C)*, **45**:516-526.

Rousseeuw, P. J. and I. Ruts. 1998. "Constructing the bivariate Tukey median," *Statistica Sinica*, **8**:827-839.

Rousseeuw, P, J., I. Ruts, and J. W. Tukey. 1999. "The bagplot: A bivariate boxplot," *The American Statistician*, **53**:382-387.

Roweis, S. T. and L. K. Saul. 2000. "Nonlinear dimensionality reduction by locally linear embedding," *Science*, **290**:2323-2326.

Salton, Gerard, Chris Buckley, and Maria Smith. 1990. 'On the application of syntactic methodologies,' *Automatic Text Analysis, Information Processing & Management*, **26**:73 - 92.

Saul, L. K. and S. T. Roweis. 2002. "Think globally, fit locally: Unsupervised learning of nonlinear manifolds," Technical Report MS CIS-02-18, University of Pennsylvania.

Schena, M., Shalon, D., R. W. Davis, and P. O. Brown. 1995. "Quantitative monitoring of gene expression patterns with a complementary DNA microarray," *Science*, **270**:497-470.

Schimek, M. G. (ed.) 2000. *Smoothing and Regression: Approaches, Computation, and Application*, New York: John Wiley & Sons.

Schwarz, G. 1978. "Estimating the dimension of a model," *The Annals of Statistics*, **6**:461-464.

Scott, A. J. and M. J. Symons. 1971. "Clustering methods based on likelihood ratio criteria," *Biometrics*, **27**:387-397.

Scott, David W. 1992. *Multivariate Density Estimation: Theory, Practice, and Visualization*, New York: John Wiley & Sons.

Sebastani, P., E. Gussoni, I. S. Kohane, & M. F. Ramoni. 2003. "Statistical Challenges in Functional Genomics," *Statistical Science*, **18**:33-70.

Seber, G. A. F. 1984. *Multivariate Observations*, New York: John Wiley & Sons.

Shepard, R. N. 1962a. "The analysis of proximities: Multidimensional scaling with an unknown distance function I," *Psychometrika*, **27**:125-140.

Shepard, R. N. 1962b. "The analysis of proximities: Multidimensional scaling with an unknown distance function II," *Psychometrika*, **27**:219-246.

Shneiderman, B. 1992. "Tree visualization with tree-maps: 2-D space-filling approach," *ACM Transactions on Graphics*, **11**:92-99.

Sieklecki, W., K. Siedlecka, and J. Sklansky. 1988. "An overview of mapping techniques for exploratory pattern analysis," *Pattern Recognition*, **21**:411-429.

Silverman, B. W. 1986. *Density Estimation for Statistics and Data Analysis*, London: Chapman and Hall.

Simonoff, J. S. 1996. *Smoothing Methods in Statistics*, New York: Springer-Verlag.

Slomka, M. 1986. "The analysis of a synthetic data set," *Proceedings of the Section on Statistical Graphics*, 113-116.

Sneath, P. H. A. and R. R. Sokal. 1973. *Numerical Taxonomy*, San Francisco: W. H. Freeman.

Solka, J. and W. L. Martinez. 2004. "Model-based clustering with an adaptive mixtures smart start," to appear in *Computing Science and Statistics*.

Späth, Helmuth. 1980. *Cluster Analysis Algorithms for Data Reduction and Classification of Objects*, New York: Halsted Press.

Steyvers, Mark. 2002. "Multidimensional scaling," in *Encyclopedia of Cognitive Science*.

Stone, Charles J. 1977. "Consistent nonparametric regression," *The Annals of Statistics*, **5**:595-645.

Strang, Gilbert. 1988. *Linear Algebra and its Applications*, Third Edition, San Diego: Harcourt Brace Jovanovich.

Strang, Gilbert. 1993. *Introduction to Linear Algebra*, Wellesley, MA: Wellesley-Cambridge Press.

Stuetzle, Werner. 1987. "Plot windows," *Journal of the American Statistical Association*, **82**:466-475.

Sturges, H. A. 1926. "The choice of a class interval," *Journal of the American Statistical Association*, **21**:65-66.

Swayne, D. F., D. Cook, and A. Buja. 1991. "XGobi: Interactive dynamic graphics in the X window system with a link to S," *ASA Proceedings of the Section on Statistical Graphics*. 1-8.

Tamayho, P., D. Slonim, J. Mesirov, Q. Zhu, S. Kitareewan, E. Dmitrovsky, E. S. Lander, and T. R. Golub. 1999. "Interpreting patterns of gene expression with self-organizing maps: Methods and application to hematopoietic differentiation," *Proceedings of the National Academy of Science*, **96**:2907-2912.

Tenenbaum, J. B., V. de Silva, and J. C. Langford. 2000. "A global geometric framework for nonlinear dimensionality reduction," *Science*, **290**:2319-2323.

Tibshirani, R., G. Walther, D. Botstein, and P. Brown. 2001. "Cluster validation by prediction strength," Technical Report, Stanford University.

Tibshirani, R., Guenther Walther, and Trevor Hastie. 2001. "Estimating the number of clusters in a data set via the gap statistic," *Journal of the Royal Statistical Society, B,* **63**:411-423.

Tiede, J. J. and M. Pagano. 1979. "The application of robust calibration to radioimmunoassay," *Biometrics,* **35**:567-574.

Timmins, D. A. Y. 1981. "Study of sediment in mesolithic middens on Oronsay," MA Thesis, University of Sheffield, Sheffield, UK.

Titterington, D. B. 1985. "Common structure of smoothing techniques in statistics," *International Statistical Review,* **53**:141-170.

Titterington, D. M., A. F. M. Smith, and U. E. Makov. 1985. *Statistical Analysis of Finite Mixture Distributions,* New York: John Wiley & Sons.

Torgerson, W. S. 1952. "Multidimensional scaling: 1. Theory and method," *Psychometrika,* **17**:401-419.

Trunk, G. 1968. "Statistical estimation of the intrinsic dimensionality of data collections," Inform. Control, **12**:508-525.

Trunk, G. 1976. "Statistical estimation of the intrinsic dimensionality of data," *IEEE Transactions on Computers,* **25**:165-171.

Tufte, E. 1983. *The Visual Display of Quantitative Information,* Cheshire, CT: Graphics Press.

Tufte, E. 1990. *Envisioning Information,* Cheshire, CT: Graphics Press.

Tufte, E. 1997. *Visual Explanations,* Cheshire, CT: Graphics Press.

Tukey, John W. 1973. "Some thoughts on alternagraphic displays," Technical Report 45, Series 2, Department of Statistics, Princeton University.

Tukey, John W. 1975. "Mathematics and the picturing of data," *Proceedings of the International Congress of Mathematicians,* **2**:523-531.

Tukey, John W. 1977. *Exploratory Data Analysis,* New York: Addison-Wesley.

Tukey, John W. 1980. "We need both exploratory and confirmatory," *The American Statistician,* **34**:23-25.

Ultsch, A. and H. P. Siemon. 1990. "Kohonen's self-organizing feature maps for exploratory data analysis," *Proceedings of the International Neural Network Conference (INNC'90),* Dordrecht, Netherlands, 305-308.

Vandervieren, E. and M. Hubert. 2004. "An adjusted boxplot for skewed distributions," *COMPSTAT 2004,* 1933-1940.

Velicer, W. F., and D. N. Jackson. 1990. "Component analysis versus common factor analysis: Some issues on selecting an appropriate procedure (with discussion)," *Journal of Multivariate Behavioral Research,* **25**:1-114.

Venables, W. N. and B. D. Ripley. 1994. *Modern Applied Statistics with S-Plus,* New York: Springer-Verlag.

Verveer, P. J. and R. P. W. Duin. 1995. "An evaluation of intrinsic dimensionality estimators," *IEEE Transactions on Pattern Analysis and Machine Intelligence,* **17**:81-86.

Vesanto, Juha. 1997. "Data mining techniques based on the self-organizing map," Master's Thesis, Helsinki University of Technology.

Vesanto, Juha. 1999. "SOM-based data visualization methods," *Intelligent Data Analysis*, **3**:111-126.

Vesanto, Juha and Esa Alhoniemi. 2000. "Clustering of the self-organizing map," *IEEE Transactions on Neural Networks*, **11**:586-6000.

Wainer, H. 1997. *Visual Revelations: Graphical Tales of Fate and Deception from Napoleon Bonaparte to Ross Perot*, New York: Copernicus/Springer-Verlag.

Wall, M. E., P. A. Dyck, and T. S. Brettin. 2001. "SVDMAN - singular value decomposition analysis of microarray data," *Bioinformatics*, **17**:566-568.

Wall, M. E., A. Rechtsteiner, and L. M. Rocha. 2003. "Chapter 5: Singular value decomposition and principal component analysis," in *A Practical Approach to Microarray Data Analysis*, D. P. Berar, W. Dubitsky, M. Granzow, eds., Kluwer: Norwell, MA.

Wand, M. P. and M. C. Jones. 1995. *Kernel Smoothing*, London: Chapman and Hall.

Ward, J. H. 1963. "Hierarchical groupings to optimize an objective function," *Journal of the American Statistical Association*, **58**:236-244.

Webb, Andrew. 1999. *Statistical Pattern Recognition*, Oxford: Oxford University Press.

Wegman, E. J. 1986. *Hyperdimensional Data Analysis Using Parallel Coordinates*, Technical Report No. 1, George Mason University Center for Computational Statistics.

Wegman, E. 1988. "Computational statistics: A new agenda for statistical theory and practice," *Journal of the Washington Academy of Sciences*, **78**:310-322.

Wegman, E. J. 1990. "Hyperdimensional data analysis using parallel coordinates," *Journal of the American Statistical Association*, **85**:664-675.

Wegman, E. J. 1991. "The grand tour in *k*-dimensions," *Computing Science and Statistics: Proceedings of the 22nd Symposium on the Interface*, 127-136.

Wegman, E. J. 2003. "Visual data mining," *Statistics in Medicine*, **22**:1383-1397.

Wegman, E. J. and D. Carr. 1993. "Statistical graphics and visualization," in *Handbook of Statistics, Vol 9*, C. R. Rao, ed., The Netherlands: Elsevier Science Publishers, 857-958.

Wegman, E. J. and Q. Luo. 1997. "High dimensional clustering using parallel coordinates and the grand tour," *Computing Science and Statistics*, **28**:361-368.

Wegman, E. J. and J. Shen. 1993. "Three-dimensional Andrews plots and the grand tour," *Proceedings of the 25th Symposium on the Interface*, 284-288.

Wegman, E. J. and J. Solka. 2002. "On some mathematics for visualizing high dimensional data," *Sankkya: The Indian Journal of Statistics*, **64**:429-452.

Wegman, E. J. and I. W. Wright. "Splines in statistics," *Journal of the American Statistical Association*, **78**:351-365.

Wegman, E. J., D. Carr, and Q. Luo. 1993. "Visualizing multivariate data," in *Multivariate Analysis: Future Directions*, C. R. Rao, ed., The Netherlands: Elsevier Science Publishers, 423-466.

Wehrens, R., L. M. C. Buydens, C. Fraley, and A. E. Raftery. 2003. "Model-based clustering for image segmentation and large datasets via sampling," Technical Report 424, Department of Statistics, University of Washington (*to appear in Journal of Classification*).

Weihs, C. 1993. "Multivariate exploratory data analysis and graphics: A tutorial," *Journal of Chemometrics*, **7**:305-340.

Weihs, C. and H. Schmidli. 1990. "OMEGA (Online multivariate exploratory graphical analysis): Routine searching for structure," *Statistical Science*, **5**:175-226.

West, M., C. Blanchette, H. Dressman, E. Huang, S. Ishida, R. Spang, H. Zuzan, J. A. Olson, Jr., J. R. Marks, and J. R. Nevins. 2001. "Predicting the clinical status of human breast cancer by using gene expression profiles," *Proceedings of the National Academy of Science*, **98**:11462-11467.

Wijk, J. J. van and H. van de Wetering. 1999. "Cushion treemaps: Visualization of hierarchical information," in: G. Wills, De. Keim (eds.), *Proceedings IEEE Symposium on Information Visualization (InfoVis'99)*, 73-78.

Wilhelm, A. F. X., E. J. Wegman, and J. Symanzik. 1999. "Visual clustering and classification: The Oronsay particle size data set revisited," *Computational Statistics*, **14**:109-146.

Wilk, M. B. and R. Gnanadesikan. 1968. "Probability plotting methods for the analysis of data," *Biometrika*, **55**:1-17.

Wilkinson, Leland. 1999. *The Grammar of Graphics*, New York: Springer-Verlag.

Wills, G. J. 1998. "An interactive view for hierarchical clustering," *Proceedings IEEE Symposium on Information Visualization*, 26-31.

Witten, I. H., A. Moffat and T. C. Bell. 1994. *Managing Gigabytes: Compressing and Indexing Documents and Images*, New York, NY: van Nostrand Reinhold.

Wolfe, J. H. 1970. "Pattern clustering by multivariate mixture analysis," *Multivariate Behavioral Research*, **5**:329-350.

Yeung, K. Y. and W. L. Ruzzo. 2001. "Principal component analysis for clustering gene expression data," *Bioinformatics*, **17**:363-774

Yeung, K. Y., C. Fraley, A. Murua, A. E. Raftery, and W. L. Ruzzo. 2001. "Model-based clustering and data transformation for gene expression data," Technical Report 396, Department of Statistics, University of Washington.

Young, Forrest W. 1985. "Multidimensional scaling," in *Encyclopedia of Statistical Sciences*, Kotz, S. and Johnson, N. K. (eds.), 649-659.

Young, F. W., and Penny Rheingans. 1991. "Visualizing structure in high-dimensional multivariate data," *IBM Journal of Research and Development*, **35**:97-107.

Young, F. W., R. A. Faldowski, and M. M. McFarlane. 1993. "Multivariate statistical visualization," in *Handbook of Statistics, Vol 9*, C. R. Rao, ed., The Netherlands: Elsevier Science Publishers, 959-998.

Young, G. and A. S. Householder. 1938. "Discussion of a set of points in terms of their mutual distances," *Psychometrika*, **3**:19-22.

Author Index

A

Adams, P. 193
Alhoniemi, E. 100
Alter, O. 42, 57
Anderberg, M. R. 157
Andrews, D. F. 209, 322, 333
Anscombe, F. J. 332
Arabie, P. 142
Asimov, D. 103, 104, 105, 106, 107, 110, 125
Asman, W. A. H. 314
Azzalini, A. 228

B

Babu, G. 192
Baeza-Yates, R. 10
Bailey, T. A. 52, 54, 158
Balasubramanian, M. 100
Banfield, J. D. 164, 168, 275
Basford, K. E. 164, 192, 193
Basseville, M. 343
Becker, R. A. 301, 308, 311, 333, 351, 353, 354
Bell, T. C. 10
Benjamini, Y. 268, 275
Bensmail, H. 187, 193
Berry, M. W. 10, 42, 43, 44, 46
Bertin, J. 332
Bhattacharjee, A. 16, 17
Bickel, P. J. 57
Biernacki, C. 193
Binder, D. A. 164
Bishop, C. M. 94, 95, 100
Blasius, J. 332
Bock, H. 151, 164
Bolton, R. J. 125
Bonner, R. 129
Borg, I. 63, 64, 68, 74, 99
Botstein, D. 42, 57, 158
Bouldin, D. W. 158

Bowman, A. W. 228
Brettin, T. S. 42
Brinkman, N. D. 352
Brown, P. O. 42, 57, 158
Browne, M. 10, 44
Bruls, D. M. 255
Bruntz, S. M. 352
Bruske. J. 57
Buckley, C. 10
Buijsman, E. 314
Buja, A. 103, 104, 105, 107, 110, 125, 333
Buta, R. 206, 352
Buydens, L. M. C. 193

C

Cabrera, J. 125
Calinski, R. 160
Campbell, J. G. 192
Carr, D. 299, 332
Cattell, R. B. 38, 49
Cavalli-Sforza, L. L. 164
Celeux, G. 168, 169, 177, 187, 193
Chakrapani, T. K. 352
Chambers, J. M. 332, 354
Charniak, E. 9
Chatfield, C. 5
Chee, M. 13
Chen, Z. 125
Chernoff, H. 293
Cho, R. J. 14
Cleveland, W. S. 198, 199, 201, 206, 209, 211, 212, 216, 218, 220, 223, 228, 269, 284, 289, 301, 308, 309, 311, 313, 314, 332, 333, 351, 352, 353, 354
Cohen, A. 255
Cook, D. 125, 333
Cook, W. J. 251
Cooper, M. C. 23, 145, 158
Costa, J. A. 57
Cottrell, M. 100
Cox, M. A. A. 61, 62, 64, 65, 68, 74, 75, 99, 343, 354
Cox, T. F. 61, 62, 64, 65, 68, 74, 75, 99, 343, 354
Cox. M. A. A. 74
Crawford, S. 125
Crile, G. 303, 351

Subject Index

A

adjacent values 271
agglomerative model-based clustering 181
alternagraphics 303
Andrews' curves 321
Andrews' function 321
ANN 90
artificial neural network 90
at-random method 107
average linkage 131

B

bag 286
bagplot 286
Batch Map, SOM 92
batch training method, SOM 92
Bayes factors 182
Bayesian Information Criterion 184
best-matching unit, SOM 91
between-class scatter matrix 160
BIC 182, 184
bigram proximity matrix 9, 10
binary tree 233
bisquare 209
BMU 91
box-and-whisker 268
box-percentile plot 275
boxplot, variable width 275
boxplots 268
brushing 308
    modes 308
    operations 308

C

canonical basis vector 105

Cartesian coordinates 319
centroid linkage 132
chaining 131
Chernoff faces 294
chi-square index 114
city block metric 339
classical scaling 64
classification likelihood 181
cluster 129
cluster Assessment 128
cluster, orientation 169
cluster, shape 169
cluster, volume 169
codebooks 91
color histogram 249
common factors 47
communality 47
complete linkage 131
component density 168
confirmatory data analysis 3
convex hull 222
cophenetic coefficient 143
coplots 309
correlation matrix 37
correlation similarity 340
cosine measure 44, 340
covariance matrix 24, 34, 168, 177, 338

D

data abstraction 128
data image 249
Data Visualization Toolbox 207, 311
degree of polynomial 199
dendrogram 133, 233
denoising text 10
dependence panels 309
depth median 287
diagonal family 171
dimensionality reduction 31
direction cosine 35
discriminant analysis 127
disparity 72